Low Power Design Essentials

Series on Integrated Circuits and Systems

Series Editor: Anantha Chandrakasan
                Massachusetts Institute of Technology
                Cambridge, Massachusetts

Low Power Design Essentials
Jan Rabaey
ISBN 978-0-387-71712-8

Carbon Nanotube Electronics
Ali Javey and Jing Kong (Eds.)
ISBN 978-0-387-36833-7

Wafer Level 3-D ICs Process Technology
Chuan Seng Tan, Ronald J. Gutmann, and L. Rafael Reif (Eds.)
ISBN 978-0-387-76532-7

Adaptive Techniques for Dynamic Processor Optimization: Theory and Practice
Alice Wang and Samuel Naffziger (Eds.)
ISBN 978-0-387-76471-9

mm-Wave Silicon Technology: 60 GHz and Beyond
Ali M. Niknejad and Hossein Hashemi (Eds.)
ISBN 978-0-387-76558-7

Ultra Wideband: Circuits, Transceivers, and Systems
Ranjit Gharpurey and Peter Kinget (Eds.)
ISBN 978-0-387-37238-9

Creating Assertion-Based IP
Harry D. Foster and Adam C. Krolnik
ISBN 978-0-387-36641-8

Design for Manufacturability and Statistical Design: A Constructive Approach
Michael Orshansky, Sani R. Nassif, and Duane Boning
ISBN 978-0-387-30928-6

Low Power Methodology Manual: For System-on-Chip Design
Michael Keating, David Flynn, Rob Aitken, Alan Gibbons, and Kaijian Shi
ISBN 978-0-387-71818-7

Modern Circuit Placement: Best Practices and Results
Gi-Joon Nam and Jason Cong
ISBN 978-0-387-36837-5

CMOS Biotechnology
Hakho Lee, Donhee Ham and Robert M. Westervelt
ISBN 978-0-387-36836-8

SAT-Based Scalable Formal Verification Solutions
Malay Ganai and Aarti Gupta
ISBN 978-0-387-69166-4, 2007

Ultra-Low Voltage Nano-Scale Memories
Kiyoo Itoh, Masashi Horiguchi and Hitoshi Tanaka
ISBN 978-0-387-33398-4, 2007

Continued after index

Jan Rabaey

Low Power Design Essentials

Springer

Jan Rabaey
Department of Electrical Engineering &
  Computer Science (EECS)
University of California
Berkeley, CA 94720
USA
jan@eecs.berkeley.edu

Additional material to this book can be downloaded from http://extras.springer.com.

ISSN 1558-9412
ISBN 978-0-387-71712-8   e-ISBN 978-0-387-71713-5
DOI 10.1007/978-0-387-71713-5

Library of Congress Control Number: 2008932280

© Springer Science+Business Media, LLC 2009
All rights reserved. This work may not be translated or copied in whole or in part without the written permission of the publisher (Springer Science+Business Media, LLC, 233 Spring Street, New York, NY 10013, USA), except for brief excerpts in connection with reviews or scholarly analysis. Use in connection with any form of information storage and retrieval, electronic adaptation, computer software, or by similar or dissimilar methodology now known or hereafter developed is forbidden.
The use in this publication of trade names, trademarks, service marks, and similar terms, even if they are not identified as such, is not to be taken as an expression of opinion as to whether or not they are subject to proprietary rights.

Printed on acid-free paper

springer.com

To Kathelijin
For so many years, my true source of support and motivation.

To My Parents
While I lost you both in the past two years, you still inspire me to reach ever further.

Preface

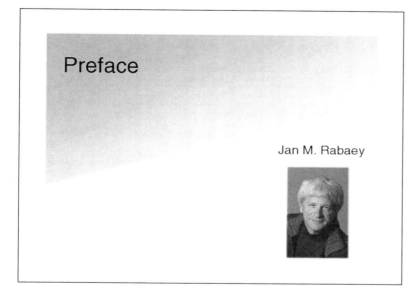

Slide 0.1

Welcome to this book titled "Low Power Design Essentials". (A somewhat more accurate title for the book would be "Low Power Digital Design Essentials", as virtually all of the material is focused on the digital integrated-circuit design domain.)

In recent years, power and energy have become one of the most compelling issues in the design of digital circuits. On one end, power has put a severe limitation on how fast we can run our circuits; at the other end, energy reduction techniques have enabled us to build ubiquitous mobile devices that can run on a single battery charge for an exceedingly long time.

Slide 0.2

You may wonder why there is a need for yet another book on low-power design, as there are quite a number of those already on the market (some of them co-authored by myself). The answer is quite simple: all these books are edited volumes, and target the professional who is already somewhat versed in the main topics of design for power or energy. With these topics becoming one of the most compelling issues in design today, it is my opinion that it is time for a book with *an educational approach*. This means building up from the basics, and exposing the different subjects in a rigorous and methodological way with consistent use of notations and definitions. Concepts are illustrated with examples using state-of-the-art technologies (90 nm and below). The book is primarily intended for use in short-to-medium length courses on low-power design. However, the format also should work well for the working professional, who wants to update her/himself on low-power design in a self-learning manner.

> **Goals of This Book**
>
> - Provide an educational perspective on low-power desgn for digital integrated circuits
> - Promote a structured design methodology for low power/energy design
> - Traverse the levels of the design hierarchy
> - Explore bounds and roadblocks
> - Provide future perspectives

This preface also presents an opportunity for me to address an issue that has been daunting low-power design for a while. Many people in the field seem to think that it is just a "bag of tricks" applied in a somewhat ad hoc fashion, that it needs a guru to get to the bottom, and that the *concept of a low-power methodology* is somewhat an oxymoron. In fact, in recent years researchers and developers have demonstrated that this need not be the case at all. One of the most important realizations over the past years is that minimum-energy design, though interesting, is not what we truly are pursuing. In general, we design in an energy–delay trade-off space, where we try to find design with the lowest energy for a given performance, or vice versa. A number of optimization and design exploration tools can be constructed that help us to traverse this trade-off space in an informed fashion, and this at all levels of the design hierarchy.

In addition to adhering to such a methodology throughout the text, we are also investigating the main *roadblocks* that we have to overcome in the coming decades if we want to keep reducing the energy per operation. This naturally leads to the question of what the *physical limits* of energy scaling might be. Wherever possible, we also venture some perspectives on the future.

> **An Innovative Format**
>
> - Pioneered in W. Sansen's book *Analog Design Essentials* (Springer)
> - PowerPoint slides present a quick outline of essential points and issues, and provide a graphical perspective
> - Side notes provide depth, explain reasonings, link topics
> - Supplemented with web-site:
>   http://bwrc.eecs.berkeley.edu/LowPowerEssentials
> - An ideal tool for focused-topic courses

Slide 0.3

Already in this preface, you observe the somewhat unorthodox approach the book is taking. Rather than choosing the traditional approach of a lengthy continuous text, occasionally interspersed with some figures, we use the reverse approach: graphics first, text as a side note. In my experience, a single figure does a lot more to convey a message than a page of text ("A picture is worth a 1000 words"). This approach was pioneered by Willy Sansen in his book *Analog Design Essentials* (also published by Springer). The first time I saw the book, I was immediately captivated by the idea. The more I looked at it, the more I liked it. Hence this book .... When browsing through it, you will notice that the slides and the notes play entirely different roles. Another advantage of the format is that the educator has basically all the lecturing material in her/his hands rightaway. Besides distributing the slideware freely, we also offer additional material and tools on the web-site of the book.

Slide 0.4

> **Outline**
>
> - **Background**
>   1. Introduction
>   2. Advanced MOS Transistors and Their Models
>   3. Power Basics
> - **Optimizing Power @ Design Time**
>   4. Circuits
>   5. Architectures, Algorithms, and Systems
>   6. Interconnect and Clocks
>   7. Memories
> - **Optimizing Power @ Standby**
>   8. Circuits and Systems
>   9. Memory
> - **Optimizing Power @ Runtime**
>   10. Circuits, Memory, and Systems
> - **Perspectives**
>   11. Ultra Low Power/ VoltageDesign
>   12. Low Power Design Methodologies and Flows
>   13. Summary and Perspectives

The outline of the book proceeds as follows: After first establishing the basics, we proceed to address power optimization in three different operational modes: design time, standby time, and run time. The techniques used in each of these modes differ considerably. Observe that we treat dynamic and static power simultaneously throughout the text – in today's semiconductor technology, leakage power is virtually on par with switching power. Hence separating them does not make much sense. In fact, a better design is often obtained if the two are carefully balanced. Finally, the text concludes with a number of general topics such as design tools, limits on power, and some future projections.

Slide 0.5

> **Acknowledgements**
>
> The contributions of many of my colleagues to this book are greatly appreciated. Without them, building this collection of slides would have been impossible. Especially, I would like to single out the inputs of the following individuals who have contributed in a major way to the book: Ben Calhoun, Jerry Frenkil, and Dejan Marković. As always, it has been an absolute pleasure working with them.
>
> In addition, a large number of people have helped to shape the book by contributing material, or by reviewing the chapters as they emerged. I am deeply indebted to all of them: E. Alon, T. Austin, D. Blaauw, S. Borkar, R. Brodersen, T. Burd, K. Cao, A. Chandrakasan, H. De Man, K. Flautner, M. Horowitz, K. Itoh, T. Kuroda, B. Nikolić, C. Rowen, T. Sakurai, A. Sangiovanni-Vincentelli, N. Shanbhag, V. Stojanović, T. Sakurai, J. Tschanz, E. Vittoz, A. Wang, and D. Wingard, as well as all my graduate students at BWRC.
>
> I also would like to express my appreciation for the funding agencies that have provided strong support to the development of low-power design technologies and methodologies. Especially the FCRP program (and its member companies) and DARPA deserve special credit.

Putting a book like this together without help is virtually impossible, and a couple of words of thanks and appreciation are in order. First and foremost, I am deeply indebted to Ben Calhoun, Jerry Frenkil, Dejan Marković, and Bora Nikolić for their help and co-authorship of some of the chapters. In addition, a long list of people have helped in providing the basic slideware used in the text, and in reviewing the earlier drafts of the book. Special gratitude goes to a number of folks who have shaped the low-power design technology world in a tremendous way – and as a result have contributed enormously to this book: Bob Brodersen, Anantha Chandrakasan, Tadahiro Kuroda, Takayasu Sakurai, Shekhar Borkar, and Vivek De. Working with them over the past decade(s) has been a great pleasure and a truly exciting experience!

Slide 0.6–0.7

Every chapter in the book is concluded with a set of references supporting the material presented in the chapter. For those of you who are truly enamored with the subject of low-power design, these slides enumerate a number of general reference works, overview papers, and visionary presentations on the topic.

Low Power Design – Reference Books

- A. Chandrakasan and R. Brodersen, *Low Power CMOS Design*, Kluwer Academic Publishers, 1995.
- A. Chandrakasan and R. Brodersen, *Low-Power CMOS Design*, IEEE Press, 1998 (Reprint **Volume**).
- A. Chandrakasan, Bowhill, and Fox, *Design of High-Performance Microprocessors*, IEEE Press, 2001.
  - Chapter 4, "Low-Voltage Technologies," by Kuroda and Sakuraipggy
  - Chapter 3, "Techniques for Leakage Power Reduction," by De, et al.
- M. Keating et al., *Low Power Methodology Manual*, Springer, 2007.
- S. Narendra and A. Chandrakasan, *Leakage in Nanometer CMOS Technologies*, Springer, 2006.
- M. Pedram and J. Rabaey, Ed., *Power Aware Design Methodologies*, Kluwer Academic Publishers, 2002.
- C. Piguet, Ed., *Low-Power Circuit Design*, CRC Press, 2005.
- J. Rabaey and M. Pedram, Ed., *Low Power Design Methodologies*, Kluwer Academic Publishers, 1995.
- J. Rabaey, A. Chandrakasan, and B. Nikolic, *Digital Integrated Circuits - A Design Perspective*, Prentice Hall, 2003.
- S. Roundy, P. Wright and J.M. Rabaey, *Energy Scavenging for Wireless Sensor Networks*, Kluwer Academic Publishers, 2003.
- A. Wang, *Adaptive Techniques for Dynamic Power Optimization*, Springer, 2008.

Low-Power Design – Special References

- S. Borkar, "Design challenges of technology scaling," *IEEE Micro*, 19 (4), p. 23–29, July–Aug. 1999.
- T. Kuroda, T. Sakurai, "Overview of low-power ULSI circuit techniques," *IEICE Trans. on Electronics*, E78-C(4), pp. 334–344, Apr. 1995.
- Journal-o fLow Power Electronics (JOLPE), http://www.aspbs.com/jolpe/
- Proceedings of the IEEE, Special Issue on Low Power Design, Apr. 1995.
- Proceedings of the ISLPED Conference (starting 1994)
- Proceedings of ISSCC, VLSI Symposium, ESSCIRC, A-SSCC, DAC, ASPDAC, DATE, ICCAD conferences

I personally had a wonderful and truly enlightening time putting this material together while traversing Europe during my sabbatical in the spring of 2007. I hope you will enjoy it as well.

Jan M. Rabaey, Berkeley, CA

Contents

1. Introduction 1
2. Nanometer Transistors and Their Models 25
3. Power and Energy Basics 53
4. Optimizing Power @ Design Time: Circuit-Level Techniques 77
5. Optimizing Power @ Design Time – Architecture, Algorithms, and Systems 113
6. Optimizing Power @ Design Time – Interconnect and Clocks 151
7. Optimizing Power @ Design Time – Memory 183
8. Optimizing Power @ Standby – Circuits and Systems 207
9. Optimizing Power @ Standby – Memory 233
10. Optimizing Power @ Runtime: Circuits and Systems 249
11. Ultra Low Power/Voltage Design 289
12. Low Power Design Methodologies and Flows 317
13. Summary and Perspectives 345

Index 357

Chapter 1
Introduction

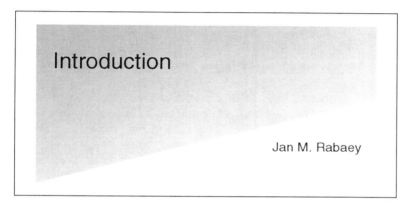

Slide 1.1

In this chapter we discuss why power and energy consumption has become one of the main (if not *the* main) design concerns in today's complex digital integrated circuits. We first aalyze the different application domains and evaluate how each has its own specific concerns and requirements, from a power perspective soon. Most projections into the future show that these concerns most likely will not go away. In fact, everything seems to indicate that they will even aggravate. Next, we evaluate technology trends – in the idle hope that technology scaling may help to address some of these problems. Unfortunately, CMOS scaling only seems to make the problem worse. Hence, design solutions will be the primary mechanism in keeping energy/power consumption in control or within bounds. Identifying the central design themes and technologies, and finding ways to apply them in a structured and methodological fashion, is the main purpose of this book. For quite some time, low-power design consisted of a collection of ad hoc techniques. Applying those techniques successfully on a broad range of applications and without too much "manual" intervention requires close integration in the traditional design flows. Over the past decade, much progress in this direction was made. Yet, the gap between low-power design technology and methodology remains.

Slide 1.2

There are many reasons why designers and application developers worry about power dissipation. One concern that has come consistently to the foreground in recent years is the need for "green" electronics. While the power dissipation of electronic components until recently was only a small fraction of the overall electrical power budget, this picture has changed substantially in the last few decades. The pervasive use of desktops and laptops has made its mark in both the office and home environments. Standby power of electronic consumer components and set-up boxes is rising rapidly such that at the time of writing this book their power drain is becoming equivalent to

Why Worry About Power?

The Tongue-in-Cheek Answer

- Total energy of Milky Way galaxy: 10^{59} J
- Minimum switching energy for digital gate (1 electron @ 100 mV): 1.6×10^{-20} J (limited by thermal noise)
- Upper bound on number of digital operations: 6×10^{78}
- Operations/year performed by 1 billion 100 MOPS computers: 3×10^{24}
- Entire energy might be consumed in **180 years**, assuming a doubling of computational requirements every year (Moore's Law).

year, the total energy of our galaxy would be exhausted in the relatively low time span of 180 years (even if we assume that every digital operation is performed at its lowest possible level). However, as Gordon Moore himself stated in his keynote address at the 2001 ISSCC conference, "No exponential is forever", adding quickly thereafter, "... but forever can be delayed".

that of a decent-size fridge. Electronics are becoming a sizable fraction of the power budget of a modern automobile. These trends will only become more pronounced in the coming decade(s).

In this slide, the growing importance of electronics as part of the power budget is brought home with a "tongue-in-cheek" extrapolation. If Moore's law would continue unabated in the future and the computational needs would keep on doubling every

Slide 1.3

Power: The Dominant Design Constraint (1)

Cost of large data centers solely determined by power bill ...

Columbia River

Google Data Center, The Dalles, Oregon

- 400 Millions of Personal Computers worldwide (Year 2000)
  - Assumed to consume 0.16 Tera (10^{12}) kWh per year
  - Equivalent to 26 nuclear power plants
- Over 1 Giga kWh per year just for cooling
  - Including manufacturing electricity

[Ref: Bar-Cohen et al., 2000]

NY Times, June 06

Google's Server Growth — 8,000; 100,000; 450,000

The subsequent slide sets evaluate the power need and trends for a number of dominant application areas of digital integrated circuits. First, the domains of computation and communication infrastructure are discussed. The advent of the Internet, combined with ubiquitous access to the network using both wired and wireless interfaces, has dramatically changed the nature of computing. Today massive data storage and computing centers operated by large companies at a number of centralized locations have absorbed a huge amount of the worldwide computational loads of both corporations and individuals. And this trend is not showing any signs of slowing down, as new server farms are being brought online at a staggering rate. Yet, this centralization comes at a price. The "computational density" of such a center, and hence the power usage, is substantial. To quote Luis Barosso from Google (a company which is one of the most prolific promoters of the remote-computation concept), the cost of a data center is determined

Introduction

solely by the monthly power bill, not by the cost of hardware or maintenance. This bill results from both the power dissipation in the electronic systems and the cost of removing the dissipated heat – that is, air conditioning. This explains why most data centers are now implanted at carefully chosen locations where power is easily available and effective cooling techniques are present (such as in the proximity of major rivers – in an eerie similarity to nuclear plants).

While data centers represent a major fraction of the power consumed in the computation and communication infrastructure, other components should not be ignored. The fast routers that zip the data around the world, as well as the wireless base stations (access points) which allow us to connect wirelessly to the network, offer major power challenges as well. Owing to their location, the availability of power and the effectiveness of cooling techniques are often limited. Finally, the distributed computing and communication infrastructure cannot be ignored either; the wired and wireless data routers in the office, plant, or home, the back-office computing servers and the desktop computers add up to a sizable power budget as well. A large fraction of the air conditioning bill in offices is due to the ever-growing computational infrastructure.

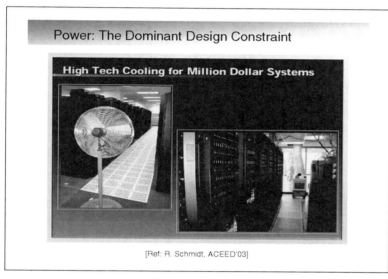

Slide 1.4

It is worth spending some time on the cooling issue. A typical computing server rack in a server farm can consume up to 20 kW. With the racks in a farm easily numbering over one hundred, power dissipation can top the 2 MW (all of which is transformed into heat). The design of the air conditioning system and the flow of air through the room and the racks is quite complicated and requires extensive modeling and analysis. The impact of an ill-designed system can be major (i.e., dramatic failure), or more subtle. In one such data center design, cool air is brought in from the floor and is gradually heated while it rises through the blades (boards) in the rack. This leads to a temperature gradient, which may mean that processors closer to the floor operate faster than the ones on the top! Even with the best air-cooling design practices, predicting the overall dynamics of the center can be hard and can lead to under-cooling. Sometimes some quick improvised fixes are the only rescue, as witnessed in these ironic pictures, provided by Roger Schmidt, a distinguished engineer at IBM and a leading expert in the engineering and engineering management of the thermal design of large-scale IBM computers.

Slide 1.5

While temperature gradients over racks can lead to performance variations, the same is true for the advanced high-performance processors of today. In the past die sizes were small enough, and activity over the die was quite uniform. This translated into a flat temperature profile at the surface of the die. With the advent of Systems-on-a-Chip (SoC), more and more diverse

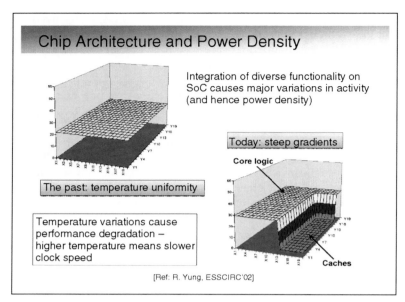

functionality is integrated in close proximity, very often with very different workloads and activity profiles. For instance, most high-performance microprocessors (or multi-core processors) integrate multiple levels of cache memories on the die, just next to the high-performance computing engine(s). As the data path of the processor is clocked at the highest speed and is kept busy almost 100% of the time, its power dissipation is substantially higher than that of the cache memories. This results in the creation of hot spots and temperature gradients over the die. This may impact the long-term reliability of the part and complicate the verification of the processor. Execution speed and propagation delay are indeed strongly dependent on temperature. With temperature gradients over the die (which may change dynamically depending upon the operation modes of the processors), simulation can now not be performed for a single temperature, as was the common practice.

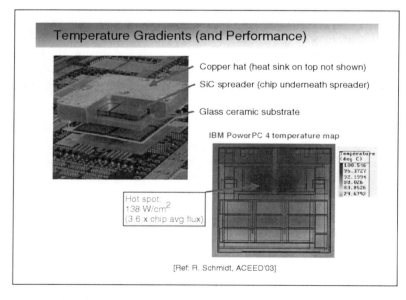

Slide 1.6
The existence of these thermal gradients is perfectly illustrated in this slide, which plots the temperature map of the IBM PowerPC 4 (a late 1990s microprocessor). A temperature difference of over 20°C can be observed between the processor core and the cache memory. Even more staggering, the heat generation at the hot spot (the data pipeline) equals almost 140 W/cm^2. This is 3.6 times the heat removal capacity of the chip cooling system. To correct for this imbalance, a complex package has to be constructed, which allows for the heat to spread over a wider area, thus improving the heat removal process. In high-performance

Introduction

components, packaging cost has become an important (if not dominating) fraction of the total cost. Techniques that help to mitigate the packaging problems either by reducing the gradients or by reducing the power density of selected sub-systems are hence essential. Structured low-power design methodologies, as advocated in this book, do just that.

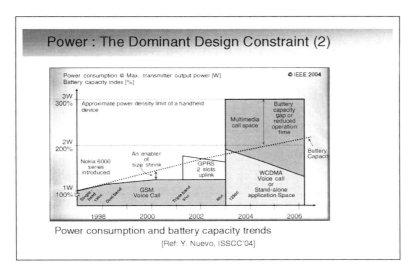
Power consumption and battery capacity trends
[Ref: Y. Nuevo, ISSCC'04]

Slide 1.7
The second reason why design for low power/energy has become so important is the emergence of mobile electronics. While mobile consumer electronics has been around for a while (FM radios, portable CD players), it is the simultaneous success of portable laptops and digital cellular phones that has driven the quest for low-energy computing and communication. In a battery-operated device, the available energy is fixed, and the rate of power consumption determines the lifetime of the battery (for non-rechargeables) or the time between recharges. Size, aspect ratio, and weight are typically set by the application or the intended device. The allowable battery size of a cellular phone typically is set to at most 4–5 cm^3, as dictated by user acceptance. Given a particular battery technology, the expected operational time of the device – cell phone users today expect multiple days of standby time and 4–5 h of talk time – in between recharges sets an upper bound on the power dissipation for the different operational modes. This is turn determines what functionality can be supported by the device, unless breakthroughs in low-power design can be accomplished. For instance, the average power dissipation limit of a cell phone is approximately 3 W, dictated by today's battery technologies. This in turn dictates whether your phone will be able to support digital video broadcasting, MP3 functionality, and 3G cellular and WIFI interconnectivity.

Slide 1.8
From this perspective, it is worthwhile to classify consumer and computing devices into a number of categories, based on their energy needs and hence functionality. In the "ambient intelligent" home of the future (a term coined by Fred Boekhorst from Philips in his ISSCC keynote in 2002), we may identify three styles of components. First, we have the "Watt nodes" (P > 1 W). These are nodes connected to the power grid, offering computational capacity of around 1 GOPS and performing functions such as computing and data serving, as well as routing and wireless access. The availability of energy, and hence computational prowess, makes them the ideal home for advanced media processing, data manipulation, and user interfaces.

The second tier of devices is called the "Milliwatt nodes" (1 mW < P < 1 W). Operating at a couple of MOPS, these represent mobile, untethered devices such as PDAs, communication devices (connecting to WANs and LANs), and wireless displays. These components are battery-powered and fall into the scope of devices discussed in the previous slide.

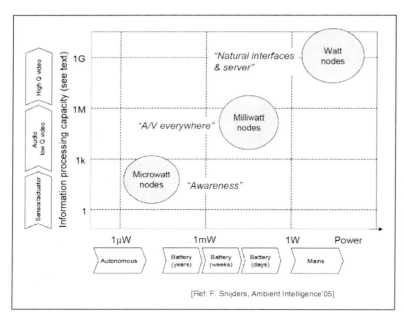

The "Microwatt nodes" represent the final category (P < 1 mW). Their function is to add awareness to the network, providing sensing functionality (temperature, presence, motion, etc.), and to transmit that data to the more capable nodes. The 1 KOPS computational capability severely limits their functionality. Given that a typical home may contain a huge number of these nodes, they have to be energy self-contained or powered using energy scavenging. Their very low power levels enable the atter. More information about this class of nodes follows in later slides.

Slide 1.9
From the above discussion, obviously the question arises: "Where is battery technology heading?" As already observed in Slide 1.7, battery capacity (i.e., the amount of energy that can be stored and delivered for a given battery volume) doubles approximately every 10 years. This represents an improvement of 3–7% every year (the slope tends to vary based on the introduction of new technologies). This growth curve lags substantially behinds Moore's law, which indicates a doubling in computational complexity every 18 months. The challenge with battery technology is that chemical processes are the underlying force, and improvements in capacity are often related to new chemicals or electrode materials. These are hard to come by. Also, the manufacturing processes for every new material take a long time to develop. Yet, an analysis of the available chemicals seems to show some huge potential. The energy density

Introduction 7

of alcohol or gasoline is approximately two orders of magnitude higher than that of lithium-polymer. Unfortunately, concerns about the effective and safe handling of these substances make it hard to exploit them in small form factors.

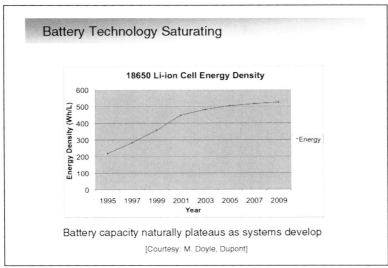

Slide 1.10
The historical trends in battery capacity actually vary quite a bit. Up to the 1980s, very little or even no progress was made – there was actually little incentive to do so, as the scope of application was quite limited. Flash lights were probably the driving application. In the 1990s, mobile applications took off. Intensive research combined with advanced manufacturing strategies changed the slope substantially, improving the capacity by a factor of four in almost a decade. Unfortunately, the process has stalled somewhat since the beginning of the 21st century. A major improvement in battery capacity can only be achieved by the introduction of new battery chemicals. It should also be observed that the capacity of a battery (that is the energy that can be extracted from it) also depends upon its discharge profile. Draining a battery slowly will deliver more energy than flash discharge. It is hence worthwhile to match the battery structure to the application at hand.

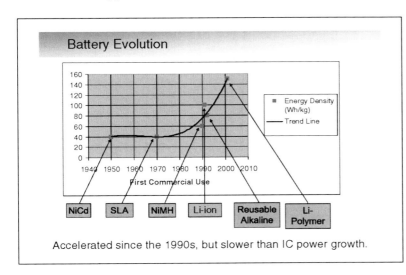

Slide 1.11
The fact that the energy delivery capacity of a battery is ultimately determined by the basic chemical properties of the materials involved is clearly illustrated in this slide. In the 1990s, the capacity of Lithium-ion batteries improved substantially. This was mostly due to better engineering: improved electrode structures, better charging technology, and advanced battery system design. This ultimately saturated as the intrinsic maximum potential of the material is being approached. Today, progress in Lithium-ion battery technology has stalled, and little improvement is foreseen in the future.

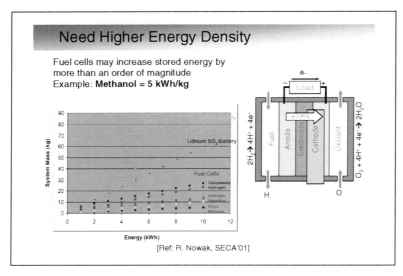

Slide 1.12
The bottom line from the presented trends is that only a dramatic change in chemicals will lead to substantial increase in battery capacity. The opportunity is clearly there. For instance, hydrogen has an energy density 4–8 times that of Lithium-ion. It is no surprise that hydrogen fuel cells are currently under serious consideration for the powering of electrical or hybrid cars. The oxidation of hydrogen produces water and electrical current as output. Fuels such as alcohol, methanol, or gasoline are even better. The challenge with these materials is to maintain the efficiency in small form factors while maintaining safety and reliability.

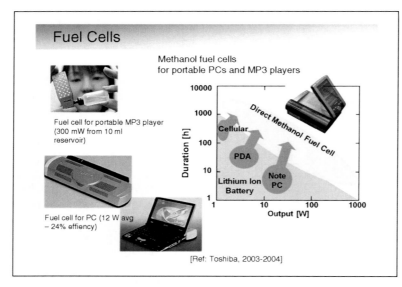

Slide 1.13
It should be of no surprise that research in this area is intensive and that major companies as well as start-ups are vying for a piece of the potentially huge cash pot. Success so far has been few and far between. Toshiba, for instance, has introduced a number of methanol fuel cells, promising to extend the operational time of your cell phone to 1000 h (i.e., 40 days!). Other companies actively exploring the fuel cell option are NEC and IBM. Yet, the technology still has to find its way into the markets. Long-term efficiency, safety, and usage models are questionable. Other candidates such as solid oxygen fuel cells (also called ceramic fuel cells) are waiting behind the curtain. If any one of these becomes successful, it could change the energy equation for mobiles substantially.

Introduction

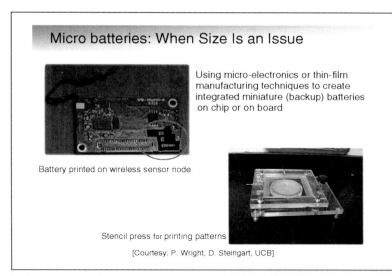

Slide 1.14

Another interesting new entry in the battery field is the "micro battery". Using technologies inherited from thin-film and semiconductor manufacturing, battery anodes and cathodes are printed on the substrate, and micromachined encapsulations are used to contain the chemicals. In this way, it is possible to print batteries on printed circuit boards (PCBs), or even embed them into integrated circuits. While the capacity of these circuits will never be large, micro batteries can serve perfectly well as backup batteries or as energy storage devices in sensor nodes. The design of the battery involves trading off between current delivery capability (number of electrodes) and capacity (volume occupied by the chemicals). This technology is clearly still in its infancy but could occupy some interesting niche in the years to come.

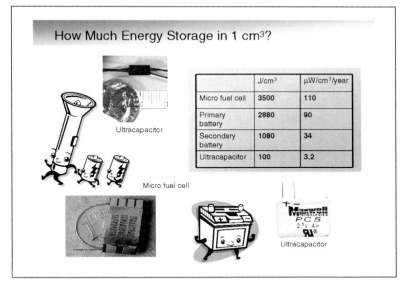

Slide 1.15

As a summary of the above discussions, it is worthwhile ordering the various energy storage technologies for mobile nodes based on their capacity (expressed in J/cm^3). Another useful metric is the average current that can be delivered over the time span of a year by a $1\,cm^3$ battery ($\mu W/cm^3/$year), which provides a measure of the longevity of the battery technology for a particular application.

Miniature fuel cells clearly provide the highest capacity. In their currently best incarnation, they are approximately three times more efficient than the best rechargeable (secondary) batteries. Yet, the advantage over non-rechargeables (such as alkaline) is at most 25%.

One alternative strategy for the temporary storage of energy was not discussed so far: the capacitor. The ordinary capacitor constructed from high-quality dielectrics has the advantage of simplicity, reliability, and longevity. At the same time, its energy density is limited. One technology that attempts to bridge the gap between capacitor and battery is the so-called supercapacitor or ultracapacitor, which is an electrochemical capacitor that has an unusually high energy density when compared to common capacitors, yet substantially lower than that of rechargeable batteries. A major advantage of (ultra)capacitors is the instantaneous availability of a high discharge current, which makes them very attractive for bursty applications. It is expected that new materials such as carbon nanotubes, carbon aerogels, and conductive polymers may substantially increase the capacity of supercapacitors in the years to come.

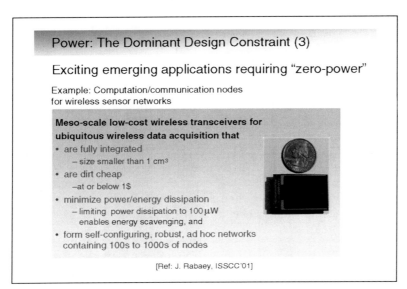

Slide 1.16
The third and final motivation behind "ultra low power" design is the emergence of a new class of frontier applications, called "zero-power electronics" or "disappearing electronics" (microwatt nodes in the Boekhorst classification). The continuing miniaturization of computing and communication components, enabled by semiconductor scaling, allows for the development of tiny wireless sensor nodes, often called motes. With sizes in the range of cubic centimeters or less, these devices can be integrated into the daily-living environment, offering a wide range of sensing and monitoring capabilities. By providing spatial and temporal information about, for instance, the environmental conditions in a room, more efficient and more effective conditioning of the room is enabled. The integrated format and the low cost make it possible to deploy large or even huge numbers of these motes. These emerging "wireless sensor networks (WSN)" have made some major inroads since their inception in the late 1990s. Energy is one of the main hurdles to be overcome, if the WSN paradigm is to be successful. Given the large number of nodes in a network, regular battery replacement is economically and practically out of question. Hence, nodes should in principle be energy self-contained for the lifetime of the application (which can be tens of years). Hence, a node should be able to operate continuously on a single battery charge, or should be capable of replenishing its energy supply by energy-scavenging techniques. As both energy storage and scavenging capacities are proportional to volume and the node size is limited, ultra low-power design is absolutely essential. In the "PicoRadio" project, launched by the author in 1998, it was determined that the average power dissipation of the node could not be larger than 100 μW.

Slide 1.17
Since the inception of the WSN concept, much progress was made in reducing the size, cost, and power dissipation of the mote. First-generation nodes were constructed from off-the-shelf components, combining generic microcontrollers, simple wireless transceivers with little power

Introduction

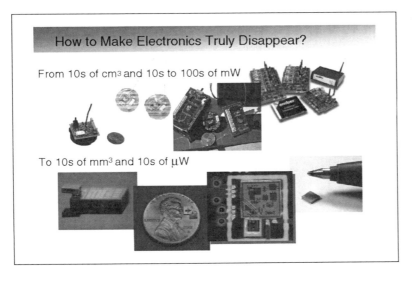

optimization, and standard sensors. The resulting motes were at least one or two orders off the stated goals in every aspect (size, cost, power).

Since then, research in miniature low-power electronics has blossomed, and has produced spectacular results. Advanced packaging technologies, introduction of novel devices (sensors, passives, and antennas), ultra low-voltage design, and intelligent power management have produced motes that are close to meeting all the stated goals. The impact of these innovations goes beyond the world of wireless sensor networks and can equally be felt in areas such as implantable devices for health monitoring or smart cards.

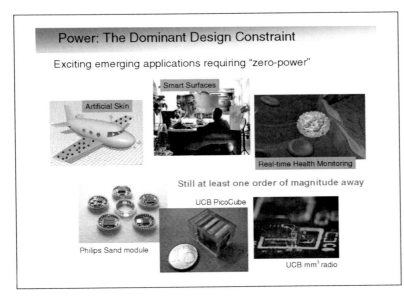

Slide 1.18
Even more, progress in ultra low-power design and extreme miniaturization may enable the emergence of a number of applications that otherwise would be completely impossible. A couple of examples may help to illustrate this. Dense networks of sensor nodes deployed on a broad surface may lead to "artificial skin", sensitive to touch, stress, pressure, or fatigue. Obvious applications of such networks would be intelligent plane wings, novel user interfaces, and improved robots. Embedding multiple sensors into objects may lead to smart objects such as intelligent tires that sense the condition of the road and adjust the driving behavior accordingly. The concept of "inject-able" health diagnostic, monitoring, and, eventually, surgery devices was suggested in the science fiction world in the 1960s (for instance, in the notorious "Fantastic Voyage" by Isaac Asimov), but it may not be fiction after all. Yet, bringing each of these applications into reality will require power and size reduction by another order of magnitude (if not two). The cubic-centimeter nodes of today should be reduced to true "dust" size

(i.e., cubic millimeter). This provides a true motivation for further exploration of the absolute boundaries of ultra low-power design, which is the topic of Chapter 11 in this book.

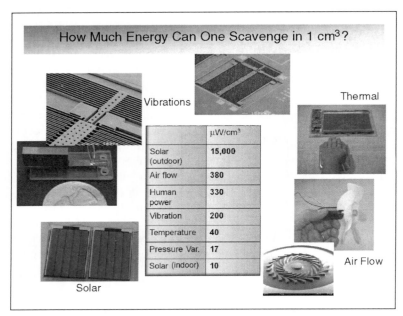

Slide 1.19

Energy scavenging is an essential component for the success of microwatt nodes. The idea is to transform the physical energy present in various sources in the environment into electrical power. Examples of the former are temperature or pressure gradients, light, acceleration, and kinetic and electromagnetic energy. In recent years, researchers both in academics and in industry have spent substantial efforts in cataloguing and metricizing the effectiveness of the various scavenging technologies [Roundy03, Paradiso05]. The efficiency of an energy harvester is best expressed by the average power provided by a scavenger of 1 cm^3, operating under various conditions. Just like with batteries, scavenging efficiency is linearly proportional to volume (or, as in the case for solar cells, to surface area).

From the table presented in the slide, it is clear that light (captured by photovoltaic cells) is by far the most efficient source of energy, especially in outdoors conditions. A power output of up to 15 mW/cm^2 can be obtained. Unfortunately, this drops by two or three orders of magnitudes when operated in ambient indoor conditions. Other promising sources of energy that are ubiquitously available are vibration, wind, and temperature and pressure gradients. The interested reader can refer to the above-mentioned reference works for more information. The main takeaway is that average power levels of around 100 W/cm^3 are attainable in many practical situations.

The discussion so far has not included some other sources of energy, magnetic and electromagnetic, that are prime targets for scavenging. Putting moving coils in a magnetic field (or having a variable magnetic field) induces current in a coil. Similarly, an antenna can capture the energy beamed at it in the form of an electromagnetic wave. This concept is used effectively for the powering of passive RF-IDs. None of these energy sources occurs naturally though, and an "energy transmitter" has to be provided. Issues such as the impact on health should be considered, if large power levels are required. Also, the overall efficiency of these approaches is quite limited.

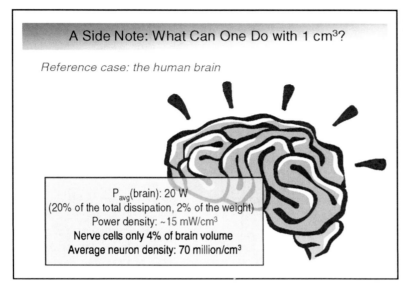

Slide 1.20

This introduction has so far focused on the various application domains of microelectronics and their power needs and constraints. In the subsequent slides, we will discuss the energy and power trends from a technology perspective, looking back at past evolutions and projecting future developments. Before doing so, one more side note is probably useful. To put the energy efficiency of microelectronic systems into perspective, it is worth comparing them with other "computational engines", in this case biological machinery (i.e., the brain).

The average power consumption of an average human brain approximately equals 20 W, which is approximately 20% of the total power dissipation of the body. This fraction is quite high, especially when considering that the brain represents only 2% of the total body mass – in fact, the ratio of power to the brain versus the total body power is a telling indicator of where the being stands on the evolutionary ladder. Again considering the average brain size (1.33 dm^3), this leads to a power consumption of 15 mW/cm^3 – similar to what could be provided by 1 cm^2 of solar cells. Active neurons only represent a small fraction of this volume (4%) – most of the rest is occupied by blood vessels, which transport energy in and heat out of the brain, and the dense interconnecting network.

Judging the energy efficiency of the brain is a totally different matter, though. Comparing the "computational complexity" of a neuron with that of a digital gate or a processor is extremely hard, if not irrelevant. The brain contains on the average 70 million neurons per cubic centimeter, each of which performs complex non-linear processing. For the interested readers, a great analysis of and comparison between electronic and neurological computing is offered in the best-selling book by Ray Kurzweil, "The Singularity Is Near."

Power Versus Energy

- Power in high-performance systems
  - Heat removal
  - Peak power and its impact on power delivery networks
- Energy in portable systems
  - Battery life
- Energy/power in "zero-power systems"
  - Energy-scavenging and storage capabilites
- Dynamic (energy) vs. static (power) consumption
  - Determined by operation modes

Slide 1.21
Before discussing trends, some words about useful metrics are necessary (more details to follow in Chapter 3). So far, we have used the terms power and energy quite interchangeably. Yet, each has its specific role depending upon the phenomena that are being addressed or the constraints of the application at hand. Average power dissipation is the prominent parameter when studying heat-removal and packaging concerns of high-performance processors. Peak power dissipation, on the other hand, is the parameter to watch when designing the complex power supply delivery networks for integrated circuits and systems.

When designing mobile devices or sensor network nodes, the type of energy source determines which property is the most essential. In a battery-powered system, the energy supply is finite, and hence energy minimization is crucial. On the other hand, the designer of an energy-scavenging system has to ensure that the average power consumed is smaller than the average power provided by the scavenger.

Finally, dividing power dissipation into dynamic (proportional to activity) and static (independent of activity) is crucial in the design of power management systems exploiting the operational modes of the system. We will see later that the reality here is quite complex and that a careful balancing between the two is one of the subtleties of advanced low-power design.

Slide 1.22
While concerns about power density may seem quite recent to most designers, the issue has surfaced a number of times in the design of (electrical) engineering systems before. Obviously heat removal was and is a prime concern in many thermodynamic systems. In the electronics world, power dissipation, and consequent high temperatures, was a main cause of unreliability in vacuum-tube computers. While bipolar computer design offered prime performance, exceeding what could be delivered by MOS implementations at that time, power density and the ensuing reliability concerns limited the amount of integration that could be obtained. The same happened with pure NMOS logic – the static current inherent in non-complimentary logic families ultimately caused semiconductor manufacturers to switch to CMOS, even though this meant an increased process complexity and a loss in performance. When CMOS was adopted as the technology-of-choice in the mid 1980s, many felt that the power problem had been dealt with effectively, and that CMOS design would enjoy a relatively trouble-free run to ever higher performance. Unfortunately, it was not to be. Already in the early 1990s, the ever-increasing clock frequencies and the emergence of new application domains brought power back to the foreground.

The charts in this slide document how the increases in heat flux in bipolar and CMOS systems mirror each other, only offset by about a decade. They make the interesting point that exponentials are hard to get around. New technologies create a fixed offset, but the exponential increases in

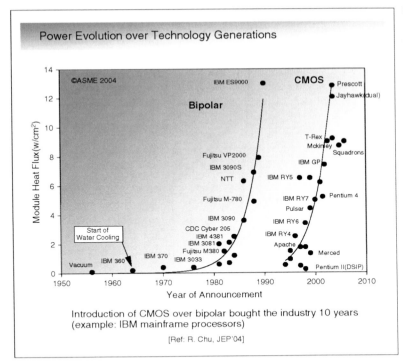

complexity – so essential to the success of the semiconductor industry – conspire to eliminate that in the shortest possible time.

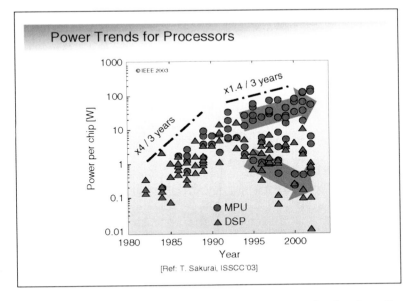

Slide 1.23
The power trends of the past are best observed by empirically sampling the leading processor designs over the years (as embodied by publications in ISSCC, the leading conference in the field) [Courtesy of T. Kuroda and T. Sakurai]. Plotting the power dissipations of microprocessors and DSPs as a function of time reveals some interesting trends. Up to the mid 1990s, the average power dissipation of a processor rose by a factor of four every three years. At that time, a discontinuity occurred. A major slowdown in the rise of power dissipation of leading-edge processors is apparent (to approximately a factor of 1.4 every three years). Simultaneously, another downward vector emerged: owing to the introduction of mobile devices, a market for lower-power lower-performance processors was materializing. One obviously wonders about the

discontinuity around 1995, the answer for which is quite simple: Owing to both power and reliability concerns, the semiconductor industry finally abandoned the idea of a supply voltage fixed at 5 V (the "fixed-voltage scaling model"), and started scaling supply voltages in correspondence with successive process nodes. Fixed-voltage scaling was an attractive proposition, as it simplified the interfacing between different components and parts, yet the power cost became unattainable. Reasoning about the precise value of the slope factors is somewhat simplified when studying *power density* rather than *total power*, as the former is independent of the actual die size.

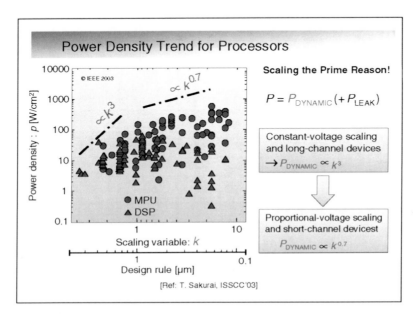

Slide 1.24

Under the assumptions of fixed-voltage scaling [see Rabaey03] and long-channel devices (more about this in Chapter 2 on devices), it is assumed that the supply voltage remains constant and the discharge current scales. Under these conditions, the clock frequency f scales between technology generations as k^2, where k is the technology scaling factor (which typically equals 1.41). The power density

$$p = CV_{DD}^2 f$$

then evolves as

$$k_p = k \times 1 \times k^2 = k^3.$$

Consider now the situation after 1995. Under the full-scaling mode, supply voltages were scaled in proportion to the minimum feature size of the technology. Also at that time, short-channel device effects such as velocity saturation (again see Chapter 2) were becoming important, causing the saturation current (i.e., the maximum discharge current) to scale approximately as $k^{-0.3}$, leading to a slowdown in the clock frequency increase to $k^{1.7}$. For the power density, this means that

$$p = CV_{DD}^2 f$$

now scales as

$$k_p = k \times (1/k)^2 \times k^{1.7} = k^{0.7},$$

Introduction

which corresponds with the empirical data. Even though this means that power density is still increasing, a major slowdown is observed. This definitely is welcome news.

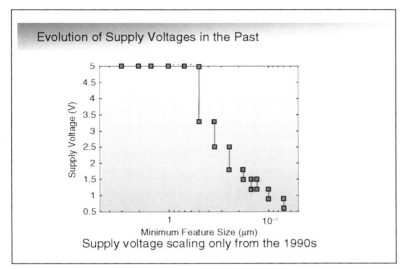

Slide 1.25
To illustrate the fact that the full scaling model was truly adopted starting around the 0.65 µm CMOS technology node, this slide plots the range of supply voltages that were (are) typically used for every generation. Up to the early 1990s, supply voltages were pretty much fixed at 5 V, dropping for the first time to 3.3 V for the 0.35 µm generation. Since then, supply voltages have by and large followed the minimum feature size.

For instance, the nominal supply voltage for the 180 nm processor equals 1.8 V; for 130 nm it is 1.3 V; and so on. Unfortunately, this trend is gradually changing for the worse again, upsetting the subtle balance between performance and power density, as will become clear in the following slides.

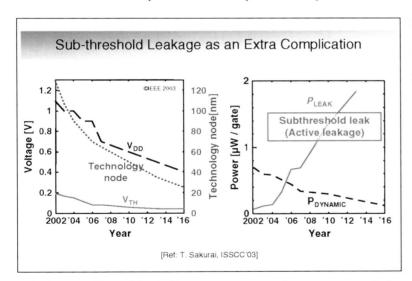

Slide 1.26
By the end of the 20th century, new storm clouds were gathering on the horizon. The then prevalent scaling model made the assumption that a certain ratio between supply voltage and threshold voltage is maintained. If not, a substantial degradation in maximum clock speed (which was generally equated to system performance) results, a penalty that the designers of that time were not willing to accept. The only plausible solution to address this challenge was to maintain a constant ratio by scaling the threshold voltages as well. This, however, posed a whole new problem. As we will discuss in detail in later chapters, the off-current of a MOS transistor (i.e., the current when the gate–source voltage is set to zero) increases exponentially with a reduction in the threshold voltage. Suddenly, static power dissipation – a problem that had gone away with the introduction of CMOS – became a forefront issue again. Projections indicated that, if left unattended, static power dissipation would overtake dynamic power sometime in the mid to late 2000s.

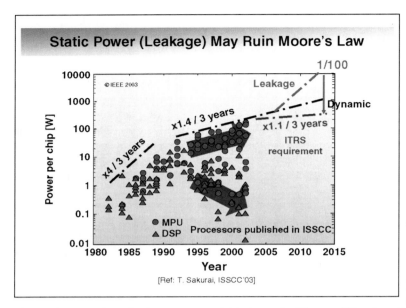

Slide 1.27

The problem was such that one was afraid that leakage might become the undoing of Moore's law. While the International Technology Roadmap for Semiconductors (ITRS) was prescribing a further slowdown in the average power dissipation (by a factor of approximately 1.1 every three years), static power dissipation potentially was registering a very rapid increase instead.

Fortunately, designers have risen to the challenge and have developed a range of techniques to keep leakage power within bounds. These will be described in detail in later chapters. Yet, static power has become a sizable fraction of the overall power budget of today's integrated circuits, and most indicators suggest that this problem will only get more severe with time.

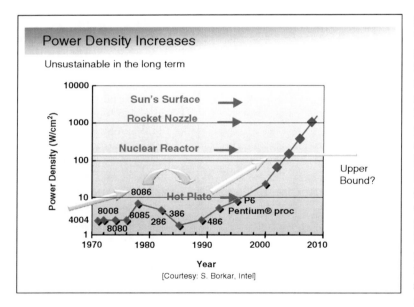

Slide 1.28

There exist very compelling reasons why a further increase in power density should be avoided at all costs. As shown in an earlier slide for the PowerPC 4, power densities on chips can become excessive and lead to degradation or failure, unless extremely expensive packaging techniques are used. To drive the point home, power density levels of some well-known processors are compared to general-world examples, such as hot plates, nuclear reactors, rocket nozzles, or even the sun's surface. Surprisingly, high-performance ICs are not that far off from some of these extreme heat sources! Classic wisdom dictates that power densities above 150 W/cm^2 should be avoided for the majority of designs, unless the highest performance is an absolute must and cost is not an issue.

Introduction 19

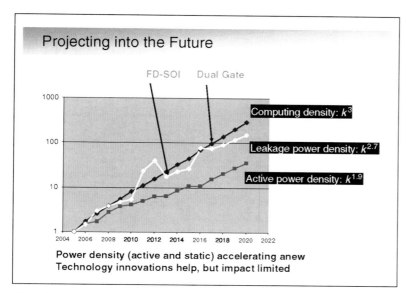

Slide 1.29

At this point, it is worth gazing into the future a bit and injecting some projections on how voltages, power, and computing densities may evolve over the coming decades. The plotted curves are based on the 2005 version of the ITRS. Obviously, unforeseen innovations in manufacturing, devices, and design technology may drastically alter the slope of some of these curves. Hence, they should be taken with a grain of salt. Yet, they help to present the dire consequences of what would happen if we do not act and identify areas where intensive research is necessary.

The first observation is that *computing density* (defined as the number of computations per unit area and time) continues to increase at a rate of k^3. This assumes that clock frequencies continue to rise linearly, which is probably doubtful considering the other trends. The *dynamic power density* is projected to accelerate anew (from $k^{0.7}$ to $k^{1.9}$). This is particularly bad news, and is mainly due to a continuing increase in clock speed combined with a slowdown in supply voltage scaling (as is plotted in the next slide). The latter is a necessity if static power dissipation is to be kept somewhat within bounds. Yet, even when accounting for a slowdown in supply- and threshold-voltage scaling, and assuming some technology and device breakthroughs such as full-depleted SOI (FD-SOI) and dual-gate transistors, static power density still grows at a rate of $k^{2.7}$. This means that leakage power if left unattended will come to dominate the power budget of most integrated circuits.

Most probably, the above scenario will not play out. Already clock frequencies of leading processors have saturated, and architectural innovations such as multi-core processing are used to maintain the expected increase in overall performance. The obtained slack can be used to reduce either dynamic or static power, or both. In addition, the heterogeneous composition of most SoCs means that different scenarios apply to various parts of the chip.

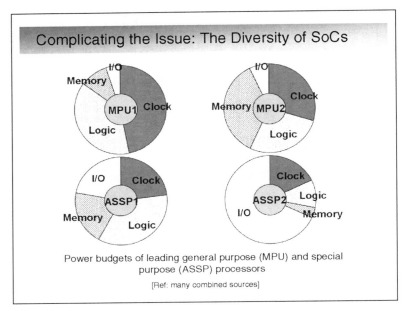

Slide 1.30
To emphasize the last argument, this slide plots the power budget of a number of microprocessors and DSPs from different companies. The distribution of power over different resources, such as computation, memory, clock, and interconnect, varies wildly. Looking forward, this trend will only accelerate. Complex SoCs for communication, media processing, and computing contain a wide variety of components with vastly different performance and activity profiles (including mixed signal, RF, and passive components). Managing the different scaling trajectories of each of these is the task of the "power management", which is the topic of Chapter 10.

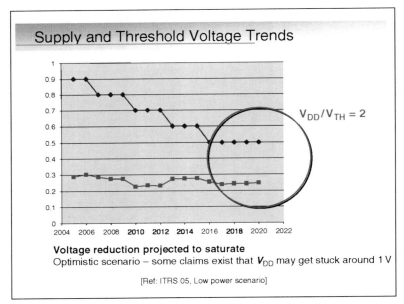

Slide 1.31
Leakage concerns put a lower bound on the threshold voltages. Barring the (improbable) event that a leakage-resistant logic family suddenly emerges, threshold voltages are unlikely to drop below 0.25 V. This severely impedes further scaling of the supply voltages. The ITRS (low-power scenario) optimistically projects that supply voltages will be reduced to 0.5 V. Getting there presents a severe challenge though. It is even doubtful whether reliable memories are feasible at all at these low voltage levels.

Innovations at the device and circuit level may come somewhat to the rescue. Transistors with higher mobility are currently researched at a number of institutions. Higher current drive means

Introduction 21

that performance can be maintained even at a low V_{DD}/V_{TH} ratio. Transistors with a sharp transition between the on and off states are another opportunity. In later chapters, we will also explore how we can design reliable and efficient circuits, even at very low voltages.

Slide 1.32

A 20 nm Scenario

Assume V_{DD} = 1.2 V
- FO4 delay < 5 ps
- Assuming no architectural changes, digital circuits could be run at 30 GHz
- Leading to power density of 20 kW/cm² (??)

Reduce V_{DD} to 0.6 V
- FO4 delay ≈ 10 ps
- The clock frequency is lowered to 10 GHz
- Power density reduces to 5 kW/cm² (still way too high)

[Ref: S. Borkar, Intel]

A simple example is often the best way to drive the arguments home. Assume a fictitious microprocessor with an architecture that is a direct transplant of current-generation processors. In a 20 nm technology, clock speeds of up to 30 GHz are theoretically plausible if the supply voltage is kept unchanged at 1.2 V. The power density however goes through the roof, even when the supply voltage is reduced to 0.6 V, and the clock frequency limited to 10 GHz.

Slide 1.33

A 20 nm Scenario (contd)

Assume optimistically that we can design FETs (Dual-Gate, FinFet, or whatever) that operate at 1 kW/cm² for FO4 = 10 ps and V_{DD} = 0.6 V [Frank, Proc. IEEE, 3/01]

- For a 2cm x 2cm high-performance microprocessor die, this means 4 kW power dissipation
- If die power has to be limited to 200 W, only 5% of these devices can be switching at any time, assuming that nothing else dissipates power.

[Ref: S. Borkar, Intel]

Let us be optimistic for a while, and assume the device innovations allow us to maintain the 10 GHz clock frequency, while reducing the power density by a factor of five. Still, a 4 cm² processor would consume 4 kW. Bringing this down to an acceptable 200 W requires that most of the devices not be switching 95% of the time, and also not leaking. A formidable challenge indeed!

This example clearly demonstrates that a drastic review of design strategies and computational architecture is necessary.

> **An Era of Power-Limited Technology Scaling**
>
> **Technology innovations offer some relief**
> – Devices that perform better at low voltage without leaking too much
>
> **But also are adding major grief**
> – Impact of increasing process variations and various failure mechanisms more pronounced in low-power design regime
>
> **Most plausible scenario**
> – Circuit- and system-level solutions essential to keep power/energy dissipation in check
> – Slow down growth in computational density and use the obtained slack to control power density increase
> – Introduce design techniques to operate circuits at nominal, not worst-case, conditions

Slide 1.34

In summary, this introductory chapter spells out the reasons why most of the innovators involved in the semiconductor industry believe that we have entered an era of power-limited scaling. This means power considerations are the primary factors determining how process, transistor, and interconnect parameters are scaled. This is a fundamental break with the past, where technology scaling was mostly guided by performance considerations. Furthermore, we do not believe that there is a "life-saving" transition – such as the one from bipolar to MOS – on its way soon. Novel devices that are currently in the lab phase hold some great promises, but only provide a limited amount of healing. In fact, the introduction of scaled devices adds an amount of suffering to the blessings (such as decreasing reliability and increasing variability). In the end, it is new design strategies and innovative computational architectures that will set the course. The main concepts underlying those will be treated in detail in the coming chapters.

> **Some Useful References ...**
>
> **Selected Keynote Presentations**
>
> - F. Boekhorst,"Ambient intelligence, the next paradigm for consumer electronics: How will it affect Silicon?," *Digest of Technical Papers ISSCC*, pp.28–31, Feb. 2002.
> - T.A.C.M. Claasen, "High speed: Not the only way to exploit the intrinsic computational power of silicon," *Digest of Technical Papers ISSCC* , pp.22–25, Feb.1999.
> - H. DeMan, "Ambient intelligence: Gigascale dreams and nanoscale realities," *Digest of Technical Papers ISSCC*, pp.29–35, Feb. 2005.
> - P.P. Gelsinger, "Microprocessors for the new millennium: Challenges, opportunities, and new frontiers," *Digest of Technical Papers ISSCC*, pp.22–25, Feb. 2001.
> - G.E. Moore, "No exponential is forever: But "Forever" can be delayed!," *Digest of Technical Papers ISSCC*, pp.20–23, Feb. 2003.
> - Y. Neuvo,"Cellular phones as embedded systems," *Digest of Technical Papers ISSCC*, pp.32–37, Feb. 2004.
> - T. Sakurai,"Perspectives on power-aware electronics," *Digest of Technical Papers ISSCC*, pp.26–29, Feb. 2003.
> - R. Yung, S.Rusu and K.Shoemaker, "Future trend of microprocessor design," *Proceedings ESSCIRC*, Sep. 2002.
>
> **Books and Book Chapters**
>
> - S. Roundy, P. Wright and J.M. Rabaey, "Energy scavenging for wireless sensor networks," Kluwer Academic Publishers, 2003.
> - F. Snijders, "Ambient Intelligence Technology: An Overview," In *Ambient Intelligence*, Ed. W. Weber et al., pp. 255–269, Springer, 2005.
> - T. Starner and J. Paradiso, "Human-Generated Power for Mobile Electronics," In *Low-Power Electronics*, Ed.C. Piguet, pp. 45-1-35, CRC Press 05.

Slide 1.35 – 1.36
Some useful references

Some Useful References (cntd)

Publications
- A. Bar-Cohen, S. Prstic, K. Yazawa and M. Iyengar. "Design and Optimization of Forced Convection Heat Sinks for Sustainable Development", Euro Conference – New and Renewable Technologies for Sustainable Development, 2000.
- S. Borkar, numerous presentations over the past decade.
- R. Chu, "The challenges of electronic cooling: Past, current and future,"*Journal of Electronic Packaging*, 126, p. 491, Dec. 2004.
- D. Frank, R. Dennard, E. Nowak, P. Solomon, Y. Taur, and P. Wong, "Device scaling limits of Si MOSFETs and their application dependencies," *Proceedings of the IEEE*, Vol 89 (3), pp. 259 –288, Mar. 2001.
- International Technology Roadmap for Semiconductors, *http://www.itrs.net/*
- J. Markoff and S. Hansell, "Hiding in Plain Sight, Google Seeks More Power", NY Times, http://www.nytimes.com/2006/06/14/technology/14search.html?r=1&oref=slogin, June 2006.
- R. Nowak, "A DARPA Perspective on Small Fuel Cells for the Military," presented at Solid State Energy Conversion Alliance (SECA) Workshop, Arlington, Mar. 2001.
- J. Rabaey et al. "PicoRadios for wireless sensor networks: the next challenge in ultra-low power design,"*Proc. 2002 IEEE ISSCC Conference*, pp. 200–201, San Francisco, Feb. 2002.
- R. Schmidt, "Power Trends in the Electronics Industry –Thermal Impacts," ACEED03, IBM Austin Conference on Energy-Efficient Design, 2003.
- Toshiba, "Toshiba Announces World's Smallest Direct Methanol Fuel Cell With Energy Output of 100 Milliwatts," http://www.toshiba.co.jp/about/press/2004_06/pr2401.htm, June 2004.

Chapter 2
Nanometer Transistors and Their Models

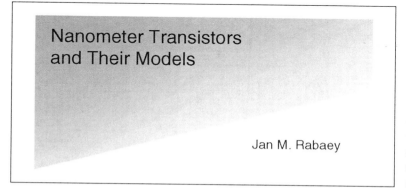

Slide 2.1

As has become apparent in Chapter 1, the behavior of the MOS transistor, when scaled into the sub-100 nm regime, is having a large impact on how and where power is consumed in the next-generation integrated circuits. Hence, any discussion on low-power design should start with a good understanding of the deep submicron MOS transistor, and an analysis of its future trends. In addition, the availability of adequate models, for both manual and computer-aided analysis, is essential. As this book emphasizes optimization, simple yet accurate models that can serve in an automated (MATLAB-style) optimization framework are introduced.

Results in this and in the coming chapters are based on the Predictive MOS models, developed by UCB and the University of Arizona, as well as industrial models spanning from 180 nm down to 45 nm. Whenever possible, MATLAB code is made available on the web site of the book.

Slide 2.2

The chapter starts with a discussion of the nanometer transistor and its behavior. Special attention is devoted to the leakage behavior of the transistor. The increasing influence of variability is analyzed next. At the end of the chapter, we evaluate some innovative devices that are emerging from the research labs and discuss their potential impact on low-power design technology.

J. Rabaey, *Low Power Design Essentials*, Series on Integrated Circuits and Systems,
DOI 10.1007/978-0-387-71713-5_2, © Springer Science+Business Media, LLC 2009

Slide 2.3

Nanometer Transistors and Their Models

- Emerging devices in the sub-100 nm regime post challenges to low-power design
  - Leakage
  - Variability
  - Reliability
- Yet also offer some opportunities
  - Increased mobility
  - Improved control (?)
- State-of-the-art low-power design should build on and exploit these properties
  - Requires clear understanding and good models

Beyond the degeneration of the on/off behavior of the MOS transistor, mentioned in Chapter 1, sub-100 nm transistors also suffer from increased variability effects, due both to manufacturing artifacts and to physical limitations. Once the feature sizes of the process technology approach the dimensions of a molecule, it is obvious that some quantum effects start to play. In addition, the reduced dimensions make the devices also prone to reliability failures such as soft errors (single-event upsets) and time-dependent degradation.

While these issues affect every MOS circuit design, their impact is more pronounced in low-power designs. Reducing power dissipation often means reducing the operational signal-to-noise margins of the circuits (for instance, by lowering the supply voltage). Effects such as variation in performance and unreliability are more apparent under these conditions. It is fair to say that today's low-power design is closely interwoven with design for variability or reliability. In this sense, low-power design often paves the way for the introduction of novel techniques that are later adopted by the general-purpose design community.

While it may seem that scaling MOS transistors down to tens of nanometers only brings bad karma, some emerging devices may actually help to reduce power density substantially in the future. Especially transistors with higher mobility, steeper sub-threshold slopes, better threshold control, and lower off-currents are attractive.

Slide 2.4

The Sub-100 nm Transistor

- **Velocity-saturated**
  - Linear dependence between I_D and V_{GS}
- **Threshold voltage V_{TH} strongly impacted by channel length L and V_{DS}**
  - Reduced threshold control through body biasing
- **Leaky**
  - Sub-threshold leakage
  - Gate leakage
- → Decreasing I_{on} over I_{off} ratio

From an operational perspective, the main characteristics of the sub-100 nm MOS transistors can be summarized as follows: a linear dependence exists between voltage and current (in the strong-inversion region); threshold is a function of channel length and operational voltages; and leakage (both sub-threshold and gate) plays a major role. Each of these issues is discussed in more detail in the following slides.

Nanometer Transistors and Their Models

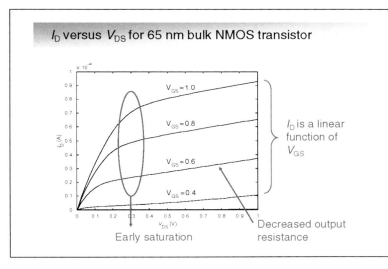

Slide 2.5

The simulated I_D versus V_{DS} behavior of a 65 nm NMOS transistor clearly demonstrates the linear relationship between I_D and V_{GS} in the saturation region. This is a result of the well-known *velocity-saturation* effect, which started to impact CMOS transistors around the 250 nm technology generation. The main impact is a reduced current drive for a given gate voltage. Of course, this means that the simple models of the past are inaccurate. To address this issue, we introduce some simplified transistor models of varying complexity and accuracy. The main goal in the context of this book is to provide the necessary tools to the circuit and systems designer to predict power and performance quickly.

Another important effect to be observed from the curves is the decrease in output resistance of the device in saturatio

Slide 2.6

Probably the most accurate model, which still allows for fast analysis and requires only a restricted set of parameters, was introduced by Taur and Ning in 1998. One important parameter in this model is the *critical electrical field* E_C, which determines the onset of velocity saturation. The problem with this model is its highly non-linear nature, which makes it hard to use in optimization programs (and hand analysis); for instance, E_C itself is a function of V_{GS}. Hence, some further simplification is desirable.

Drain Current Under Velocity Saturation

$$I_{DSat} = v_{Sat} W C_{ox} \frac{(V_{GS} - V_{TH})^2}{(V_{GS} - V_{TH}) + E_C L}$$

- Good model, could be used in hand or MATLAB analysis

$$I_{DSat} = \frac{W}{L} \frac{\mu_{eff} C_{ox}}{2} V_{DSat} (V_{GS} - V_{TH})$$

with $\quad V_{DSat} = \dfrac{(V_{GS} - V_{TH}) E_C L}{(V_{GS} - V_{TH}) + E_C L}$

[Ref: Taur-Ning, '98]

Slide 2.7

The "unified model" of the MOS transistor was introduced in [Rabaey03]. A single non-linear equation suffices to describe the transistor in the saturation and linear regions. The main simplification in this model is the assumption that velocity saturation occurs at a fixed voltage V_{DSat}, independent of the value of V_{GS}. The main advantages of the model are its elegance and simplicity. A total of only five parameters are needed to describe the transistor: k', V_{TH}, V_{DSat}, λ and γ. Each of these can be empirically derived using curve-fitting with respect to the actual device plots. Observe that these parameters are *purely empirical*, and have no or little relation to traditional physical device parameters such as the channel-length modulation λ.

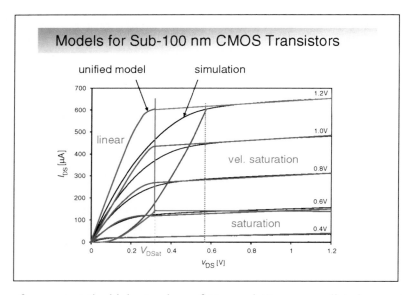

Slide 2.8

Simplicity comes at a cost however. Comparing the I–V curves produced by the model to those of the actual devices (BSIM-4 SPICE model), a large discrepancy can be observed for intermediate values of V_{DS} (around V_{DSat}). When using the model for the derivation of propagation delays (performance) of a CMOS gate, accuracy in this section of the overall operation region is not that crucial. What is most important is that the values of current at the highest values of V_{DS} and V_{GS} are predicted correctly – as these predominantly determine the charge and discharge times of the output capacitor. Hence, the propagation delay error is only a couple of percents, which is only a small penalty for a major reduction in model complexity.

Nanometer Transistors and Their Models

Alpha Power Law Model

- Alternate approach, useful for hand analysis of propagation delay

$$I_{DS} = \frac{W}{2L}\mu C_{ox}(V_{GS} - V_{TH})^\alpha$$

- Parameter α is between 1 and 2.
- In 65–180 nm CMOS technology $\alpha \sim 1.2$–1.3

- This is not a physical model
- Simply empirical:
  - Can fit (in minimum mean squares sense) to a variety of α's, V_{TH}
  - Need to find one with minimum square error – fitted V_{TH} can be different from physical

[Ref: Sakurai, JSSC'90]

Slide 2.9
Even simpler is the *alpha model*, introduced by Sakurai and Newton in 1990, which does not even attempt to approximate the actual I–V curves. The values of α and V_{TH} are purely empirical, chosen such that the propagation delay of a digital gate, approximated by $t_p = \frac{kV_{DD}}{(V_{DD}-V_{TH})^\alpha}$, best resembles the propagation delay curves obtained from simulation. Typically, curve-fitting techniques such as the minimum-mean square (MMS) are used. Be aware that these do not yield unique solutions and that it is up to the modeler to find the ones with the best fit.

Owing to its simplicity, the alpha model is the corner stone of the optimization framework discussed in later chapters.

Output Resistance

- Drain current keeps increasing beyond the saturation point
- Slope in I–V characteristics caused by:
  - Channel-length modulation (CLM)
  - Drain-induced barrier lowering (DIBL).
- The simulations show approximately linear dependence of I_{DS} on V_{DS} in saturation (modeled by λ factor)

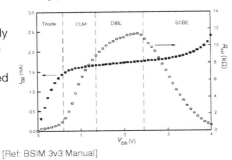

[Ref: BSIM 3v3 Manual]

Slide 2.10
Beyond velocity saturation, reducing the transistor dimensions also lowers the output resistance of the device in saturation. This translates into reduced noise margins for digital gates. Two principles are underlying this phenomenon: (1) channel-length modulation – which was also present in long-channel devices – and (2) drain-induced barrier lowering (DIBL). The latter is a deep-submicron effect and is related to a reduction of the threshold voltage as a function of the drain voltage. DIBL primarily impacts leakage (as discussed later), yet its effect on output resistance is quite sizable as well. SCBE (Substrate Current Body Effect) only kicks in at voltages higher than the typical operation regime, and its impact is hence not that important.

Fortunately, the relationship between drain voltage and current proves to be approximately linear, and is adequately modeled with a single parameter λ.

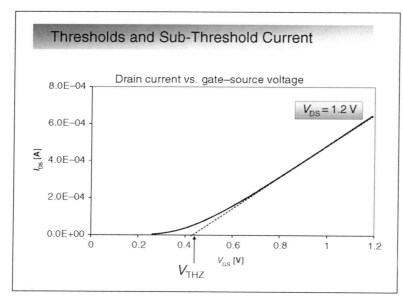

Slide 2.11
With the continuing reduction of the supply voltages, scaling of the threshold voltage is a necessity as well, as illustrated at length in Chapter 1. Defining the actual threshold voltage of a transistor is not simple, as many factors play and measurements may not be that straightforward. The "physics" definition of the threshold voltage is the value of V_{GS} that causes strong inversion to occur underneath the gate. This is however impossible to measure. An often-used empirical approach is to derive V_{TH} from the I_D–V_{GS} plot by linearly extrapolating the current in the saturation region (see plot). The cross-point with the zero-axis is then defined as V_{TH} (also called V_{THZ}). Another approach is the "constant-current" (CC) technique, which defines the threshold voltage as the point where the drain–source current drops below a fixed value (I_{D0}), scaled appropriately with respect to the (W/L) ratio of the transistor. The choice of I_{D0} is however quite arbitrary. Hence, in this book we use the extrapolation technique, unless otherwise mentioned.

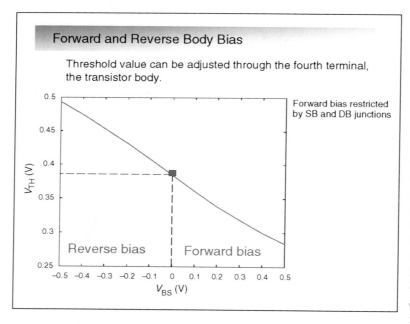

Slide 2.12
The threshold voltage is unfortunately not a constant parameter, but is influenced by a number of operational parameters. The foremost is the body-bias or back-bias effect, where the fourth terminal of the transistor (the bulk or well voltage) serves as an extra control knob. The relationship between V_{TH} and V_{SB} is well-known, and requires the introduction of one extra device parameter, the *body-effect parameter* γ. Observe that body-biasing can be used either to increase (reverse bias) or to decrease (forward bias) the threshold voltage. The forward-biasing effect is limited in its scope, as the

source–bulk diode must remain in reverse-bias conditions (that is $V_{SB} > -0.6$ V). If not, current is directly injected into the body from the source, effectively killing the gain of the transistor. For the 130 nm technology, a 1 V change in V_{SB} changes the threshold voltage by approximately 200 mV.

The beauty of the body-biasing effect is that it allows for a dynamic adjustment of the threshold voltage during operation, thus allowing for the compensation of process variations or a dynamic trade-off between performance and leakage.

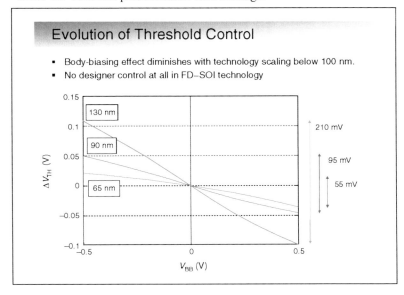

Slide 2.13
Regrettably, scaling of device technology is gradually eroding the body-biasing effect. With the doping levels in the channel increasing, changes in the bias voltage have little effect on the onset of strong inversion. This is clearly illustrated in this slide, which plots the impact of body biasing for three technology nodes.

Emerging technologies, such as fully-depleted SOI (in which the body of the transistor is floating), even do away completely with the body biasing. This development is quite unfortunate, as this takes away one of the few parameters a designer can use to actively control leakage effects.

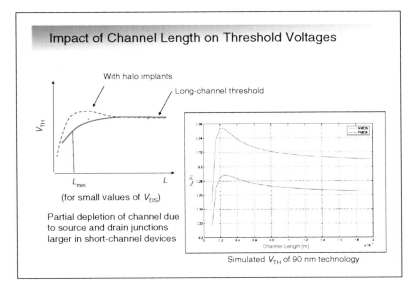

Slide 2.14
Channel length is another parameter that influences the threshold voltage. For very short channels, the depletion regions of the drain (and source) junctions themselves deplete a sizable fraction of the channel. Turning the transistor on becomes easier, thus causing a reduction in the threshold voltage. To offset this effect, device engineers add some extra "halo implants", which cause the threshold to peak around the nominal value of the channel length. While this is beneficial in general, it also increases the sensitivity of the threshold

voltage with respect to channel-length variations. For instance, it may happen that the channel lengths of a particular wafer batch are consistently below the nominal value. This causes the thresholds to be substantially below the expected value, leading to faster, but leaky, chips.

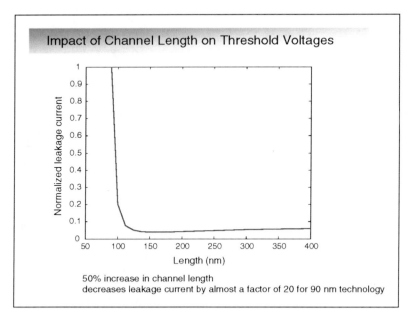

Slide 2.15
Designers vying for large threshold values with relatively small variations often size their transistors above the nominal channel length. This obviously comes at a penalty in area. The impact can be quite substantial. In a 90 nm technology, leakage currents can be reduced by an order of magnitude by staying away from the minimum channel lengths. Just a 10% increase already reaps major benefits. This observation has not escaped the attention of designers of leakage-sensitive modules, such as SRAM memories.

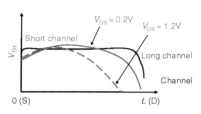

Slide 2.16
Drain voltage is another variable that has a sizable impact on the threshold voltage. The DIBL effect was already mentioned in the context of the output resistance of short-channel devices. As the drain voltage increases, the depletion region of the junction between the drain and the channel increases in size and extends under the gate, effectively lowering the threshold voltage (this is a hugely simplified explanation, but it catches the main tenet). The most negative feature of DIBL effect is that it turns the threshold voltage into a signal-dependent variable. For all practical purposes, it is fair to assume that V_{DS} changes the threshold in a linear fashion, with λ_d being the proportionality factor.

Nanometer Transistors and Their Models

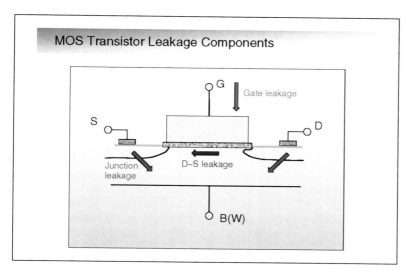

Slide 2.17

Quite a number of times in the introduction, we have alluded to the increasing effects of "leakage" currents in the nanometer MOS transistor. An ideal MOS transistor (at least from a digital perspective) should not have any currents flowing into the bulk (or well), should not conduct any current between drain and source when off, and should have an infinite gate resistance. As indicated in the accompanying slide, a number of effects are causing the contemporary devices to digress from this ideal model.

Leakage currents, flowing through the reverse-biased source–bulk and drain–bulk *pn* junctions, have always been present. Yet, the levels are so small that their effects could generally be ignored, except in circuitry that relies on charge storage such as DRAMs and dynamic logic.

The scaling of the minimum feature sizes has introduced some other leakage effects that are far more influential and exceed junction leakage currents by 3–5 orders of magnitude.

Most important are the sub-threshold drain–source and the gate leakage effects, which we will discuss in more detail.

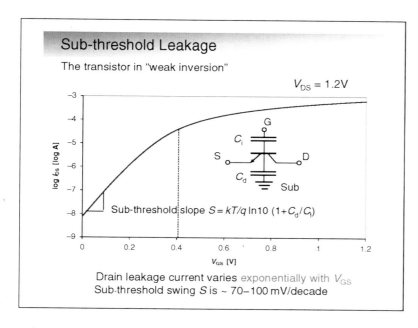

Slide 2.18

In earlier slides, we have alluded to a relationship between the value of the threshold voltage V_{TH} and (sub-threshold) leakage. When the gate voltage of a transistor is lowered below the threshold voltage, the transistor does not turn off instantaneously. In fact, the transistor enters the so-called "sub-threshold regime" (or weak inversion). In this operation mode, the drain–source current becomes an exponential function of V_{GS}. This is clearly observed from the I_D–V_{GS} curves, if the current is plotted on a logarithmic scale.

The exponential dependence is best explained by the fact that under these conditions the MOS transistor behaves as a bipolar device (*npn* for an NMOS) with its base coupled to the gate through a capacitive divider. We know that for an ideal bipolar transistor, the base current relates to the base–emitter voltage as $I_{CE} = e^{\frac{V_{BE}}{(kT/q)}}$ where k is the Boltzmann constant and T the absolute temperature. The so-called *thermal voltage* (kT/q) equals approximately 25 mV at room temperature. For an ideal bipolar transistor, every increase in V_{BE} by 60 mV [= 25 mV x ln(10)] increases the collector current by a factor of 10!

In the weak inversion mode of the MOS transistor, the exponential is somewhat deteriorated by the capacitive coupling between gate and channel (base). Hence, the sub-threshold current is best modeled as $I_{DS} = e^{\frac{V_{GS}}{n(kT/q)}}$ where n is the slope factor ranging around 1.4–1.5 for modern technologies. The net effect of this degradation is that, for the current to drop by one order of magnitude in the sub-threshold region, the reduction in V_{GS} needed is not of 60 mV, but more like 70–100 mV. Obviously, for an ideal switch we would hope that the current drops immediately to zero when V_{GS} is lowered below V_{TH}.

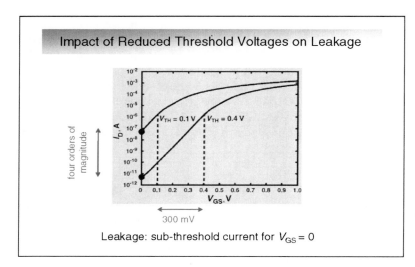

Slide 2.19

The growing importance of sub-threshold leakage can now be understood. If the threshold voltage is set, for example, at 400 mV, the leakage current drops by five orders of magnitude between $V_{GS} = V_{TH}$ and $V_{GS} = 0$ (assuming a sub-threshold swing of approximately 80 mV/decade). Assume now that the threshold voltage is scaled to 100 mV to maintain performance under reduced supply voltage conditions. The leakage current at $V_{GS} = 0$ for this low-threshold transistor will be approximately four orders of magnitude higher than that for the high-threshold device, or the *leakage current goes up exponentially with a linear reduction in threshold voltage*. This serves as another example of the impact of exponential relations.

Slide 2.20

> **Sub-threshold Current**
>
> - Sub-threshold behavior can be modeled physically
>
> $$I_{DS} = 2n\mu C_{ox} \frac{W}{L} \left(\frac{kT}{q}\right)^2 e^{\frac{V_{GS}-V_{TH}}{nkT/q}} \left(1 - e^{\frac{-V_{DS}}{kT/q}}\right) = I_S e^{\frac{V_{GS}-V_{TH}}{nkT/q}} \left(1 - e^{\frac{-V_{DS}}{kT/q}}\right)$$
>
> where n is the *slope factor* (≥ 1, typically around 1.5) and $I_S = 2n\mu C_{ox} \frac{W}{L} \left(\frac{kT}{q}\right)^2$
>
> - Very often expressed in base 10
>
> $$I_{DS} = I_S \, 10^{\frac{V_{GS}-V_{TH}}{S}} \left(1 - 10^{\frac{-nV_{DS}}{S}}\right) \quad \approx 1 \text{ for } V_{DS} > 100 \text{ mV}$$
>
> where $S = n \left(\frac{kT}{q}\right) \ln(10)$, the *sub-threshold swing*, ranging between 60 mV and 100 mV

Since sub-threshold leakage is playing such a dominant role in the nanometer design regime, it is quite essential to have good models available. One of the (few) advantages of the sub-threshold operational regime is that physical modeling is quite possible, and that the basic expressions of the drain current as a function of V_{GS} or V_{DS} can be easily derived.

Slide 2.21

> **Sub-threshold Current – Revisited**
>
> - **Drain-Induced Barrier Lowering (DIBL)**
>   - Threshold reduces approximately linearly with V_{DS}
>
> $$V_{TH} = V_{TH0} - \lambda_d V_{DS}$$
>
> - **Body-Biasing Effect**
>   - Threshold reduces approximately linearly with V_{BS}
>
> $$V_{TH} = V_{TH0} - \gamma_d V_{BS}$$
>
> Leading to:
>
> $$I_{DS} = I_S \, 10^{\frac{V_{GS}-V_{TH0}+\lambda_d V_{DS}+\gamma_d V_{BS}}{S}} \left(1 - 10^{\frac{-nV_{DS}}{S}}\right)$$
>
> Leakage is an exponential function of drain and bulk voltages

The simple model of the previous slide does not cover two effects that dynamically modulate the threshold voltage of the transistor: DIBL and body biasing. While these effects influence the strong-inversion operational mode of the transistor (as discussed earlier), their impact is felt far more in the sub-threshold mode owing to the exponential relation between drain current and threshold voltage. The current model is easily adjusted to include these effects with the addition of two parameters: λ_d and γ_d.

Slide 2.22

Especially DIBL turns out to have a huge impact on the sub-threshold leakage of the nanometer CMOS transistor. Assume, for instance, an NMOS transistor in the off-mode ($V_{GS} = 0$). The sub-threshold current of the transistor is now strongly dependent on the applied V_{DS}. For instance, for the device characteristics shown in the slide, raising V_{DS} from 0.1 V to 1 V increases the leakage current by a factor of 10 (while in an ideal device it should stay approximately flat). This creates

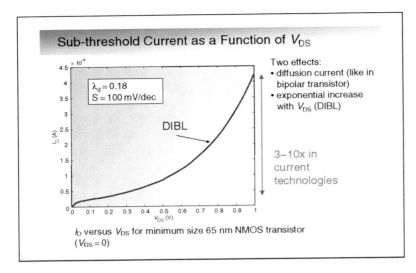

both a challenge and an opportunity, as it means that leakage becomes strongly data-dependent. Leaving this unchecked may lead to substantial problems. At the same time, it offers the innovative designer an extra parameter to play with.

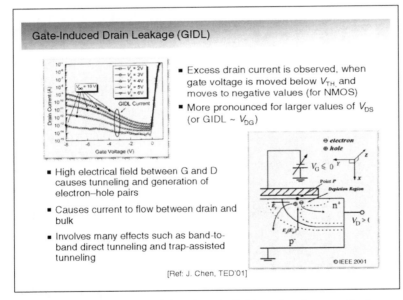

Slide 2.23

In addition, the current flowing through the drain in the off-state is influenced by the "gate-induced drain leakage" (GIDL) effect. While one would expect the drain current to drop continuously when reducing V_G below V_{TH} for a given drain voltage V_D, the inverse is actually true. Especially at negative values of V_G, an increase in drain current is observed. This is the result of a combination of effects such as band-to-band tunneling and trap-assisted tunneling. A high value of the electric field under the gate/drain overlap region [as occurring for low values of V_G (0 V or lower) and high V_D] causes deep depletion and an effective thinning of the depletion width of the drain–well junction. This effectively leads to electron–hole pair creation and an accompanying drain-to-bulk current. The effect is proportional to the applied value of V_{DG}. The impact of GIDL is mostly felt in the off-state of the transistor with $V_{GS} = 0$. The upward bending of the drain current curve causes an effective increase of the leakage current.

It should be noted that the GIDL effect is substantially larger in NMOS than in PMOS transistors (by about two orders of magnitude). Also observe that the impact of GIDL is quite small for typical supply voltages, which are at 1.2 V or lower.

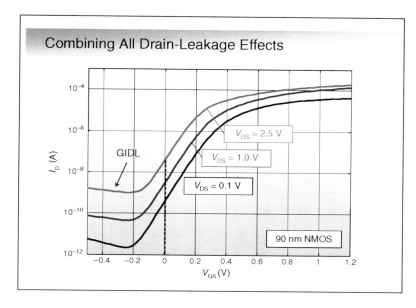

Slide 2.24

The combined effect of all drain leakage components is best illustrated by plotting I_D versus V_{GS} for different values of V_{DS} (as shown in the slide for a 90 nm NMOS transistor). Most important from a leakage perspective is the current at $V_{GS} = 0\,\text{V}$. For low values of V_{DS}, the drain current is set by the sub-threshold current for the nominal V_{TH} (as well as the drain–well junction leakage current, which is ignorable). When raising V_{DS}, DIBL reduces V_{TH} and causes a substantial increase in leakage current. For instance, increasing V_{DS} from 0.1 to 1.0 V causes the drain current to increase by a factor of almost 8.

The GIDL effect can clearly be observed for values of V_{GS} smaller than -0.1 V. However, even for V_{DS} at a very high value of 2.5 V, the impact at $V_{GS} = 0$ is still ignorable. GIDL hence plays a minor role in most of today's designs.

It is worth contemplating the overall picture that emerges from this. For a minimum-sized device in a low-leakage technology with a V_{TH} around 0.35 V, the drain leakage hovers around 1 nA at room temperature. This amounts to a total leakage current of approximately 0.1 A for a design with a hundred million gates (or equivalent functions). This value increases substantially at higher temperatures (which is the standard operating condition), increases linearly with the device width, and rises exponentially with a reduction in threshold voltage. Designs with standby leakage currents of multiple Amperes are hence very plausible and real, unless care is taken to stop the bleeding.

Slide 2.25

While sub-threshold currents became an issue with the introduction of the 180 nm technology node, another leakage effect is gaining importance once technology scales below the 100 nm level – that is, *gate leakage*. One of the attractive properties of the MOS transistor has always been its very high (if not infinite) input resistance. In contrast, a finite base current is inherent to the structure of the bipolar transistor, making the device unattractive for usage in complex digital designs.

To maintain the current drive of the transistor while scaling its horizontal dimensions, general scaling theory prescribes that the gate oxide (SiO_2) thickness is scaled as well. Once however the oxide thickness becomes of the order of just a few molecules, some significant obstacles emerge. And this is exactly what happens with the sub-100 nm transistors, as is illustrated by the cross-section SEM picture of a 65 nm MOS transistor with an oxide thickness of 1.2 nm. It is clear that the oxide is barely a couple of molecules thick.

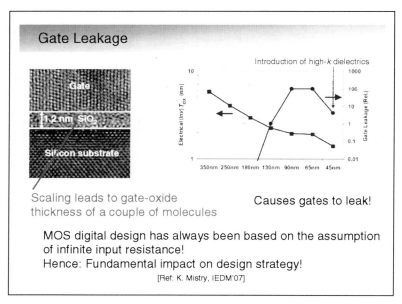

While posing some obvious limits on scaling, the very thin oxides also cause a reduction in the gate resistance of the transistor, as current starts to leak through the dielectric. This trend is clearly illustrated in the chart, which shows the evolution of the gate thickness and the gate leakage over various technology generations at Intel. From 180 nm to 90 nm, the gate leakage current increased by more than four orders of magnitude. The reasons behind the leveling and subsequent drop in subsequent generations will become clear in the following slides. Observe also that gate leakage increases strongly with temperature.

Unlike sub-threshold currents, which primarily cause an increase in standby power, gate currents threaten some fundamental concepts used in the design of MOS digital circuits.

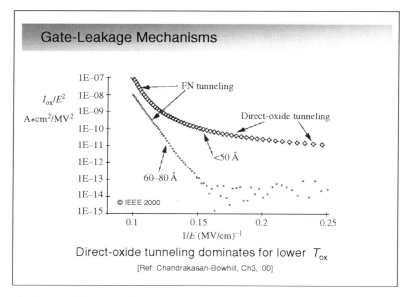

Slide 2.26
Gate leakage finds its source in two different mechanisms: Fowler–Nordheim (FN) tunneling, and direct-oxide tunneling. FN tunneling is an effect that has been effectively used in the design of non-volatile memories, and is already quite substantial for oxide thickness larger than 6 nm. Its onset requires high electric-field strengths, though. With reducing oxide thicknesses, tunneling starts to occur at far lower field strengths. The dominant effect under these conditions is direct-oxide tunneling.

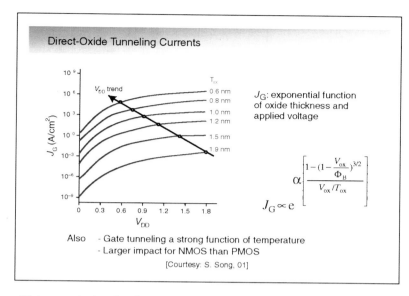

Slide 2.27

In this slide, the dependence of the direct-oxide tunneling current is plotted as a function of the applied voltage and the SiO_2 thickness. The leakage current is shown to vary exponentially with respect to both of these parameters. Hence, even though we are scaling the supply voltages with successive process generations, the simultaneous scaling of the oxide thickness causes the gate leakage current density to continuously increase. This trend clearly threatens the further scaling of MOS technology, unless some innovative process technology solutions emerge.

A first approach to address the challenge is to stop or slow down the scaling of the oxide thickness, while continuing the scaling of the other critical device dimensions. This negatively impacts the obtainable current density and reduces the performance benefit that typically comes with technology scaling. Yet, even considering these negative implications, this is exactly what most semiconductor companies did when moving to the 65 nm node (as is apparent in Slide 2.25). This should however be considered a temporary therapy, accompanying the mastering of some quite substantial device innovations such as high-k gate dielectrics and high-mobility transistors.

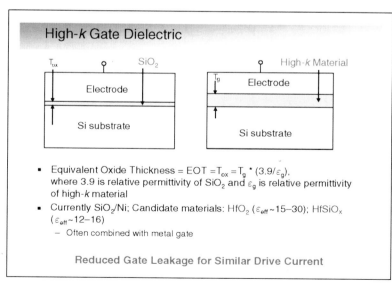

Slide 2.28

The MOS transistor current is proportional to the *process transconductance* parameter $k' = \mu C_g = \mu \varepsilon / t_g$. To increase k' through scaling, one must either find a way to increase the mobility of the carriers or increase the gate capacitance (per unit area). The former requires a fundamental change in the device structure (to be discussed later). With the traditional way of increasing the gate capacitance (i.e., scaling T_g) running out of steam, the only remaining option is to look for gate dielectrics with a higher permittivity ε – the so-called high-k

dielectrics. Replacing SiO$_2$ with a "high-k" material yields the same effect as scaling the thickness, while keeping gate leakage under control.

Device technologists have introduced a metric to measure the effectiveness of novel dielectrics: the "*equivalent oxide thickness*" or EOT, which equals $T_g \times (\varepsilon_{ox}/\varepsilon_g)$.

Introducing new gate materials is however not a trivial process change, and requires a complete redesign of the gate stack. In fact, most dielectric materials under consideration today require a metal gate electrode, replacing the traditional polysilicon gate. Incorporating major changes of this type into a high-yield manufacturing process takes time and substantial investments. This explains why the introduction of high-k dielectrics into production processes was postponed a number of times. Major semiconductor companies such as IBM and Intel have now adopted hafnium oxide (HfO$_2$) as the dielectric material of choice for their 45 nm and 32 nm CMOS processes in combination with a metal gate electrode. The relative permittivity of HfO$_2$ equals 15–30, compared to 3.9 for SiO$_2$. This is equivalent to between two and three generations of technology scaling, and should help to address gate leakage at least for a while. The resulting drop in gate leakage current for the Intel 45 nm processor is apparent in the chart on Slide 2.25.

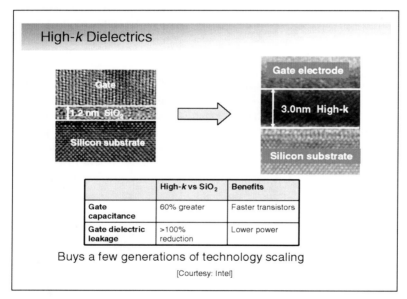

Slide 2.29
The advantages offered by high-k gate dielectrics are quite clear: faster transistors and/or reduced gate leakage.

Slide 2.30

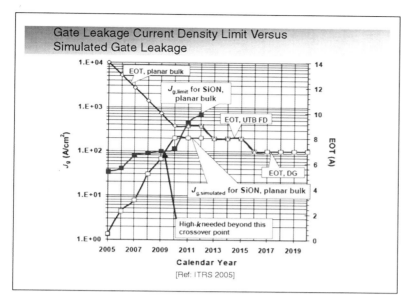

The expected evolution of gate leakage and gate materials is best summarized by this chart, extracted from the International Technology Roadmap on Semiconductors (2005). By analyzing the maximum allowable leakage current density (obviously, this number is disputable – what is allowable depends upon the application domain), it is concluded that the step to high-k dielectrics is necessary by around 2009 (the 45 nm technology node). Combined with some other device innovations such as FD-SOI and dual-gate (more about these later in this Chapter), this may allow for the EOT to scale to around 0.7 nm (!), while keeping the gate leakage current density at approximately $100\,\text{A/cm}^2$ (or $1\,\mu\text{A}/\mu\text{m}^2$).

Slide 2.31

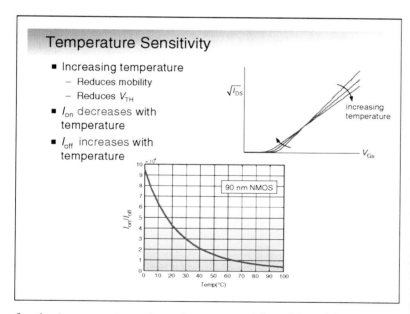

The influence of temperature on the leakage behavior of a transistor was mentioned a number of times before. In general, it can be assumed that the on-current of a transistor reduces (slightly) with an increase in temperature. The decrease in threshold voltage is not sufficient to offset the decrease in carrier mobility. The threshold reduction on the other hand has an exponential impact on the leakage current. Hence, higher temperatures are detrimental for the I_{on} versus I_{off} ratio as demonstrated for a 90 nm NMOS transistor. Increasing the temperature from 0 to 100°C reduces the ratio by almost 25. This is mostly due to the increase in leakage current (by a factor of 22), but also to slight decrease in on-current (10%).

Variability

- Scaled device dimensions leading to increased impact of variations
  - Device physics
  - Manufacturing
  - Temporal and environmental
- Impacts performance, power (mostly leakage) and manufacturing yield
- More pronounced in low-power design due to reduced supply/threshold voltage ratios

Slide 2.32
The topic of variability rounds out the discussion of the nanometer transistor and its properties. It has always been the case that transistor parameters such as the geometric dimensions or the threshold voltage are not deterministic. When sampled between wafers, within a wafer, or even over a die, each of these parameters exhibits a statistical nature. In the past, the projection of the parameter distributions onto the performance space yielded quite a narrow distribution. This is easily understandable. When the supply voltage is 3 V and the threshold is at 0.5 V, a 25 mV variation in the threshold has only a small impact on the performance and leakage of the digital module. However, when the supply voltage is at 1 V and the threshold at 0.3 V, the same variation has a much larger impact.

So, in past generation processors it was sufficient to evaluate a design over its worst-case corners (FF, SS, FS, SF) in addition to the nominal operation point to determine the yield distributions. Today, this is not sufficient, as the performance distributions have become much wider, and a pure worst-case analysis leads to wasteful design and does not give a good yield perspective either.

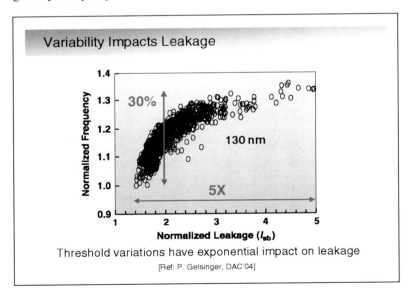

Threshold variations have exponential impact on leakage
[Ref: P. Gelsinger, DAC'04]

Slide 2.33
While variations influence the high-performance design regime, their impact is far more pronounced in the low-power design arena. First of all, the prediction of leakage currents becomes hard. The sub-threshold current is an exponential function of the threshold voltage, and each variation in the latter is amplified in a major way in leakage fluctuations. This is illustrated very well in the performance–leakage distribution plot (for 130 nm technology). When sampled over a large number of dies (and wafers), gate performance varies over 30%, while the leakage current fluctuates by a factor of 5. Observe that the leakiest designs are also the ones with the highest performance (this should be no surprise).

Other reasons why variations play a more pronounced role in low-power design will emerge in the subsequent chapters. However, they can be summarized in the following observation: In general, low-power designs operate at lower supply voltages, lower V_{DD}/V_{TH}, and smaller signal-to-noise ratios; these conditions tend to amplify the importance of parameter variations.

Variability Sources and Their Time Scales

- **Physical**
  - Changes in characteristics of devices and wires.
  - Caused by IC manufacturing process, device physics & wear-out (electro-migration).
  - Time scale: 10^9 s (years).
- **Environmental**
  - Changes in operational conditions (modes), V_{DD}, temperature, local coupling.
  - Caused by the specifics of the design implementation.
  - Time scale: 10^{-6} to 10^{-9} s (clock tick).

Slide 2.34
Process variations are not the only cause behind the variability in the performance parameters of a design (such as delay and power dissipation). It actually originates from a broad set of causes with very different temporal characteristics. In a broad sense, we can classify them into physical, manufacturing, environmental, and operational categories. It is probably fair to state that manufacturing variations – that is, fluctuations in device and interconnect parameters caused by the manufacturing process – are dominant in today's designs. However, with device dimensions approaching the molecular scale, statistical quantum-mechanical effects start to play a role, as the "law of large numbers" starts to be less applicable. Environmental and operational conditions are closely related. While operating a circuit, some "external" parameters such as the supply voltage, the operating temperature, and the coupling capacitance may change dynamically as a result of environmental conditions or the activity profile of the design.

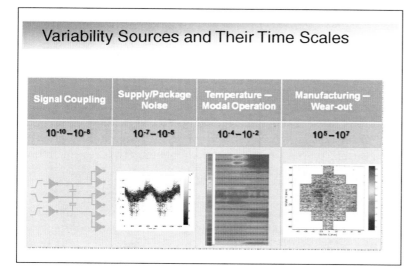

Slide 2.35
When trying to create design solutions to address the variability concerns, it pays to understand the nature and the behavior of the sources of variation, as these will ultimately determine what design techniques can be effective in eliminating or reducing the impact. The most important statistical parameters of concern are the temporal and spatial correlations. If a parameter has a strong spatial correlation (i.e., all

devices in the neighborhood show the same trend), a solution such as global tuning proves to be effective. The same is true in the time domain. Very strong temporal correlations (i.e., a device parameter is totally predictable or may not even change over time) can again be addressed by one-time or slow adaptation.

In this slide, we have classified the different sources of variations from a temporal perspective. At the slow extreme of the spectrum are manufacturing variations, which last for the lifetime of the product. Almost similar from a lifetime perspective, but entirely different in nature, are variations caused by wear-out, which manifest themselves only after a very long time of operation (typically years). Examples of such sources are electro-migration, hot-electron degradation, and negative-bias temperature instability (NBTI). Next on the time scale are slow operational or environmental conditions. The temperature gradients on a die vary slowly (in the range of milliseconds), and changes are typically the result of alterations in the operation mode of the system. An example of such is putting a module to sleep or standby mode after a time of intensive computation. Other variations happen at a much faster time scale such as the clock period or even a single signal transition. Their very dynamic nature does not leave room for adaptive cancellation, and circuit techniques such as shielding are the only way to eliminate their impact.

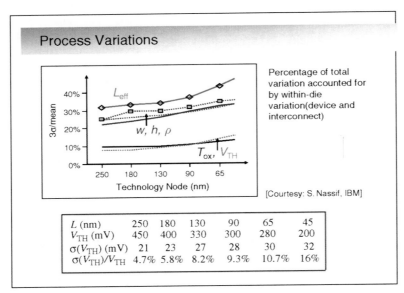

Slide 2.36
Process and manufacturing variations are probably of the most concern. The evolutionary trend is clear: virtually all technology parameters such as transistor length, width, oxide thickness, and interconnect resistivity show an increasing variability over time (as measured by the ratio of standard deviation over the mean value). Although each of these parameters is important on its own, the resulting impact on the threshold voltage is what counts most from a digital-design perspective. As shown in the table, the threshold variability is rising from 4% to 16% while evolving from 250 nm to 45 nm CMOS technologies. One may assume that this variation primarily results from the increasing deviations in channel length, since the V_{TH} is quite sensitive to variations in L around the critical dimension (remember the halo implants). The resulting impact on both performance and power metrics is quite substantial.

Slide 2.37
Since the lengths of neighboring transistors tend to be similarly affected by deviations in the manufacturing process, one would assume that the threshold voltages of closely spaced transistors should be strongly correlated. This conclusion holds especially for > 100 nm technology nodes, where strong systematic trends in thresholds of local neighborhoods can be observed.

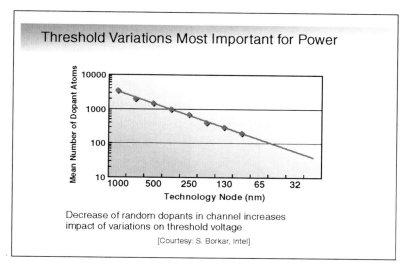

Decrease of random dopants in channel increases impact of variations on threshold voltage

[Courtesy: S. Borkar, Intel]

However, the observation becomes less true with continued scaling, when deviations in another device parameter, channel doping, start to become an issue. As shown in the graph, the number of dopant atoms, which is a discrete number, drops below 100 for transistor dimensions smaller than 100 nm. The exact number of dopants in the channel is a random variable, and can change from transistor to transistor. We may hence expect that the correlation in threshold voltages between neighboring transistors will reduce substantially in future technology generations.

The main takeaway from this discussion on process variations is that most device and design parameters will feature broader distributions over time, and that this is primarily caused by variations in the transistor threshold. While these variations tend to be mostly systematic today, we may expect larger random components in the future.

Device and Technology Innovations

- Power challenges introduced by nanometer MOS transistors can be partially addressed by new device structures and better materials
  - Higher mobility
  - Reduced leakage
  - Better control
- However ...
  - Most of these techniques provide only a one (or two) technology generation boost
  - Need to be accompanied by circuit and system level methodologies

Slide 2.38

One may wonder whether these many and profound challenges may make design in the nanometer regime all but impossible. This is a very valid question indeed, which has kept many semiconductor company executives awake at night over the past years.

While reflecting, it pays to keep the following considerations in mind. Over the years, designers have proven to be quite ingenious, and they have come up over and over again with new design technologies and methodologies to address emerging challenges and roadblocks. We can be confident that this will continue to happen in the future (this is what this book is about, after all). At the same time, device engineers are not sitting still either. On the drawing board are a number of device structures that may help to address some, if not all, of the concerns raised in this chapter. For a designer, it is important to be aware of what may be coming down the device pipeline and plan accordingly.

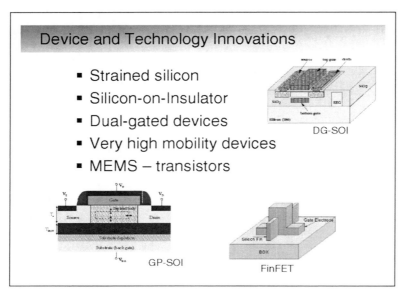

Slide 2.39

The devices introduced in the coming slides present any of the following features: higher mobility, better threshold control, or faster sub-threshold current roll-off.

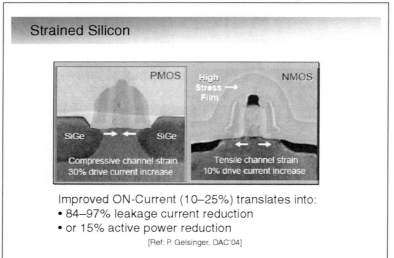

Slide 2.40

The concept of strained silicon was introduced by IBM to increase the mobility in traditional CMOS transistors. From the 65 nm generation onward, it is used almost universally by all semiconductor manufacturers. The generic idea is to create a layer of silicon (typically in the transistor channel), in which the silicon atoms are stretched (or strained) beyond their normal inter-atomic distance.

A generic way to create strain is to put a layer of silicon over a substrate of silicon germanium (SiGe). As the atoms in the silicon layer align with the atoms in the silicon–germanium layer, where the atoms are further apart, the links between the silicon atoms become stretched – thereby leading to strained silicon. Moving the atoms further apart reduces the atomic forces that interfere with the movement of electrons through the transistors, resulting in higher mobility.

The practical realization may differ between manufacturers. The slide illustrates one strategy, as employed by Intel. To stretch the silicon lattice, Intel deposits a film of silicon nitride over the whole transistor at a high temperature. Because silicon nitride contracts less than silicon as it cools, it locks the silicon lattice beneath it in place with a wider spacing than it would normally adopt. This improves electron conduction by 10%. For PMOS transistors, the silicon is compressed. This is accomplished by carving trenches along opposite ends of the channel. These are filled with silicon germanium, which has a larger lattice size than silicon alone and so compresses the regions nearby. This improves hole conduction by 25%.

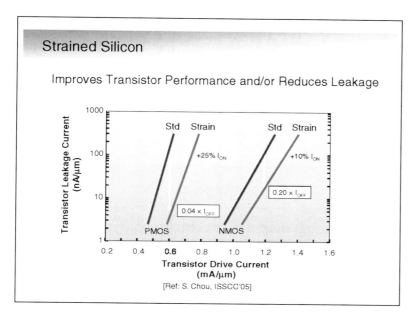

Slide 2.41
The higher mobility may be used to increase the performance. From a power perspective, a better approach is to use the higher mobility to obtain the same performance with either a higher threshold voltage (reducing leakage), or with a reduced V_{DD}/V_{TH} ratio, as is illustrated in this slide.

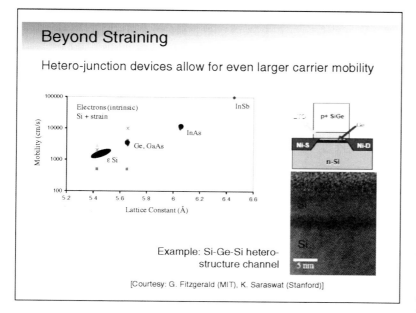

Slide 2.42
Straining is only one first step toward higher mobility. Materials such as Ge and GaAs are known to have an intrinsic electron mobility that is substantially above what Si can offer. Researchers at various locations are exploring the potential of so-called hetero-devices that combine Si with other materials such as Ge, offering the potential of carriers that are 10 times as mobile, while still relying on traditional Si technology. An example of such a device is the Si-Ge-Si heterostructure developed at Stanford (this is only one example of the many structures being investigated). While these high-mobility devices will need quite some time before making it to the production line (if ever), they offer a clear glimpse at the potential for further improvement.

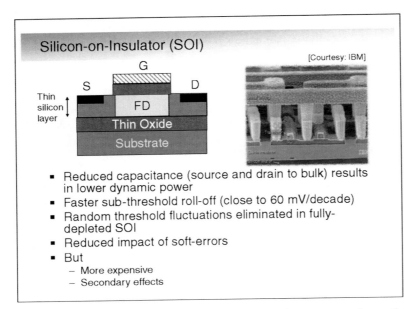

Slide 2.43

Silicon-on-Insulator (SOI) is a technology that has been "on the horizon" for quite a long time, yet it never managed to really break ground, though with some exceptions here and there. An SOI MOS transistor differs from a "bulk" device in that the channel is formed in a thin layer of silicon deposited above an electrical insulator, typically silicon dioxide.

Doing so offers some attractive features. First, as drain and source diffusions extend all the way down to the insulator layer, their junction capacitances are substantially reduced, which translates directly into power savings. Another advantage is the higher sub-threshold slope factor (approaching the ideal 60 mV/decade), reducing leakage. Finally, the sensitivity to soft errors is reduced owing to the smaller collection efficiency, leading to a more reliable transistor. There are some important negatives as well. The addition of the SiO_2 layer and the thin silicon layer increases the cost of the substrate material, and may impact the yield as well. In addition, some secondary effects should be noted. The SOI transistor is essentially a three-terminal device without a bulk (or body) contact, and a "body" that is floating. This effectively eliminates body biasing as a threshold-control technique. The floating transistor body also introduces some interesting (ironically speaking...) features such as hysteresis and state-dependency.

Device engineers differentiate between two types of SOI transistors: partially-depleted (PD-SOI) and fully-depleted (FD-SOI). In the latter, the silicon layer is so thin that it is completely depleted under nominal transistor operation, which means that the depletion/inversion layer under the gate extends all the way to the insulator. This has the advantage of suppressing some of the floating-body effects, and an ideal sub-threshold slope is theoretically achievable. From a variation perspective, the threshold voltage becomes independent of the doping in the channel, effectively eliminating a source of random variations (as discussed in Slide 2.37). FD-SOI requires the depositing of extremely thin silicon layers (3–5 times thinner than the gate length!).

Nanometer Transistors and Their Models

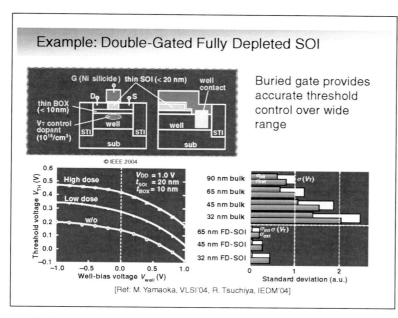

Slide 2.44

The FD-SOI device architecture can be further extended with an extra feature that reinstates threshold control through a fourth terminal. A buried gate below the SiO_2 insulator layer helps to control the charge in the channel, and thus also the threshold voltage. As shown in these graphs (published by Hitachi), the buried-gate concept pretty much reinstates the idea of body biasing as a viable design option. The reduced impact of random doping variations on the threshold voltage, as is typical in FD-SOI, is also illustrated.

Slide 2.45

The FinFET (called a trigate transistor by Intel) is an entirely different transistor structure that actually offers some properties similar to the ones offered by the device presented in the previous slide. The term FinFET was coined by researchers at the University of California at Berkeley to describe a non-planar, double-gated transistor built on an SOI substrate. The distinguishing characteristic of the FinFET is that the controlling gate is wrapped around a thin silicon "fin", which forms the body of the device. The dimensions of the fin determine the effective channel length of the device. The device structure has shown the potential to scale the channel length to values that are hard, if not impossible, to accomplish in traditional planar devices. In fact, operational transistors with channel lengths down to 7 nm have been demonstrated.

In addition to a suppression of deep submicron effects, a crucial advantage of the device is again increased control, as the gate wraps (almost) completely around the channel.

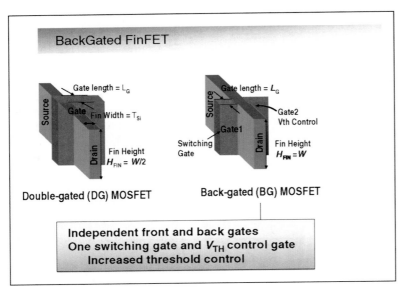

Slide 2.46

This increased two-dimensional control can be exploited in a number of ways. In the dual-gated device, the fact that the gate is controlling the channel from both sides (as well as the top) leads to increased process transconductance. Another option is to remove the top part of the gate, leading to the back-gated transistor. In this structure, one of the gates acts as the standard control gate, whereas the other is used to manipulate the threshold voltage. In a sense, this device offers similar functionality as the buried-gate FD-SOI transistor discussed earlier. Controlling the work functions of the two gates through the selection of appropriate type and quantity of the dopants helps to maximize the range and sensitivity of the control knobs.

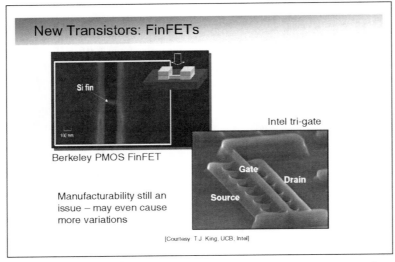

Slide 2.47

The fact that the FinFET and its cousins are dramatically different devices compared to your standard bulk MOS transistor is best-illustrated with these pictures from Berkeley and Intel. The process steps that set and control the physical dimensions are entirely different. Although this creates new opportunities, it also brings challenges, as the process steps involved are vastly different. The ultimate success of the FinFET depends greatly upon how these changes can be translated into a scalable, low-cost and high-yield process – some formidable question, indeed! Also unclear at this time is how the adoption of such a different structure impacts variability, as critical dimensions and device parameters are dependent upon entirely different process steps.

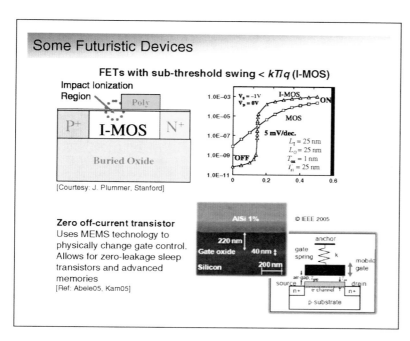

Slide 2.48

It is worth pointing out that the devices described here represent by no means the complete spectrum of new transistors and switching devices that are currently being explored. In fact, the number of options that are emerging from the research labs these days is quite extraordinary, and the excitement is palpable. Most of these will probably die with a whimper, while other ones are still decades out in terms of true applicability. Of the latter, carbon-nanotube (CNT) transistors seem to present some true potential, but the jury is still out.

When looking from a power angle, some device structures emerging from the research labs merit some special attention. The I-MOS transistor uses substantially different mechanisms, such as impact ionization, to produce a transistor with a sub-threshold slope substantially below 60 mV/decade. This opens the door for a switch with close-to-ideal characteristics. The availability of such a device would allow operation at supply voltages that are substantially lower than what we can allow today.

Another entirely new device would allow for an almost complete elimination of leakage current in standby mode: Using MEMS (Micro-electromechanical systems) technology, the suspended-gate MOSFET (SG-MOS) physically moves the actual gate up and down depending upon the applied gate voltage. In the down-position this device resembles a traditional transistor. Moving the gate into the up-position is physically equivalent to mechanically turning off the switch, effectively squelching all leakage current. The availability of such a device would come extremely handy in the design of low-standby power components.

Slide 2.49

For the circuit designer, there are some important takeaways from this chapter. Scaling into the nanometer regime has some profound impact on the behavior of the CMOS transistor, both in the ON and in the OFF modes. Simple models that capture the behavior of the transistor in both modes are available, and will help us in later chapters to build effective analysis and optimization frameworks. A profound awareness of the device characteristics and the ability to adapt to its varying properties will prove to be essential tenets in low-power design in the nanometer era.

Summary

- Plenty of opportunity for scaling in the nanometer age
- Deep-submicron behavior of MOS transistors has substantial impact on design
- Power dissipation mostly influenced by increased leakage (SD and gate) and increasing impact of process variations
- Novel devices and materials will ensure scaling to a few nanometers

Slide 2.50

Some references...

References

Books and Book Chapters
- A. Chandrakasan, W. Bowhill, and F. Fox (eds.), "Design of High-Performance Microprocessor Circuits", IEEE Press 2001
- J. Rabaey, A. Chandrakasan, and B. Nikolic, "Digital Integrated Circuits: A Design Perspective," 2nd ed, Prentice Hall 2003.
- Y. Taur and T.H. Ning, *Fundamentals of Modern VLSI Devices*, Cambridge University Press, 1998.

Articles
- N. Abele, R. Fritschi, K. Boucart, F. Casset, P. Ancey, and A.M. Ionescu, "Suspended-Gate MOSFET: Bringing New MEMS Functionality into Solid-State MOS Transistor," Proc. Electron Devices Meeting, 2005. IEDM Technical Digest. IEEE International, pp. 479–481, Dec. 2005
- BSIM3V3 User Manual, http://www.eecs.berkeley.edu/Pubs/TechRpts/1998/3486.html
- J.H. Chen et al., "An analytic three-terminal band-to-band tunneling model on GIDL in MOSFET," *IEEE Trans. On Electron Devices*, 48(7), pp. 1400–1405, July 2001.
- S. Chou, "Innovation and Integration in the Nanoelectronics Era," Digest ISSCC 2005, pp. 36–38, February 2005.
- P. Gelsinger, "Giga-scale Integration for Tera-Ops Performance," 41st DAC Keynote, DAC, 2004, (www.dac.com)
- X. Huang et al., "Sub 50-nm FinFET PMOS," International Electron Devices Meeting Technical Digest, p. 67 Dec. 5–8, 1999
- International Technology Roadmap for Semiconductors, http://www.itrs.net/
- H. Kam et al., "A new nano-electro-mechanical field effect transistor (NEMFET) design for low-power electronics," IEDM Tech. Digest, pp. 463–466, Dec. 2005.
- K. Mistry et al., "A 45nm Logic Technology with High-k+Metal Gate Transistors, Strained Silicon, 9 Cu Interconnect Layers, 193 nm Dry Patterning, and 100% Pb-free Packaging," Proceedings, IEDM, p. 247, Washington, Dec. 2007.
- Predictive Technology Model (PTM), http://www.eas.asu.edu/~ptm/
- T. Sakurai and R. Newton, "Alpha-power law MOSFET model and its applications to CMOS inverter delay and other formulas," *IEEE Journal of Solid-State Circuits*, 25(2), 1990.
- R. Tsuchiya et al., "Silicon on thin BOX: a new paradigm of the CMOSFET for low-power high-performance application featuring wide-range back-bias control," Proceedings IEDM 2004, pp. 631–634, Dec. 2004.
- M. Yamaoka et al., "Low power SRAM menu for SOC application using Yin-Yang-feedback memory cell technology," Digest of Technical Papers VLSI Symposium, pp. 288–291, June 2004.
- W. Zhao, Y. Cao, "New generation of predictive technology model for sub-45nm early design exploration," *IEEE Transactions on Electron Devices*, 53(11), pp. 2816–2823, November 2006

Chapter 3
Power and Energy Basics

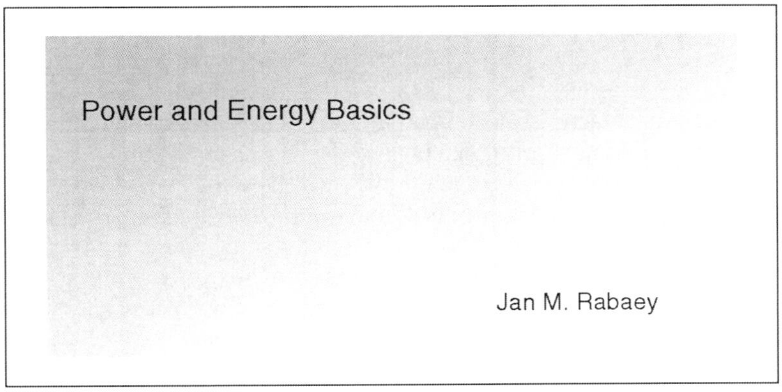

Slide 3.1

The goal of this chapter is to derive clear and unambiguous definitions and models for all of the design metrics relevant in the low-power design domain. Anyone with some training and experience in digital design is probably already familiar with a majority of them. If you are one of them, you should consider this chapter as a review. However, we recommend that everyone at least browse through the material, as some new definitions, perspectives, and methodologies are offered. In addition, if one truly wants to tackle the energy problem, it is essential to have an in-depth understanding of the causes of energy dissipation in today's advanced digital circuits.

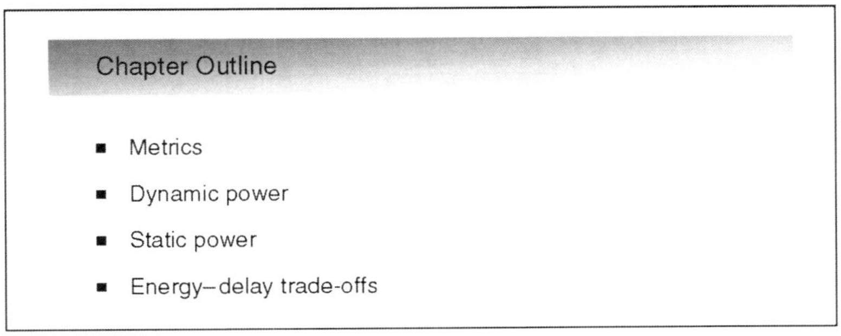

Slide 3.2

Before discussing the various sources of power dissipation in modern digital integrated circuits, it is worth spending some time evaluating the metrics typically used to evaluate the quality of a circuit or design. Unambiguous definitions are essential if one wants to provide fair comparisons. The rest of this chapter divides the sources of power roughly along the lines of dynamic and static power. At the end of the chapter, we make the point that optimization for power or energy alone rarely makes sense. Design for low power is most often a trade-off process, performed primarily in the energy-delay space. Realizing this goes a long way in setting up the foundations for an effective power-minimization design methodology.

J. Rabaey, *Low Power Design Essentials*, Series on Integrated Circuits and Systems,
DOI 10.1007/978-0-387-71713-5_3, © Springer Science+Business Media, LLC 2009

> **Metrics**
>
> - Delay (s):
>   - Performance metric
> - Energy (Joule)
>   - Efficiency metric: effort to perform a task
> - Power (Watt)
>   - Energy consumed per unit time
> - Power*Delay (Joule)
>   - Mostly a technology parameter – measures the efficiency of performing an operation in a given technology
> - Energy*Delay = Power*Delay^2 (Joule s)
>   - Combined performance and energy metric – figure of merit of design style
> - Other Metrics: Energy-Delay^n(Joule s^n)
>   - Increased weight on performance over energy

Slide 3.3
The basic design metrics – propagation delay, energy, and power – are well-known to anyone with a digital design experience. Yet, they may not be sufficient. In today's design environment, where both *delay* and *energy* play on an almost equal base, optimizing for only one parameter rarely makes sense. For instance, the design with the minimum propagation delay in general takes an exorbitant amount of energy, and, vice versa, the design with the minimum energy is unacceptably slow. Both represent extremes in a rich optimization space, where many other optimal operational points exist. Hence some other metrics of potential interest have been defined, such as the *energy–delay* product, which puts an equal weight on both parameters. In fact, the normalized energy–delay products for a number of optimized general-purpose designs fall consistently within a narrow range. While being an interesting metric, the energy–delay product of an actual design only tells us how close the design is to a perfect balance between performance and energy efficiency. In real designs, achieving that balance may not necessarily be of interest. Typically, one metric is assigned greater weight – for instance, energy is minimized for a given maximum delay or delay is minimized for a given maximum energy. For these off-balance situations, other metrics can be defined such as *energy–delay^n*. Though interesting, these derived metrics however are rarely used, as they lead to optimization for only one target in the overall design space.

It is worth at this point to recap the definition of propagation delay: it is measured as the time difference between the 50% transition points of the input and output waveforms. For modules with multiple inputs and outputs, we typically define the propagation delay as the worst-case delay over all possible scenarios.

> **Where Is Power Dissipated in CMOS?**
>
> - Active (Dynamic) power
>   - (Dis)charging capacitors
>   - Short-circuit power
>   - Both pull-up and pull-down *on* during transition
> - Static (leakage) power
>   - Transistors are imperfect switches
> - Static currents
>   - Biasing currents

Slide 3.4
Power dissipation sources can be divided in two major classes: *dynamic* and *static*. The difference between the two is that the former is proportional to the activity in the network and the switching frequency, whereas the latter is independent of both. Until recently, dynamic power vastly outweighed static power. With the emergence of leakage as a major power component

Power and Energy Basics

though, both should now be treated on an equal footing. Biasing currents for "analog" components such as sense amplifiers or level converters strictly fall under the static power-consumption class, but originate from a design choice rather than a device deficiency.

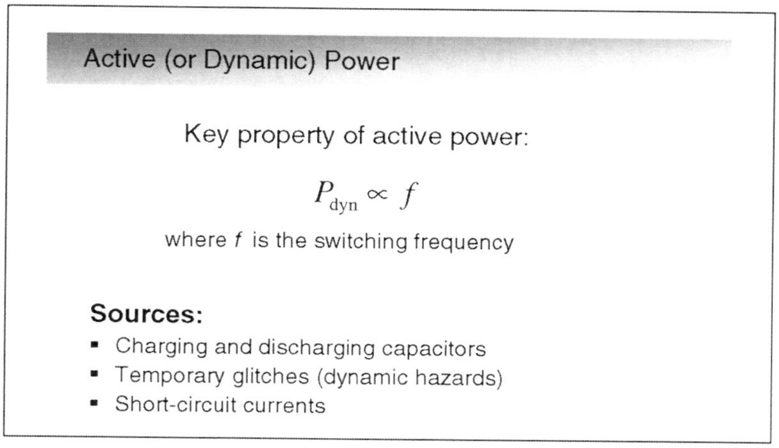

Active (or Dynamic) Power

Key property of active power:

$$P_{dyn} \propto f$$

where f is the switching frequency

Sources:
- Charging and discharging capacitors
- Temporary glitches (dynamic hazards)
- Short-circuit currents

Slide 3.5

As mentioned, dynamic power is proportional to the switching frequency. The charging and discharging of capacitances is and should be the main source of dynamic power dissipation – as these operations are at the core of what constitutes MOS digital circuit design. The other contributions (short-circuit currents and dynamic hazards or glitches) are parasitic effects and should be made as small as possible.

Charging Capacitors

Applying a voltage step

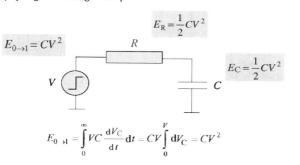

$E_{0\rightarrow 1} = CV^2$

$E_R = \frac{1}{2}CV^2$

$E_C = \frac{1}{2}CV^2$

$$E_{0\rightarrow 1} = \int_0^\infty VC\frac{dV_C}{dt}dt = CV\int_0^V dV_C = CV^2$$

Value of R does not impact energy!

Slide 3.6

The following equation is probably the most important one you will encounter in this book: to charge a capacitance C by applying a voltage step V, an amount of energy equal to CV^2 is taken from the supply. Half of that energy is stored on the capacitor; the other half is dissipated as heat in the resistance of the charging network. During discharge the stored energy in turn is dissipated as heat as well. Observe that the resistance of the networks does not enter the equation.

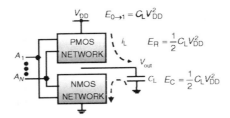

Slide 3.7
This model applies directly to a digital CMOS gate, where the PMOS and NMOS transistors form the resistive charge and discharge networks. For the sake of simplicity, the total capacitance of the network is lumped into the output capacitance of the gate.

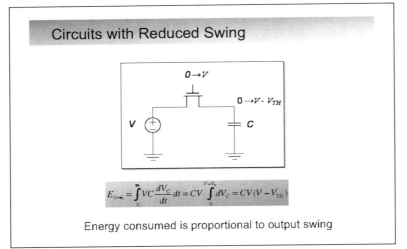

Slide 3.8
More generically, we can compute the energy it takes to charge a capacitance from a voltage V_1 to a voltage V_2. Using similar math, we derive that this requires from the supply an amount of energy equal to $CV_2(V_2-V_1)$. This equation will come in handy for a number of special circuits. One example is the NMOS pass-transistor chain. It is well-known that the maximum voltage at the end of such as chain is one threshold voltage below the supply [Rabaey03]. Using the afore-derived equation, we find that the energy dissipation in this case equals $CV_{DD}(V_{DD}-V_{TH})$, and is proportional to the swing at the output. In general, reducing the swing in a digital network results in a linear reduction in energy consumption.

Power and Energy Basics

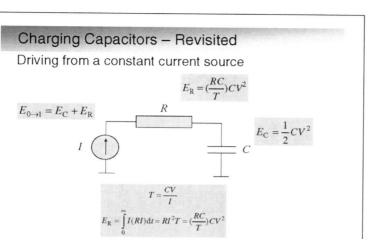

Slide 3.9
So far, we have assumed that charging a capacitor always requires an amount of energy equal to CV^2. This is true only when the driving waveform is a voltage step. It is actually possible to reduce the required energy by choosing other waveforms. Assume, for instance, that a current source with a fixed current I is used instead. Under those circumstances, the energy consumed in the resistor is reduced to $(RC/T)CV^2$ where T is the charging time, and the output voltage rises linearly with time. Observe that the resistance of the network plays a role under these circumstances. From this, it appears that the dissipation in the resistor can be reduced to very small values, if not zero, by charging the capacitor very slowly (i.e., by reducing I).

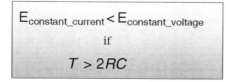

Slide 3.10
In fact, the current-driven scenario results in an actual energy reduction over the voltage-driven approach for $T > 2RC$. As a reference, the time it takes for the output of the voltage-driven circuit to move between 0% and 90% points equals $2.3RC$. Hence, the current-driven circuit is more energy-efficient than the voltage-driven one as long as it is slower.

For this scheme to work, the same approach should be used to discharge the capacitor, and the charge flowing through the source should be recovered. If not, the energy gained is just wasted in the source.

The idea of "energy-efficient" charging gained a lot of attention in the 1990s. However, the inferior performance and the complexity of the circuitry ensured that the ideas remained confined to the academic world. With the prospect of further voltage scaling bottoming out, these concepts may gain some traction anew (some more about this in Chapter 13).

Slide 3.11

Charging Capacitors

Driving using a sine wave (e.g., from resonant circuit)

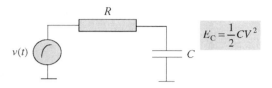

Energy dissipated in resistor can be made arbitrarily small if frequency $\omega \ll 1/RC$
(output signal in phase with input sinusoid)

Charging a capacitor using a current source is only one option. Other voltage and current waveforms can be imagined as well. For instance, assume that the input voltage waveform is a sinusoid rather than a step. A first-order analysis shows that this circuit outperforms the voltage-step approach, for sinusoid frequencies ω below $1/RC$. The easiest way to come to this conclusion is to evaluate the circuit in the frequency domain. The RC network is a low-pass filter with a single pole at $\omega_p = 1/RC$. It is well-known that for frequencies much smaller than the pole, the output sinusoid has the same amplitude and the same phase as those of the input waveform. In other words, no or negligible current is flowing through the resistor, and hence little power is dissipated. The attractive feature of the sinusoidal waveforms is that these are easily generated by resonant networks (such as LC oscillators). Again, with some notable exceptions such as power regulators, sinusoidal charging has found little industrial following.

Slide 3.12

Dynamic Power Consumption

Power = Energy per transition × Transition rate

$$= C_L V_{DD}^2 f_{0 \to 1}$$

$$= C_L V_{DD}^2 f p_{0 \to 1}$$

$$= C_{switched} V_{DD}^2 f$$

- Power dissipation is data dependent – depends on the switching probability, $P_{0 \to 1}$
- Switched capacitance $C_{switched} = p_{0 \to 1} C_L = \alpha C_L$
  (α is called the switching activity factor)

This brings us back to the generic case of the CMOS inverter. To translate the derived energy per operation into power, it must be multiplied with the rate of power-consuming transitions $f_{0 \to 1}$. The unit of the resulting metric is Watt (= Joules/sec). This translation leads right away to one of the hardest problems in power analysis and optimization: it requires knowledge of the "activity" of the circuit. Consider a circuit with a clock frequency f. The probability that a node will make a 0-to-1 transition at a given clock tick is given by αf, where $0 \le \alpha \le 1$ is the activity factor at that node. As we discuss in the following slides, α is a function of the circuit topology and the activity of the input signals. The accuracy of power estimation depends largely upon how well the activity is known – which is most often not very much.

Power and Energy Basics

The derived expression can be expanded for a complete module by summing over all nodes. The average power is then expressed as $(\alpha C)V^2 f$. Here αC is called the *effective capacitance* of the module, and equals the average amount of capacitance that is being charged in the module every clock cycle.

Slide 3.13

Impact of Logic Function

Example: Static two-input NOR gate

A	B	Out
0	0	1
0	1	0
1	0	0
1	1	0

Assume signal probabilities
$p_{A=1} = 1/2$
$p_{B=1} = 1/2$

Then transition probability
$p_{0 \to 1} = p_{out=0} \times p_{out=1}$
$= 3/4 \times 1/4 = 3/16$

If inputs switch every cycle

$\alpha_{NOR} = 3/16$

NAND gate yields similar result

Let us, for instance, derive the activity of a two-input NOR gate (which defines the topology of the circuit). Assume that each input has an equal probability of being a 1 or a 0, and that the probability of a transition at a clock tick is 50–50 as well, ensuring an even distribution between states. With the aid of the truth table we derive that the probability of a $0 \to 1$ transition (or the activity) equals 3/16. More generally, the activity at the output node can be expressed as a function of the 1-probabilities of the inputs A and B: $\alpha_{NOR} = p_A p_B (1 - p_A p_B)$.

Slide 3.14

Impact of Logic Function

Example: Static two-input XOR gate

A	B	Out
0	0	1
0	1	0
1	0	0
1	1	0

Assume signal probabilities
$p_{A=1} = 1/2$
$p_{B=1} = 1/2$

Then transition probability
$p_{0 \to 1} = p_{out=0} \times p_{out=1}$
$= 1/2 \times 1/2 = 1/4$

If inputs switch every cycle

$p_{0 \to 1} = 1/4$

A similar analysis can be performed for an XOR gate. The observed activity is a bit higher (1/4).

Slide 3.15

These results can be generalized for all basic gates.

Transition Probabilities for Basic Gates

As a function of the input probabilities

	$p_{0 \to 1}$
AND	$(1 - p_A p_B) p_A p_B$
OR	$(1 - p_A)(1 - p_B)(1 - (1 - p_A)(1 - p_B))$
XOR	$(1 - (p_A + p_B - 2 p_A p_B))(p_A + p_B - 2 p_A p_B)$

Activity for static CMOS gates
$\alpha = p_0 p_1$

Slide 3.16

The topology of the logic network has a major impact on the activity. This is nicely illustrated by comparing the activity of NAND (NOR) and XOR gates as a function of fan-in. The output-transition probability of a NAND gate goes asymptotically to zero. The probability of the output being a 0 is indeed becoming smaller with increasing fan-in. An example of such a network is a memory-address decoder. On the other hand, the activity of an XOR network is independent of fan-in. This does not bode well for the power dissipation of modules such as large en(de)cryption and coding functions, which primarily consist of XORs.

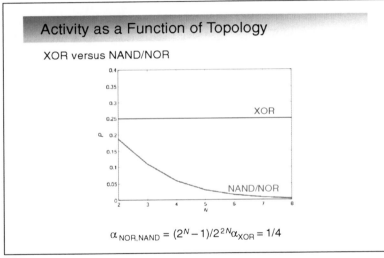

Activity as a Function of Topology

XOR versus NAND/NOR

$\alpha_{NOR, NAND} = (2^N - 1)/2^{2N} \quad \alpha_{XOR} = 1/4$

Slide 3.17

One obvious question is how the choice of logic family impacts activity and power dissipation. Some interesting global trends can be observed. Consider, for instance, the case of dynamic logic. The only power-consuming transitions in pre-charged logic occur when the output evaluates to 0, after which it has to be recharged to a high in the next pre-charge cycle. Hence, the activity factor α is equal to the probability of the output being equal to 0. This means that the *activity is always higher in dynamic logic* (compared to static), independent of the function. This does not mean per se that the power dissipation of dynamic logic is higher, as the effective capacitance is the product of

Power and Energy Basics

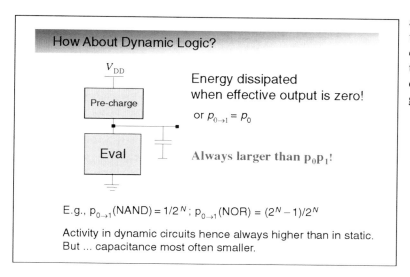

activity and capacitance, the latter being smaller in dynamic logic. In general though, the higher activity outweighs the capacitance gain.

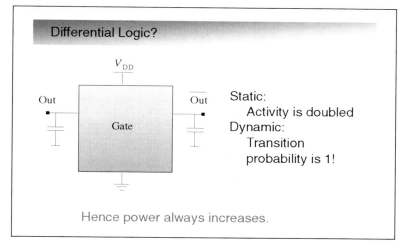

Slide 3.18
Another interesting logic family is *differential logic*, which may seem attractive for very low-voltage designs due to its increased signal-to-noise ratio. Differential implementations come unfortunately with an inherent disadvantage from a power perspective: not only is the overall capacitance higher, the activity is higher as well (for both static and dynamic implementations). The only positive argument is that differential implementation reduces the number of gates needed for a given function, and thus reduces the length of the critical path.

Slide 3.19
As activity is such an important parameter in the analysis of power dissipation, it is worthwhile spending some time on how to evaluate the activity of more complex logic networks. One may wonder whether it is possible to develop a "static power analyzer" along the lines of the "static timing analyzer". The latter evaluates the propagation delay of a logic network analyzing only the topology of the network without any simulation (hence the name "static"). The odds for successful

Evaluating Power Dissipation of Complex Logic

- Simple idea: start from inputs and propagate signal probabilities to outputs

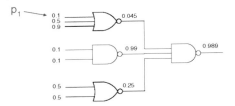

- But:
  - Reconvergent fan-out
  - Feedback and temporal/spatial correlations

static power analysis seem favorable at a first glance. Consider, for instance, the network shown on the slide, and assume that the 1- and 0-probabilities of the primary input signals are known. Using the basic gate expressions presented earlier, the output signal probabilities can be computed for the first layer of gates starting from the primary inputs. This process is then repeated until the primary outputs are reached.

This process seems fairly straightforward indeed. However, there is a catch. For the basic gate equations to be valid, the inputs must be statistically independent. In probability theory, to say that two events are *independent* intuitively means that the occurrence of one event makes it neither more nor less probable that the other occurs. While this assumption is in general true for the network of the slide (assuming obviously that all the primary input signals are independent), it unfortunately rarely holds in actual circuits.

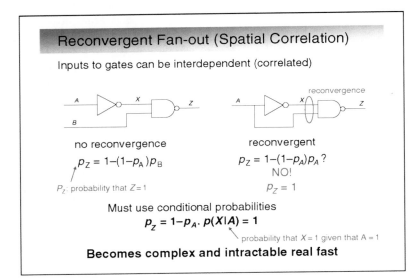

Slide 3.20

Even if the primary inputs to a logic network are independent, the signals may become correlated or "colored", while they propagate through the logic network. This is best illustrated with a simple example, which showcases the impact of a network property called *reconvergent fan-out*. In the rightmost circuit, the inputs to the NAND gate Z are not independent, but are both functions of the same input signal A. To compute the output probabilities of Z, the expression derived earlier for a NAND gate is no longer applicable, and conditional probabilities need to be used. *Conditional probability* is the probability of some event A, given the occurrence of some other event B. Conditional probability is expressed as $p(A|B)$, and is read as "the probability of A, given B". More specifically, one can derive that $p(A|B) = p(A \cap B)/p(B)$, assuming that $p(B) \neq 0$.

While propagating these conditional probabilities through the network is theoretically possible, you may guess that the complexity of doing so for complex networks rapidly becomes unmanageable – and that indeed is the case.

Power and Energy Basics

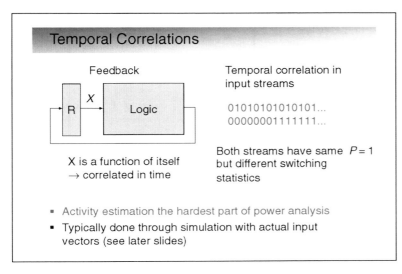

Slide 3.21

The story gets complicated even further by the occurrence of *temporal correlations*. A signal shows temporal correlation if a data value in the signal stream is dependent upon previous values in the stream. Temporal correlations are the norm in sequential networks, as any signal in the network is typically a function of its previous values owing to the existence of feedback network. In addition, primary input signals as well may show temporal dependence. For example, in a digitized speech signal any sample value is dependent upon the previous values.

All these arguments help to illustrate that static activity analysis is a very hard problem indeed, and actually all but impossible. Hence, power analysis tools either rely on simulations of actual signal traces to derive the signal probabilities or make simplifying assumptions – for instance, it is assumed that the input signals are independent and purely random. This is discussed in more detail in Chapter 12. In the following chapters, we will most often assume that activity of a module in its typical operation mode can be characterized by an independent parameter α.

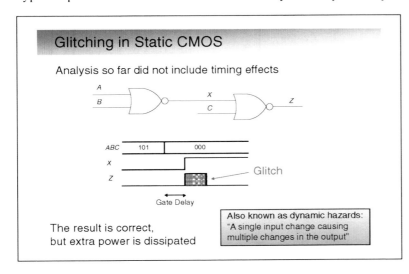

Slide 3.22

So far, we have assumed that the dynamic power dissipation solely results from the charging (and discharging) of capacitances in between clock events. Some additional sources of dynamic power dissipation (i.e., proportional to the clock frequency) should be considered. Though (dis)-charging of capacitors is essential to the operation of a CMOS digital circuit, *dynamic hazards* and *short-circuit currents* are not. They should be considered as parasitic, and be kept to an absolute minimum.

A *dynamic hazard* occurs when a single input change causes multiple transitions at the output of a gate. These events, also known as "glitches", are obviously wasteful, as a capacitor is charged and/or discharged without having an impact on the final result. In the analysis of the transition probabilities of complex logic circuits, presented in the earlier slides, glitches did not appear, as the propagation delays of the individual gates were ignored – all events were assumed to be instantaneous. To detect the occurrence of dynamic hazards a detailed timing analysis is necessary.

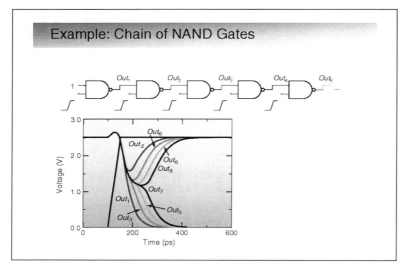

Slide 3.23

A typical example of the effect of glitching is illustrated in this slide, which shows the simulated response of a chain of NAND gates with all inputs going simultaneously from 0 to 1. Initially, all the outputs are 1, as one of the inputs was 0. For this particular transition, all the odd bits must transition to 0, while the even bits remain at the value of 1. However, owing to the finite propagation delay, the even output bits at the higher bit positions start to discharge, and the voltage drops. When the correct input ripples through the network, the output ultimately goes high. The glitch on the even bits causes extra power dissipation beyond what is required to strictly implement the logic function. Although the glitches in this example are only partial (i.e., not from rail to rail), they contribute significantly to the power dissipation. Long chains of gates often occur in important structures such as adders and multipliers, and the glitching component can easily dominate the overall power consumption.

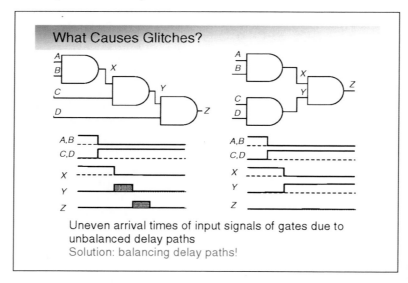

Slide 3.24

The occurrence of glitching in a circuit is mainly due to a mismatch in the path lengths in the network. If all input signals of a gate change simultaneously, no glitching occurs. On the other hand, if input signals change at different times, a dynamic hazard may develop. Such a mismatch in signal timing is typically the result of different path lengths with respect to the primary inputs of the network. This is illustrated in this slide, where two equivalent, but topologically different, realizations of the function $F = A.B.C.D$ are analyzed. Assume that the AND gate has a unit delay. The leftmost network suffers from glitching as a result of the disparity between the arrival times of the input signals for gates Y and Z. For example, for gate Z, input D settles at time 0, whereas input Y only settles at time 2. Redesigning the network so that all arrival times are identical can dramatically reduce the number of superfluous transitions, as shown in the rightmost network.

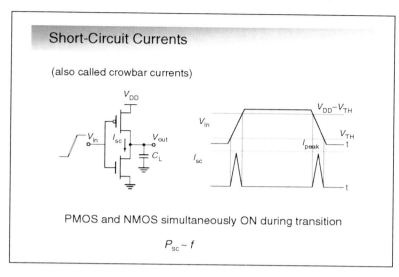

Slide 3.25
So far, it was assumed that the NMOS and PMOS transistors of a CMOS gate are never ON simultaneously. This assumption is not entirely correct, as the finite slope of the input signal during switching causes a direct current path between V_{DD} and GND for a short period of time. The extra power dissipation due to these "short-circuit" or "crowbar" currents is proportional to the switching activity, similar to the capacitive power dissipation.

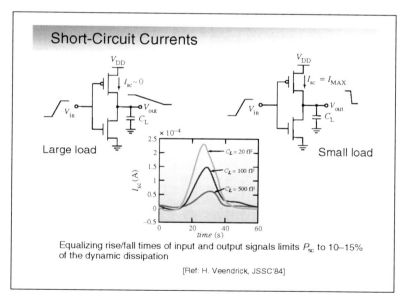

Slide 3.26
The peak value of the short-circuit current is also a strong function of the ratio between the slopes of the input and output signals. This relationship is best illustrated by the following simple analysis: Consider a static CMOS inverter with a 0→1 transition at the input. Assume first that the load capacitance is very large, so that the output fall time is significantly larger than the input rise time (left side). Under those circumstances, the input moves through the transient region before the output starts to change. As the source–drain voltage of the PMOS device is approximately zero during that period, the device shuts off without ever delivering any current. The short-circuit current is close to zero. Consider now the reverse case (right side), where the output capacitance is very small, and the output fall time is substantially smaller than the input rise time. The drain–source voltage of the PMOS device equals V_{DD} for most of the transition time, guaranteeing a maximal

short-circuit current. This clearly represents the worst-case condition. The conclusions of this intuitive analysis are confirmed by the simulation results.

This analysis may lead to the (faulty) conclusion that the short-circuit dissipation is minimized by making the output rise/fall time substantially larger than the input rise/fall time. On the other hand, making the output rise/fall time too large slows down the circuit, and causes large short-circuit currents in the connecting gates. A more practical rule that optimizes the power consumption in a global way, can be formulated: *The power dissipation due to short-circuit currents is minimized by matching the rise/fall times of the input and output signals.* At the overall circuit level, this means that rise/fall times of all signals should be kept constant within a range. Equalizing the input and output transition times of a gate is not the optimum solution for the individual gate, but keeps the overall short-circuit current within bounds (maximum 10–15% of the total dynamic power dissipation). Observe also that the impact of short-circuit current is reduced when we lower the supply voltage. In the extreme case, when $V_{DD} < V_{THn} + |V_{THp}|$, the short-circuit dissipation is completely eliminated, because the devices are never ON simultaneously.

Slide 3.27

As the short-circuit power is proportional to the clock frequency, it can be modeled as an equivalent capacitor: $P_{sc} = C_{sc}V_{DD}^2 f$, which then can be lumped into the output capacitance of the gate. Be aware however that C_{SC} is a function of the input and output transition times.

Modeling Short-Circuit Power

- Can be modeled as capacitor

$$C_{SC} = k(a\frac{\tau_{in}}{\tau_{out}} + b)$$

a, b: technology parameters
k: function of supply and threshold voltages, and transistor sizes

$$E_{SC} = C_{SC}V_{DD}^2$$

Easily included in timing and power models

Slide 3.28

Although dynamic power traditionally has dominated the power budget, static power has become an increasing concern when scaling below 100 nm. The main reasons behind this have been discussed at length in Chapter 2. Sub-threshold drain–source leakage, junction leakage, and gate leakage all play important roles, but in contemporary design it is the sub-threshold leakage that is the main cause of concern.

Transistors Leak

- Drain leakage
  - Diffusion currents
  - Drain-induced barrier lowering (DIBL)
- Junction leakage
  - Gate-induced drain leakage (GIDL)
- Gate leakage
  - Tunneling currents through thin oxide

Power and Energy Basics

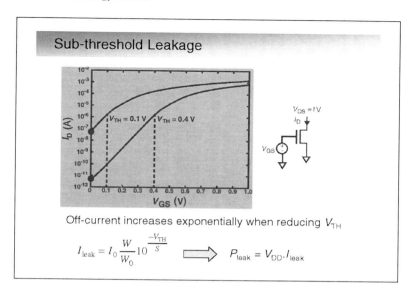

Slide 3.29
In Chapter 2, it was pointed out that the main reason behind the increase in drain–source leakage is the gradual reduction of the threshold voltage forced by the lowering of the supply voltages. Any reduction in threshold voltage causes the leakage current to grow exponentially. The chart illustrating this is repeated for the purpose of clarity.

Slide 3.30
An additional factor is the increasing impact of the DIBL effect. Combining the equations for sub-threshold leakage and the influence of DIBL on V_{TH}, an expression for the leakage power of a gate can be derived. Observe the exponential dependence of leakage power upon both V_{TH} and V_{DD}.

Slide 3.31
The dependence of the leakage current on the applied drain–source voltage creates some interesting side effects in complex gates. Consider, for example, the case of a two-input NAND gate where the two NMOS transistors in the pull-down network are turned off. If the off-resistance of NMOS transistors would be fixed, and not a function of the applied voltage, one would expect that the doubling of the resistance by putting two transistors in series would halve the leakage current (compared to a similar-sized inverter).

An actual analysis shows that the reduction in leakage is substantially larger. When the pull-down chain is off, node M settles to an intermediate voltage, set by balancing the leakage currents of transistors M1 and M2. This reduces the drain–source voltage of both transistors (especially of transistor M2), which translates into a substantial reduction in the leakage currents

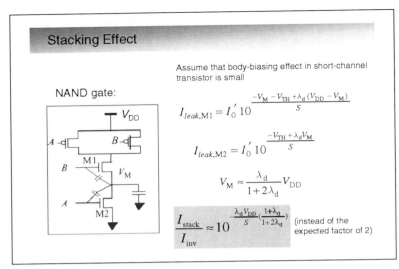

due to DIBL. In addition, the gate–source voltage of transistor M1 becomes negative, resulting in an exponential reduction of the leakage current. This is further augmented by the reverse body-biasing, which raises the threshold of M1 – this effect is only secondary though.

Using the expressions for the leakage currents derived earlier, we can determine the voltage value of the intermediate node, V_M, and derive an expression for the leakage current as a function of the DIBL factor λ_d and the sub-threshold swing S. The resulting equation shows that the reduction in leakage current obtained by stacking transistors is indeed larger than the linear factor one would initially expect. This is called the *stacking effect*.

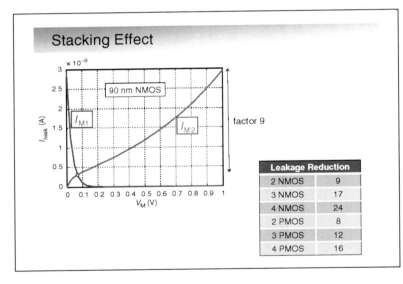

Slide 3.32

The mechanics of the stacking effect are illustrated with the example of two stacked NMOS transistors (as in the NAND gate of Slide 3.31) implemented in a 90 nm technology. The currents through transistors M1 and M2 are plotted as a function of the intermediate voltage V_M. The actual operation point is situated at the crossing of the two load lines. As can be observed, the drain–source voltage of M2 is reduced from 1 V to 60 mV, resulting in a ninefold reduction in leakage current. The negative V_{GS} of 60 mV for transistor M1 translates into a similar reduction.

The impact of the stacking effect is further detailed in the table, which illustrates the reduction in leakage currents for various stack sizes in 90 nm technology. The leakage reductions for both NMOS and PMOS stacks are quite impressive. The impact is somewhat smaller for the PMOS chains, as the DIBL effect is smaller for those devices. The stacking effect will prove to be a powerful tool in the fight against static power dissipation.

Power and Energy Basics

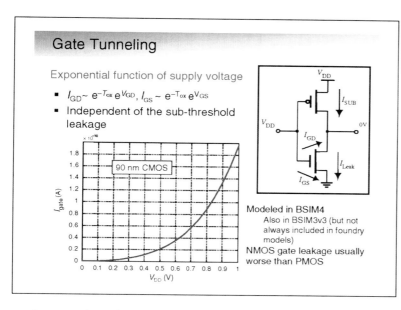

Slide 3.33

While sub-threshold currents dominate the static power dissipation, other leakage sources should not be ignored. Gate leakage is becoming significant in the sub-100 nm era. Gate leakage currents flow from one logical gate into the next one, and hence have a profoundly different impact on gate operation compared to sub-threshold currents. Whereas the latter can be reduced by increasing threshold voltages, the only way to reduce the gate leakage component is to decrease the voltage stress over the gate dielectric – which means *reducing voltage levels*.

Similar to sub-threshold leakage, gate leakage is also an exponential function of the supply voltage. This is illustrated by the simulation results of a 90 nm CMOS inverter. The maximum leakage current is around 100 pA, which is an order of magnitude lower than the sub-threshold current. Yet, even for these small values, the impact can be large, especially if one wants to store a charge on a capacitor for a substantial amount of time (such as in DRAMs, charge pumps, and even dynamic logic). Remember also that the gate leakage is an exponential function of the dielectric thickness.

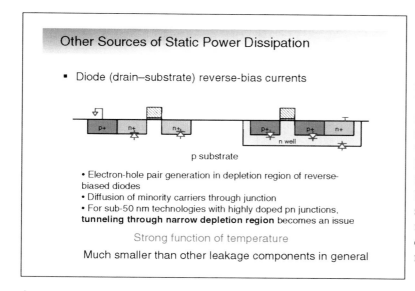

Slide 3.34

Finally, junction leakage, though substantially smaller than the previously mentioned leakage contributions, should not be ignored. With the decreasing thickness of the depletion regions owing to the high doping levels, some tunneling effects may become pronounced in sub-50 nm technology nodes. The strong dependence upon temperature must again be emphasized.

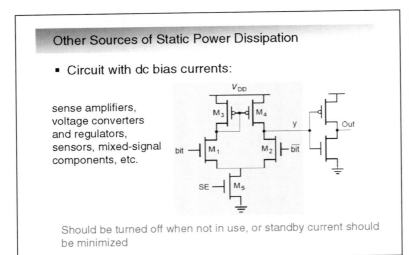

Slide 3.35
A majority of the state-of-the-art digital circuits contain a number of analog components. Examples of such circuits are sense amplifiers, reference voltages, voltage regulators, level converters, and temperature and leakage sensors. One property of each of these circuits is that they need a bias current for correct operation. These currents can become a sizable part of the total static power budget. To reduce their contribution, two mechanisms can be used:

(1) Trade off performance for current – Reducing the bias current of an analog circuit, in general, impacts its performance. For instance, the gain and slew rate of an amplifier benefit, from a higher bias current.
(2) Power management – some analog components need to operate only for a fraction of the time. For example, a sense amplifier in a DRAM or SRAM memory only needs to be ON at the end of the read cycle. Under those conditions the static power can be substantially reduced by turning off the bias when not in use. While being most effective, this technique does not always work as some bias or reference networks need to be ON all the time, or their start-up time would be too long to be practical.

In short, every analog circuit should be examined very carefully, and bias current and ON time should be minimized. The "a bias should never be on when not used" principle rules.

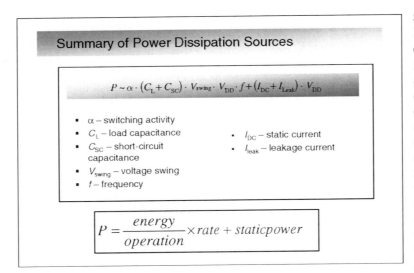

Slide 3.36
From all the preceding discussions, a global expression for the power dissipation of a digital circuit can be derived. The two major components, the dynamic and static dissipation, are easily recognized. An interesting perspective on the relationship between the two is obtained by realizing that a given computation (such as a multiplication or the execution of an instruction on a processor) is best characterized by its energy cost. Static dissipation, on the other hand, is best captured as a power quantity. To determine the

Power and Energy Basics

relative balance between the two, the former must be translated into power by multiplying it with its execution rate, or, in other words, the activity. Hence, precise knowledge of the activity is essential if one wants to estimate the overall power dissipation. Note: in a similar way, multiplying the static power with the time period leads to a global expression for energy.

The Traditional Design Philosophy

- Maximum performance is primary goal
  - Minimum delay at circuit level
- Architecture implements the required function with target throughput, latency
- Performance achieved through optimum sizing, logic mapping, architectural transformations
- Supplies, thresholds set to achieve maximum performance, subject to reliability constraints

Slide 3.37
The growing importance of power minimization and containment is revolutionizing design as we know it. Methodologies that were long-accepted have to be adjusted, and established design flows modified. Although this trend was visible already a decade ago in the embedded design world, it was only recently that it started to upend a number of traditional beliefs in the high-performance design community. Ever-higher clock frequencies were the holy grail of the microprocessor designer. Though architectural optimizations played a role in the performance improvements demonstrated over the years, reducing the clock period through technology scaling was responsible for the largest fraction.

Once the architecture was selected, the major function of the design flow was to optimize the circuitry through sizing, technology mapping, and logical transformations so that the maximum performance was obtained. Supply and threshold voltages were selected in advance to guarantee top performance.

CMOS Performance Optimization

- Sizing: Optimal performance with equal fan-out per stage

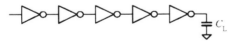

- Extendable to general logic cone through "logical effort"
- Equal effective fan-outs ($g_i C_{i+1}/C_i$) per stage
- Example: memory decoder

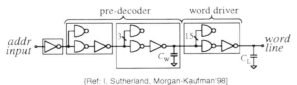

[Ref: I. Sutherland, Morgan-Kaufman'98]

Slide 3.38
This philosophy is best-reflected in the popular "logical effort"-based design optimization methodology. The delay of a circuit is minimized if the "effective fan-out" of each stage is made equal (and set to a value of approximately 4). Though this technique is very powerful, it also *guarantees* that power consumption is maximal! In the coming chapters, we will reformulate the logical-effort methodology to bring power into the equation.

Slide 3.39

That the circuit optimization philosophy of old can no longer be maintained is best illustrated by this simple example (after Shekhar Borkar from Intel). Assume a microprocessor design implemented in a given technology. Applying a single technology scaling step reduces the critical dimensions of the chip by a factor of 0.7. General scaling, which reduces the voltage by the same factor, increases the clock frequency by a factor of 1.41. If we take into account the fact that the die size typically increases (actually, used to increase is a better wording) by a factor of 14% between generations, the total capacitance of the die increases by a factor of $(1/0.7) \times 1.14^2 = 1.86$. (This simplified analysis assumes that all the extra transistors are used to good effect). The net effect is that the power dissipation of the chip increases by a factor of 1.3.

However, microprocessor designers tend to push harder than that. Over the past decades, processor frequency increased by a factor of 2 between technology generations. The extra performance improvement was obtained by circuit optimizations, such as a reduction in the logical depth. Maintaining this rate of improvement now pushes the power dissipation up by a factor of 1.8.

The situation gets even worse when the slowdown in supply voltage scaling is taken into account. Reducing the supply voltage even by a factor of 0.85 means that the power dissipation now rises by 270% from generation to generation. As this is clearly unacceptable, a change in design philosophy was the only option.

Slide 3.40

This revised philosophy backs off from the "maximum performance at all cost" theory, and abandons the notion that clock frequency is equivalent to performance. The "design slack" that results from a less-than-maximum clock speed in a new technology can now be used to keep dynamic and static power within bounds. Performance increase is still possible, but now comes mostly from architectural optimizations – sometimes, but not always, at the expense of extra die area. Design now becomes a *trade-off exercise between speed and energy* (or power).

Power and Energy Basics

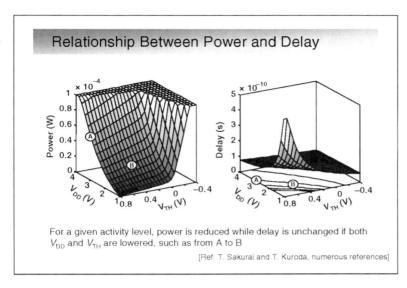

Slide 3.41
This trade-off is wonderfully illustrated by this set of, by now legendary, charts. Originated by T. Kuroda and T. Sakurai in the mid 1990s, the graphs plot power and (propagation) delay of a CMOS module as a function of the supply and threshold voltages – two parameters that were considered to be fixed in earlier years. The opposing nature of optimization for performance and power becomes obvious – the highest performance happens to occur exactly where power dissipation peaks (high V_{DD}, low V_{TH}). Another observation is that the same performance can be obtained at a number of operational points with vastly different levels of power dissipation. The existence of these "equal-delay" and "equal-power" curves proves to be an important optimization instrument when trading off in the delay–power (or energy) space.

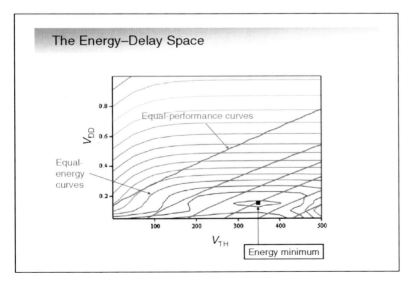

Slide 3.42
Contours of identical performance or energy are more evident in the two-dimensional plots of the delay and the (average) energy per operation as functions of supply and threshold voltages. The latter is obtained by multiplying the average power (as obtained using the expressions of Slide 3.36) by the length of the clock period. Similar trends as shown in the previous slide can be observed. Particularly interesting is that a point of minimum energy can be located. Lowering the voltages beyond this point makes little sense as the leakage energy dominates, and the performance deteriorates rapidly.

Be aware that this set of curves is obtained for one particular value of the activity. For other values of α, the balance between static and dynamic power shifts, and so do the trade-off curves. Also, the curves shown here are for fixed transistor sizes.

Slide 3.43

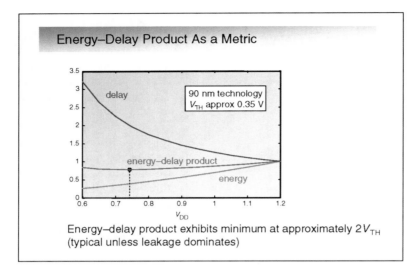

A further simplification of the graphs is obtained by keeping the threshold voltage constant. The opposing trends between energy and delay when reducing the supply voltage are obvious. One would expect that the product of the two (the energy–delay product or EDP) to show a minimum, which it does. In fact, it turns out that for CMOS designs, the minimum value of the EDP occurs approximately around two times the device threshold. In fact, a better estimate is $3V_{TH}/(3-\alpha)$ (with α the fit parameter in the alpha delay model – not to be confused with the activity factor). For $\alpha = 1.4$, this translates to 1.875 V_{TH}. Although this is an interesting piece of information, its meaning should not be over-estimated. As mentioned earlier, the EDP metric is only useful when equal weight is placed on delay and energy, which is rarely the case.

Slide 3.44

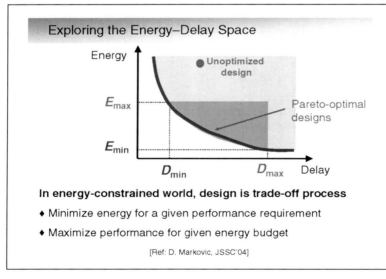

The above charts amply demonstrate that design for low power is a trade-off process. We have found that the best way to capture the duality between performance and energy efficiency is the *energy–delay curves*. Given a particular design and a set of design parameters, it is possible to derive a pareto-optimal curve that for every delay value gives the minimum attainable energy and vice versa. This curve is the best characterization of the energy and performance efficiency of a design. It also helps to redefine the design problem from "generate the fastest possible design" into a two-dimensional challenge: given a maximum delay, minimize the energy, or, given the maximum energy, find the design with the minimum delay.

We will use energy–delay curves extensively in the coming chapters. In the next chapter, we provide effective techniques to derive the energy–delay curves for a contemporary CMOS design.

Slide 3.45

Summary

- Power and energy are now primary design constraints
- Active power still dominating for most applications
  - Supply voltage, activity and capacitance the key parameters
- Leakage becomes major factor in sub-100 nm technology nodes
  - Mostly impacted by supply and threshold voltages
- Design has become energy–delay trade-off exercise!

In summary, we have analyzed in detail the various sources of power dissipation in today's CMOS digital design, and we have derived analytical and empirical models for all of them. Armed with this knowledge, we are ready to start exploring the many ways of reducing power dissipation and making circuits energy-efficient. One of the main lessons at the end of this story is that *there is no free lunch*. Optimization for energy most often comes at the expense of extra delay (unless the initial design is sub-optimal in both, obviously). Energy–delay charts are the best way to capture this duality.

Slide 3.46

References

- D. Markovic, V. Stojanovic, B. Nikolic, M.A. Horowitz and R.W. Brodersen, "Methods for true energy–performance optimization," *IEEE Journal of Solid-State Circuits*, 39(8), pp. 1282–1293, Aug. 2004.
- J. Rabaey, A. Chandrakasan and B. Nikolic, *Digital Integrated Circuits: A Design Perspective*," 2nd ed, Prentice Hall 2003.
- T. Sakurai, "Perspectives on power-aware electronics," *Digest of Technical Papers ISSCC*, pp. 26–29, Feb. 2003.
- I. Sutherland, B. Sproull and D. Harris, "Logical Effort", Morgan Kaufmann, 1999.
- H. Veendrick, "Short-circuit dissipation of static CMOS circuitry and its impact on the design of buffer circuits," *IEEE Journal of Solid-State Circuits*, SC-19(4), pp. 468–473, 1984.

Some references . . .

Chapter 4
Optimizing Power @ Design Time – Circuit-Level Techniques

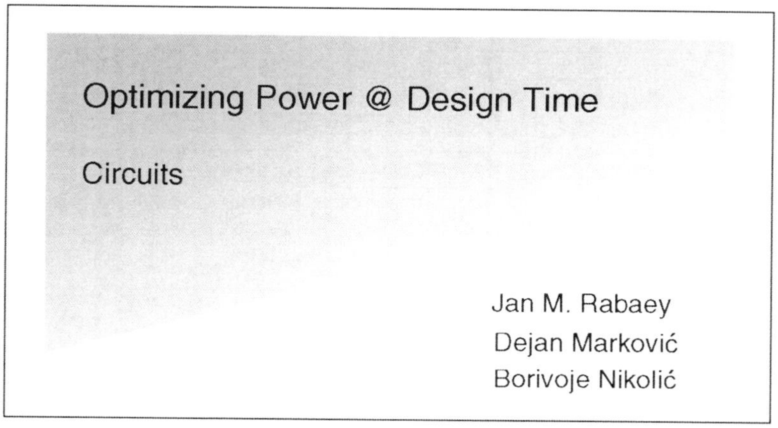

Optimizing Power @ Design Time
Circuits
Jan M. Rabaey
Dejan Marković
Borivoje Nikolić

Slide 4.1

With the sources of power dissipation in modern integrated circuits well understood, we can start to explore the various sorts of power reduction techniques. As is made clear in the beginning of the chapter, power or energy minimization can be performed at many stages in the design process and may address different targets such as dynamic or static power. This chapter focuses on techniques for power reduction at design time and at circuit level. Practical questions often expressed by designers are addressed: whether gate sizing or choice of supply voltage yields larger returns in terms of power–delay; how many supplies are needed; what the preferred ratio of discrete supplies to thresholds is; etc. As was made clear at the end of the previous chapter, all optimizations should be seen in the broader light of an energy–delay trade-off. To help guide this process, we introduce a unified sensitivity-based optimization framework. The availability of such a framework makes it possible to compare in an unbiased way the impact of various parameters such as gate size and supply and threshold voltages on a given design topology. The results serve as the foundation for optimization at the higher levels of abstraction, which is the focus of later chapters.

J. Rabaey, *Low Power Design Essentials*, Series on Integrated Circuits and Systems,
DOI 10.1007/978-0-387-71713-5_4, © Springer Science+Business Media, LLC 2009

Chapter Outline

- Optimization framework for energy–delay trade-off
- Dynamic-power optimization
  - Multiple supply voltages
  - Transistor sizing
  - Technology mapping
- Static-power optimization
  - Multiple thresholds
  - Transistor stacking

Slide 4.2

The chapter starts with the introduction of a unified energy–delay optimization framework, constructed as an extension of the powerful logical-effort approach, which was originally constructed to target performance optimization. The developed techniques are then used to evaluate the effectiveness and applicability of design-time power reduction techniques at the circuit level. Strategies to address both dynamic and static power are considered.

Energy/Power Optimization Strategy

- For given function and activity, an **optimal operation point** can be derived in the energy–performance space
- Time of optimization depends upon activity profile
- Different optimizations apply to active and static power

	Fixed Activity	Variable Activity	No Activity – Standby
Active	Design time	Run time	Sleep
Static	Design time	Run time	Sleep

Slide 4.3

Before embarking on any optimization, we should recall that the power and energy metrics are related, but that they are by no means identical. The link between the two is the activity, which changes the ratio between the dynamic and static power components, and which may vary dynamically between operational states. Take, for instance, the example of an adder. When the circuit is operated at its maximum speed and inputs are changing constantly and randomly, the dynamic power component dominates. On the other hand, when the activity is low, static power rules. In addition, the desired performance of the adder may very well vary over time as well, further complicating the optimization trajectory.

It will become apparent in this chapter that different design techniques apply to the minimization of dynamic and static power. Hence it is worth classifying power reduction techniques based on the activity level, which is a dynamically varying parameter as discussed before. Fortunately, there exists a broad spectrum of optimizations that can be readily applied at *design time*, either because they are independent of the activity level or because the module activity is fixed and known in advance. These "design-time" design techniques are the topic of the next four chapters. In general though, activity and performance requirements vary over time, and the minimization of power/energy under these circumstances requires techniques that adapt to the prevailing conditions. These are called "run-time" optimizations. Finally, one operational condition requires special attention: the case where the system is idle (or is in "standby"). Under such circumstances, the dynamic power component approaches zero, and leakage power dominates. Keeping the static power within bounds under such conditions requires dedicated design techniques.

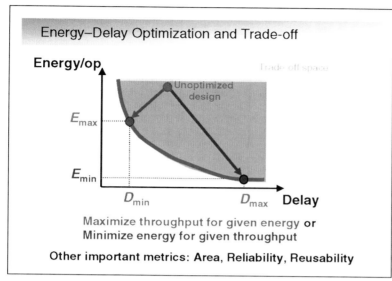

Slide 4.4
At the end of the previous chapter, it was argued that design optimization for power and/or energy requires trade-offs, and that energy and delay represent the major axes of the trade-off space. (Other metrics such as area or reliability play a role as well, but are only considered as secondary factors in this book.) This naturally motivates the use of energy–delay (E–D) space as the coordinate system in which designers evaluate the effectiveness of their techniques.

By changing the various independent design parameters, each design maps onto a constrained region of the energy–delay plane. Starting from a non-optimized design, we want to either speed up the system while keeping the design under the power cap (indicated by E_{max}), or minimize energy while satisfying the throughput constraint (D_{max}). The optimization space is bounded by the *optimal energy–delay curve*. This curve is optimal (for the given set of design parameters), because all other achievable points either consume more energy for the same delay or have a longer delay for the same energy. Although finding the optimal curve seems quite simple in this slide, in real life it is far more complex. Observe also that any optimal energy–delay curve assumes *a given activity level*, and that changes in activity may cause the curve to shift.

Slide 4.5
The problem is that there are many sets of parameters to adjust. Some of these variables are continuous, like transistor sizes, and supply and threshold voltages. Others are discrete, like different logic styles, topologies, and micro-architectures. In theory, it should be possible to consider all parameters at the same time, and to define a single optimization problem. In practice, we have learned that the complexity of the problem becomes overwhelming, and that the resulting designs (if the process ever converges) are very often sub-optimal.

Hence, design methodologies for integrated circuits rely on some important concepts to help manage complexity: abstraction (hiding the details) and hierarchy (building larger entities through a composition of smaller ones). The two most often go hand-in-hand. The abstraction stack of a typical digital IC design flow is shown in this slide. Most design parameters are, in general, confined to and selected in a single layer of the stack only. For instance, the choice between different instruction sets is a typical micro-architecture optimization, while the choice between devices with different threshold voltages is best performed at the circuit layer.

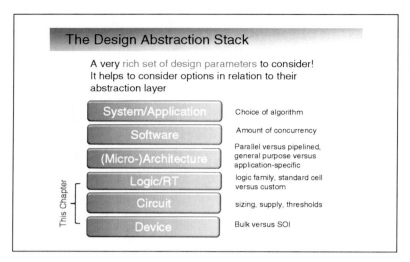

Layering, hence, is the preferred technique to manage complexity in the design optimization process.

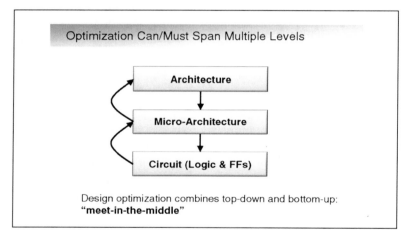

Slide 4.6
The layered approach may give the false impression that optimizations within different layers are independent of each other. This is definitely not the case. For instance, the choice of the threshold voltages at the circuit layer changes the shape of the optimization space at the logical or architectural layers. Similarly, introducing architectural transformations such as pipelining may increase the size of the optimization space at the circuit level, thus leading to larger potential gains. Hence, optimizations may and must span the layers.

Design optimization in general follows a "meet-in-the-middle" formulation: specifications and requirements are propagated from the highest abstraction layer downward (top-down), and constraints are propagated upward from the lowest abstraction later (bottom-up).

Slide 4.7
Continuous design parameters such as supply voltages and transistor sizes give rise to a continuous optimization space and a single optimal energy–delay curve. Discrete parameters, such as the choice between different adder topologies, result in a set of optimal boundary curves. The overall optimum is then defined by their composite.

For example, topology B is better in the energy-performance sense for large target delays, whereas topology A is more effective for shorter delays.

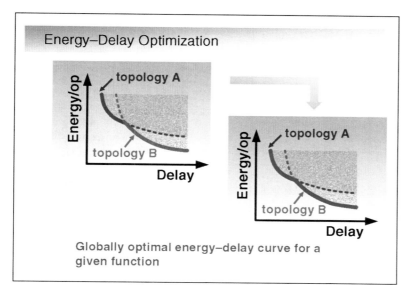

One of the goals of this chapter is to demonstrate how we can quickly search for this global optimum, and based on that, build an understanding of the scope and effectiveness of the different design parameters.

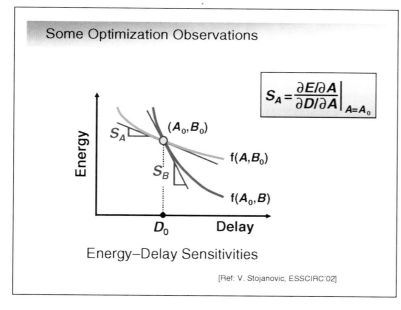

Slide 4.8

Given an appropriate formulation of the energy and delay as a function of the design parameters, any optimization program can be used to derive the optimal energy–delay curve. Most of the optimizations and design explorations in this text were performed using various modules of the MATLAB program [Mathworks].

Yet, though relying on automated optimization is very useful to address large problems or to get precise results quickly, some analytical techniques often come in handy to judge the effectiveness of a given parameter, or to come to a closed-form solution.

The *energy–delay sensitivity* is a tool that does just that: it presents an effective way to evaluate the effectiveness of changes in various design variables. It relies on simple gradient expressions that quantify the profitability of a design modification: how much change in energy and delay results from tuning one of the design variables. Consider, for instance, the operation point (A_0, B_0), where A and B are the design variables being studied. The sensitivity to each of the variables is simply the slope of the curve obtained by a small change in that variable. Observe that the sensitivities are negative owing to the nature of energy–delay trade-off (when we compare sensitivities in the rest of the text, we will use their absolute values – a larger absolute value indicates a higher potential for energy reduction). For example, variable B has higher energy–delay sensitivity at point (A_0, B_0) than the variable A. Changing B hence yields a larger potential gain.

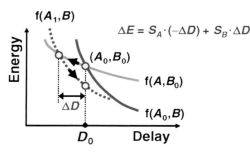

Slide 4.9
The optimal energy–delay curve as defined earlier is a *pareto-optimal curve* (a notion borrowed from economics). An assignment or operational point in a multi-dimensional search is pareto-optimal if improving *on one metric by necessity means hurting another*.

An interesting property of a pareto-optimal point is that the sensitivities to all design variables must be equal. This can be understood intuitively. If the sensitivities are not equal, the difference can be exploited to generate a no-loss improvement. Consider, for instance, the example presented here, where we strive to minimize the energy for a given delay D_0. Using the "lower-energy-cost" variable A, we first create some timing slack ΔD at a small expense in energy ΔE (proportional to A's E–D sensitivity). From the new operation point (A_1, B_0), we can now use "higher-energy-cost" variable B to achieve an overall energy reduction as indicated by the formula. The fixed point in the optimization is clearly reached when all sensitivities are equal.

Slide 4.10
In the rest of the chapter, we primarily focus on the circuit and logic layers. Let us first focus on the *active component* of power dissipation, or, in light of the E–D trade-off perspective, active energy dissipation. The latter is a product of switching activity at the output of a gate, load capacitance at the output, logic swing, and supply voltage. The simple guideline for energy reduction is therefore to reduce each of the terms in the product expression. Some variables, however, are more efficient than others.

The largest impact on active energy is effected seemingly through supply voltage scaling, because of its quadratic impact on power (we assume that the logic swing scales accordingly). All other terms have linear impact. For example, smaller transistors have less capacitance. Switching activity mostly depends on the choice of circuit topology.

For a fixed circuit topology, the most interesting trade-off exists between supply voltage and gate sizing, as these tuning knobs affect both energy and performance. Threshold voltages play a secondary role in this discussion as they impact performance without influencing dynamic energy.

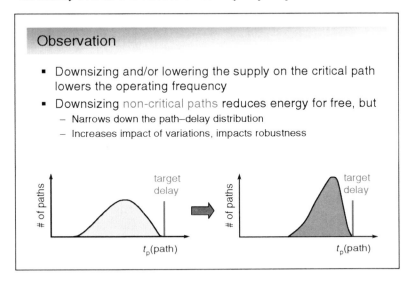

Slide 4.11
Throughout this discussion, it is useful to keep in mind that the optimizations in the E–D space also impact other important design metrics that are not captured here, such as area or reliability. Take, for example, the relationship between transistor sizing and circuit reliability. Trimming the gates on the non-critical paths saves power without a performance penalty – and hence seems to be a win-win operation. Yet in the extreme case, this results in all paths becoming critical (unless a minimum gate size constraint is reached, of course). This effect is illustrated in the slide. The downsizing of non-critical gates narrows the delay distribution and moves the average closer to the maximum delay. This makes this design vulnerable to process-variation effects and degrades its reliability.

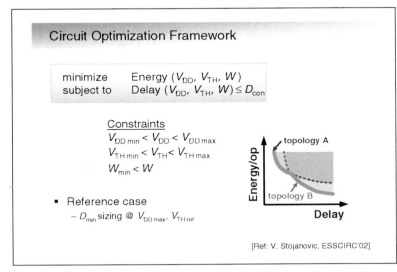

Slide 4.12
To evaluate fully the impact of the design variables in question, that of supply and threshold voltages and gate size on energy and performance, we need to construct a simple and effective, yet accurate, optimization framework. The search for a globally optimal energy–delay curve for a given circuit topology and activity level is formulated as an optimization problem:

Minimize energy subject to a delay constraint and bounds on the range of the optimization variables (V_{DD}, V_{TH}, and W).

Optimization is performed with respect to a *reference design*, sized for minimum delay at the nominal supply and threshold voltages as specified for the technology (e.g., $V_{DD} = 1.2\,V$ and $V_{TH} = 0.35\,V$ for a 90 nm process). This reference point is convenient, as it is well-defined.

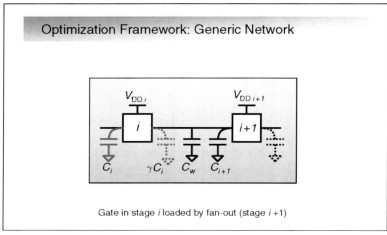

Slide 4.13

The core of the framework consists of effective models of delay and energy as a function of the design parameters. To develop the expressions, we assume a generic circuit configuration as illustrated in the slide. The gate under study is at the i-th stage of a logical network, and is loaded by a number of gates in stage $i+1$, which we have lumped into a single equivalent gate. C_w represents the capacitance of the wire, which we will assume to be proportional to the fan-out (this is a reasonable assumption for a first-order model).

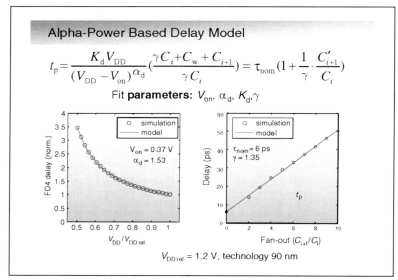

Slide 4.14

The *delay modeling* of the complex gate i proceeds in two steps. First, we derive the delay of an inverter as a function of supply voltage, threshold, and fan-out; Next, we expand this to more complex gates.

The delay of an inverter is expressed using a simple linear delay model, based on the *alpha-power law* for the drain current (see Chapter 2). Note that this model is based on curve-fitting. The parameters V_{on} and α_d are intrinsically related, yet not equal, to the transistor threshold and the velocity saturation index. K_d is another fit parameter and relates to the transconductance of the process (amongst others). The model fits SPICE simulated data quite nicely, across a range of supply voltages, normalized to the nominal supply voltage (which is 1.2 V for our 90 nm CMOS technology). Observe that this model is only valid if the supply voltage exceeds the threshold voltage by a reasonable amount. (This constraint will be removed in Chapter 11, where we present a modified model that extends into the sub-threshold region.)

The fan-out $f = C_{i+1}/C_i$ represents the ratio of the load capacitance divided by the gate capacitance. A small modification allows for the inclusion of the wire capacitance (f'). γ is another technology-dependent parameter, representing the ratio between the output and input capacitance of a minimum-sized unloaded inverter.

Slide 4.15

Combined with Logical-Effort Formulation

For Complex Gates

$$t_p = \tau_{\text{nom}}\left(p_i + \frac{f_i g_i}{\gamma}\right)$$

- Parasitic delay p_i – depends upon gate topology
- Electrical effort $f_i \approx S_{i+1}/S_i$
- Logical effort g_i – depends upon gate topology
- Effective fan-out $h_i = f_i g_i$

[Ref: I. Sutherland, Morgan-Kaufman'99]

The other part of the model is based on the *logical-effort* formulation, which extends the notion to complex gates. Using the logical-effort notation, the delay can be expressed simply as a product of the process-dependent time constant τ_{nom} and a unitless delay, $p_i + f_i g_i/\gamma$, in which g is the logical effort that quantifies the relative ability of a gate to deliver current, f is the ratio of the total output to input capacitance of the gate, and p represents the delay component due to the self-loading of the gate. The product of the logical effort and the electrical effort is called the effective fan-out h. Gate sizing enters the equation through the fan-out factor $f = S_{i+1}/S_i$.

Slide 4.16

Dynamic Energy

$$E_{\text{dyn}} = (\gamma C_i + C_w + C_{i+1}) \cdot V_{DD,i}^2 = C_i(\gamma + f'_i) \cdot V_{DD,i}^2$$

$$C_i = K_e S_i \quad f'_i = (C_w + C_{i+1})/C_i = S'_{i+1}/S_i$$

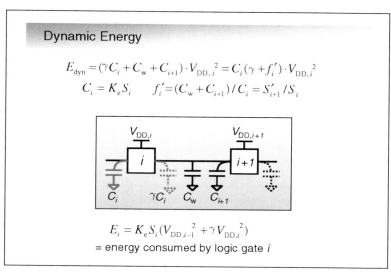

$$E_i = K_e S_i (V_{DD,i-1}^2 + \gamma V_{DD,i}^2)$$
= energy consumed by logic gate i

For the time being, we only consider the switching energy of the gate. In this model, $f'_i C_i$ is the total load at the output, including wire and gate loads, and γC_i is the self-loading of the gate. The total energy stored on these capacitances is the energy taken out of the supply voltage in stage i.

Now, if we change the size of the gate in stage i, it affects only the energy stored on the input capacitance and parasitic capacitance of that gate. E_i hence is defined as the energy that the gate at stage i contributes to the overall energy dissipation.

Slide 4.17

Optimizing Return on Investment (ROI)

Depends on Sensitivity ($\partial E/\partial D$)

- Gate Sizing

$$\frac{\partial E/\partial S_i}{\partial D/\partial S_i} = -\frac{E_i}{\tau_{nom}(h_i - h_{i-1})}$$

∞ for equal h (D_{min})

- Supply Voltage

$$\frac{\partial E/\partial V_{DD}}{\partial D/\partial V_{DD}} = -\frac{E}{D} \cdot \frac{2 \cdot (1 - \frac{V_{on}}{V_{DD}})}{\alpha_d - 1 + \frac{V_{on}}{V_{DD}}}$$

max at V_{DD}(max) (D_{min})

As mentioned, sensitivity analysis provides intuition about the profitability of optimization. Using the models developed in the previous slides, we can now derive expressions for the sensitivities to some of the key design parameters.

The formulas indicate that the largest potential for energy savings is at the minimum delay, D_{min}, which is obtained by equalizing the effective fan-out of all stages, and setting the supply voltage at the maximum allowable value. This observation intuitively makes sense: at minimum delay, the delay cannot be reduced beyond the minimum achievable value, regardless of how much energy is spent. At the same time, the potential of energy savings through voltage scaling decreases with reducing supply voltages: E decreases, while D and the ratio V_{on}/V_{DD} increase.

The key point to realize is that optimization primarily exploits the tuning variable with the largest sensitivity, which ultimately leads to the solution where all sensitivities are equal. You will see this concept at work in a number of examples.

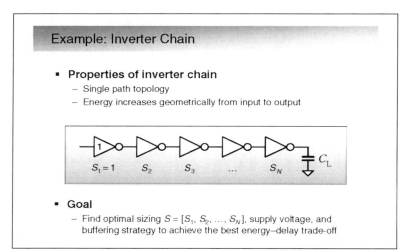

Slide 4.18

We use a number of well-known circuit topologies to illustrate the concepts of circuit optimization for energy. The examples differ in the amount of off-path loading and path reconvergence. By analyzing how these properties affect the energy profile, we may come to some general principles related to the impact of the various design parameters. More precisely, we study the (well-understood) inverter chain and the tree adder – as these examples differ widely in the number of paths and path reconvergence.

Let us begin with the inverter chain. The goal is to find the optimal sizing, the supply voltages, and the number of stages that result in the best energy–delay trade-off.

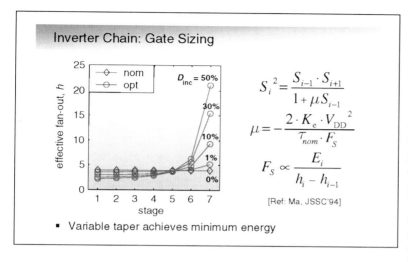

Slide 4.19

The inverter chain has been the focus of a lot of attention, as it is a critical component in digital design, and some clear guidelines about optimal design can be derived in closed form. For minimum delay, the fan-out of each stage is kept constant, and each subsequent stage is up-sized with a constant factor. This means that the energy stored per stage increases geometrically toward the output, with the largest energy stored in the final load.

In a first step, we consider solely transistor sizing. For a given delay increment, the optimum size of each stage, which minimizes energy, can be derived. The sensitivities derived in Slide 4.17 already give a first idea on what may unfold: the sensitivity to gate sizing is proportional to the energy stored on the gate, and is inversely proportional to the difference in effective fan-outs. What this means is that, for equal sensitivity in all stages, the difference in the effective fan-outs of a gate must increase in proportion to the energy stored on the gate, indicating that the difference in the effective fan-outs should increase exponentially toward the output.

This result was already analytically derived by Ma and Franzon [Ma, JSSC'94], who showed that a tapered staging is the best way to combine performance and energy efficiency. One caveat: At large delay increments, a more efficient solution can be found by reducing the number of stages — this was not included as a design parameter in this first-order optimization, in which the topology was kept unchanged.

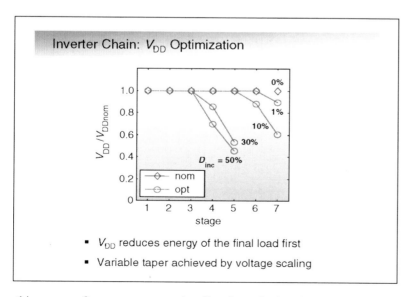

Slide 4.20

Let us now consider the potential of supply voltage scaling. We assume that each stage can be run at a different voltage. As in sizing, the optimization tackles the largest consumers – the final stages – first by scaling their supply voltages. The net effect is similar to a "virtual" tapering. An important difference between sizing and supply reduction is that sizing does not affect the energy stored in the final output load C_L. Supply reduction, on the other hand, lowers this source of energy consumption first, by reducing the supply voltage of the gate that drives the load. As (dis)charging C_L is the largest source of energy consumption, the impact of this is quite profound.

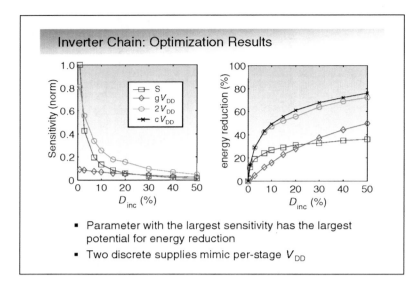

- Parameter with the largest sensitivity has the largest potential for energy reduction
- Two discrete supplies mimic per-stage V_{DD}

Slide 4.21
Now, how good can all this be in terms of energy reduction? In the graphs, we present the results of various optimizations performed on the inverter chain: sizing, reducing the global V_{DD}, two discrete V_{DD}s, and a customizable V_{DD} per stage. For each of these cases, the sensitivity and the energy reduction are plotted as functions of the delay increment (over D_{min}). The prime observation is that increasing the delay by 50% reduces the energy dissipation by more than 70%. Again, it is shown that for any value of the delay increment, the parameter with the largest sensitivity has the largest potential for energy reduction. For example, at small delay increments sizing has the largest sensitivity (initially infinity), so it offers the largest energy reduction. Its potential however quickly falls off. At large delay increments, it pays to scale the supply voltage of the entire circuit, achieving the sensitivity equal to that of sizing at around 25% excess delay. The largest reductions can be obtained by custom voltage scaling. Yet, two discrete voltages are almost as good, and are a lot simpler from an implementation perspective.

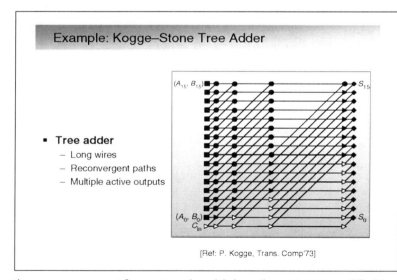

- Tree adder
  - Long wires
  - Reconvergent paths
  - Multiple active outputs

[Ref: P. Kogge, Trans. Comp'73]

Slide 4.22
An inverter chain has a particularly simple energy distribution, which grows geometrically until the final stage. This type of profile drives the optimization (for both sizing and supply) to focus on the final stages first. However, most practical circuits have a more complex energy profile.

An interesting counterpart is formed by the tree adder, which features long wires, large fan-out variations, reconvergent fan-out, and multiple active outputs qualified by paths of various logic depths. We have selected a popular instance of such an adder, the Kogge–Stone version, for our study [Kogge'93, Rabaey'03]. The overall architecture of the adder consists of a number of propagate/generate functions at the inputs (identified by the squares), followed by carry-merge operators (circles). The final-sum outputs are generated through XOR functions (diamonds). To balance the delay paths, buffers (triangles) are inserted in many of the paths.

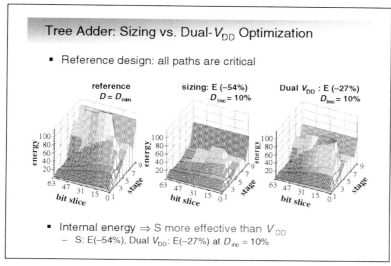

Slide 4.23

The adder topology is best understood in a two-dimensional plane. One axis is formed by the different bit slices N (we are considering a 64-bit adder for this example), whereas the other is formed by the consecutive gate stages. As befits a tree adder, the number of stages equals $\log_2(N) + M$, where M is the extra stages for propagate/generate and the final XOR functionality. The energy of an internal node is best understood when plotted with respect to this two-dimensional topology.

As always, we start from a reference design that is optimized for minimum delay, and we explore how we can trade off energy and delay starting from that point. The initial sizing makes all paths in the adder equal to the critical path. The first figure shows the energy map for the minimum delay. Though the output nodes are responsible for a sizable fraction of the energy consumption, a number of internal nodes (around stage 5) dominate.

The large internal energy increases the potential for energy reduction through gate sizing. This is illustrated by the case where we allow for a 10% delay increase. We have plotted the energy distribution resulting from sizing, as well as from the introduction of two discrete supply voltages. The former results in 54% reduction in overall energy, whereas the latter only (!) saves 27%.

This result can be explained as follows. Given the fact that the dominant energy nodes are internal, sizing allows each of these nodes to be attacked individually without too much of a global impact. In the case of dual supplies, one must be aware that driving a high-voltage node from a low-voltage node is hard. Hence the preferable assignment of low-voltage nodes is to start from the output nodes and to work one's way toward the input nodes. Under these conditions, we have already sacrificed a lot of delay slack on low-energy intermediate nodes before we reach the internal high-energy nodes. In summary, supply voltages cannot be randomly assigned to nodes. This makes the usage of discrete supply voltages less effective in modules with high internal energy.

Slide 4.24

We can now put it all together, and explore the tree adder in the energy–delay space. Each of the design parameters (V_{DD}, V_{TH}, S) is analyzed separately and in combination with the others. (Observe that inclusion of the threshold voltage as a design parameter only makes sense when the leakage energy is considered as well – how this is done is discussed later in the chapter).

A couple of interesting conclusions can be drawn:

- Through circuit optimization, we can reduce the energy consumption of the adder by a factor of 10 by doubling the delay.

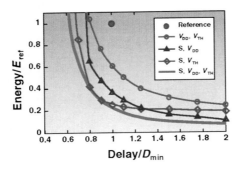

- Exploiting only two out of the three variables yields close to the optimal gain. For the adder, the most effective parameters are sizing and threshold selection. At the reference design point, sizing and threshold reduction feature the largest and the smallest sensitivities, respectively. Hence, this combination has the largest potential for energy reduction along the lines demonstrated in Slide 4.8.
- Finally, circuit optimization is most effective in a small region around the reference point. Expanding beyond that region typically becomes too expensive in terms of energy or delay cost for small gains, yielding a reduced return on investment.

Slide 4.25

So far, we have studied the theoretical impact of circuit optimization on energy and delay. In reality, the design space is more constrained. Choosing a different supply or threshold voltage for every gate is not a practical option. Transistor sizes come in discrete values, as determined by the available design library. One of the fortunate conclusions emerging from the preceding studies is that a couple of well-chosen discrete values for each of the design parameters can get us quite close to the optimum.

Let us first consider the practical issues related to the use of multiple supply voltages – a practice that until recent was not common in digital integrated circuit design at all. It impacts the layout strategy and complicates the verification process (as will be discussed in Chapter 12). In addition, generating, regulating, and distributing multiple supplies are non-trivial tasks.

A number of different design strategies exist with respect to the usage of multiple supply voltages. The first is to assign the voltage at the block/macro level (the so-called voltage island approach). This makes particular sense in case some modules have higher performance/activity requirements than others (for instance, a processor's data path versus its memory). The second and more general approach is to allow for voltage assignment all the way down to the gate level ("custom voltage assignment"). In general, this means that gates on the non-critical paths are assigned a lower supply voltage. Be aware that having signals at different voltage levels requires the insertion of level converters. It is preferable if these are limited in number (as they consume extra energy) and occur only at the boundaries of the modules.

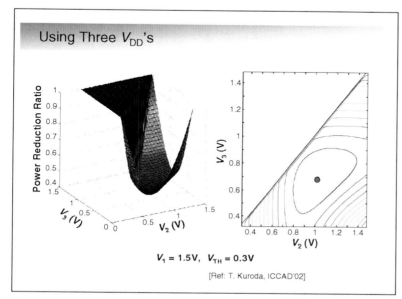

Slide 4.26

With respect to multiple supply voltages, one cannot help wondering about the following question: If multiple supply voltages are employed, how many discrete levels are sufficient, and what are their values? This slide illustrates the potential of using three discrete voltage levels, as was studied by Tadahiro Kuroda [Kuroda, ICCAD'02]. Supply assignment to the individual logic gates is performed by an optimization routine that minimizes energy for a given clock period. With the main supply fixed at 1.5 V, providing a second and third supply yields a nearly twofold power reduction ratio.

A number of useful observations can be drawn from the graphs:

- The power minimum occurs for $V_2 \approx 1$ V and $V_3 \approx 0.7$ V.
- The minimum is quite shallow. This is good news, as this means that small deviations around this minimum (as caused, for instance, by IR drops) will not have a big impact.

The question now is how much impact on power each additional supply carries.

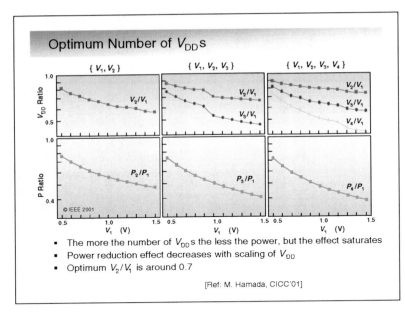

- The more the number of V_{DD}s the less the power, but the effect saturates
- Power reduction effect decreases with scaling of V_{DD}
- Optimum V_2/V_1 is around 0.7

[Ref: M. Hamada, CICC'01]

Slide 4.27

In fact, the marginal benefits of adding extra supplies quickly bottom out. Although adding a second supply yields big savings, the extra reductions obtainable by adding a third or a fourth are marginal. This makes sense, as the number of (non-critical) gates that can benefit from the additional supply shrinks with each iteration. For example, the fourth supply works only with non-critical path gates close to the tail of the delay distribution. Another observation is that the power savings obtainable from using multiple supplies reduce with the scaling of the main supply voltage (for a fixed threshold).

Lessons: Multiple Supply Voltages

- Two supply voltages per block are optimal
- Optimal ratio between the supply voltages is 0.7
- Level conversion is performed on the voltage boundary, using a level-converting flip-flop (LCFF)
- An option is to use an asynchronous level converter
  - More sensitive to coupling and supply noise

Slide 4.28

Our discussion on multiple discrete supply voltages can be summarized with a number of rules-of-thumb:

- The largest benefit is obtained by adding a second supply.
- The optimal ratio between the discrete supplies is approximately 0.7.
- Adding a third supply provides an additional 5–10% incremental savings. Going beyond that does not make much sense.

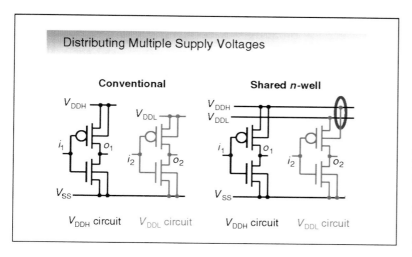

Slide 4.29

Distribution of multiple supply voltages requires careful examination of the floorplanning strategy. The conventional way to support multiple V_{DD}'s (two in this case) is to place gates with different supplies in different wells (e.g., low-V_{DD} and high-V_{DD}). This approach does not require a redesign of the standard cells, but comes with an area overhead owing to the necessary spacing between n-wells at different voltages. Another way to introduce the second supply is to provide two V_{DD} rails for every standard cell, and selectively route the cells to the appropriate supply. This "shared n-well" approach also comes with an area overhead owing to the extra voltage rail. Let us further analyze both techniques to see what kind of system-level trade-offs they introduce.

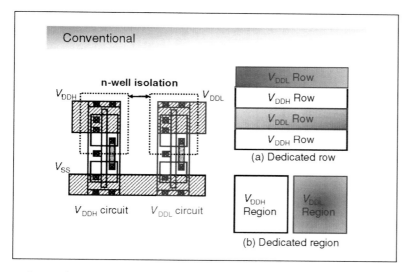

Slide 4.30

In the conventional dual-voltage approach, the most straightforward method is to cluster gates with the same supply (scheme b). This scheme works well for the "voltage island" model, where a single supply is chosen for a complete module. It does not very well fit the "custom voltage assignment" mode, though. Logic paths consisting of both high-V_{DD} and low-V_{DD} cells incur additional overhead in wire delay due to long wires between the voltage clusters. The extra wire capacitance also reduces the power savings. Maintaining spatial locality of connected combinational logic gates is essential.

Another approach is to assign voltages per row of cells (scheme a). Both V_{DDL} and V_{DDH} are routed only to the edge of the rows, and a special standard cell is added that selects between the two voltages (obviously, this applies only to the standard-cell methodology). This approach suits the "custom voltage assignment" approach better, as the per-row assignment provides a smaller granularity and the overhead of moving between voltage domains is smaller.

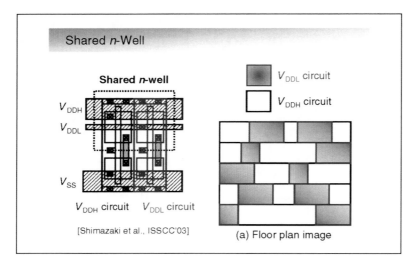

Slide 4.31
The most versatile approach is to redesign standard cells, and have both V_{DDL} and V_{DDH} rails inside the cell ("shared n-well"). This approach is quite attractive, because we do not have to worry about area partitioning – both low-V_{DD} and high-V_{DD} cells can be abutted to each other. This approach was demonstrated on a high-speed adder/ALU circuit by Shimazaki et al [Shimazaki, ISSCC'03]. However, it comes with a per-cell area overhead. Also, low-V_{DD} cells experience reverse body biasing on the PMOS transistors, which degrades their performance.

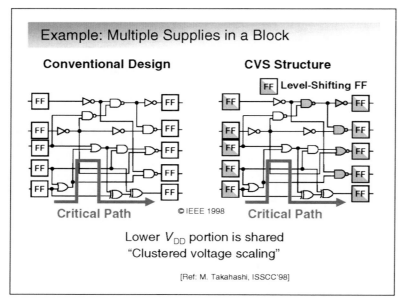

Slide 4.32
Level conversion is another important issue in designing with multiple discrete supply voltages. It is easy to drive a low-voltage gate from a high-voltage one, but the opposite transition is hard owing to extra leakage, degraded signal slopes, and performance penalty. It is hence worthwhile minimizing the occurrence of low-to-high connections.

As we will see in the next few slides, low-to-high level conversion is best accomplished using positive feedback – which is naturally present in flip-flops and registers. This leads to the following strategy: Every logical path starts at the high-voltage level. Once a path transitions to the low voltage, it never switches back. The next up-conversion happens in flip-flops. Supply voltage assignment starts from critical paths and works backward to find non-critical paths where the supply voltage can be reduced. This strategy is illustrated in the slide. The conventional design on the left has all gates operating at the nominal supply (the critical path is highlighted). Working backward from the flip-flops, non-critical paths are gradually converted to the low voltage until they become critical (gray-shaded gates operate at V_{DDL}). This technique of grouping is called *"clustered voltage scaling"* (CVS).

Slide 4.33

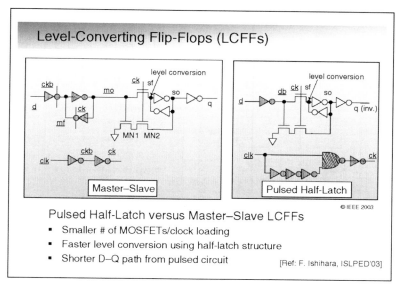

As the level-converting flip-flops play a crucial role in the CVS scheme, we present a number of flip-flops that can do level conversion and maintain good speed.

The first circuit is based on the traditional master–slave scheme, with the master and slave stages operating at the low and high voltages, respectively. The positive feedback action in the slave latch ensures efficient low-to-high level conversion. The high voltage node *sf* is isolated from low-voltage node *mo* by the pass-transistor, gated by the low-voltage signal *ck*.

The same concept can also be applied in an edge-triggered flip-flop, as shown in the second circuit (called the pulse-based half-latch). A pulse generator derives a short pulse from the clock edge, ensuring that the latch is enabled only for a very short time. This circuit has the advantage of being simpler.

Slide 4.34

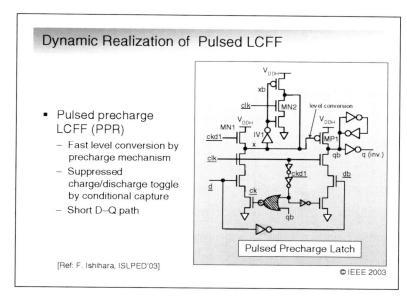

Dynamic gates with NMOS-only evaluation transistors are naturally suited for operation with reduced logic swing, as the input signal does not need to develop a full high-V_{DD} swing to drive the output node to logic zero. The reduced swing only results in a somewhat longer delay. A dynamic structure with implicit level conversion is shown in the figure.

Observe that level conversion is also possible in an asynchronous fashion. A number of such non-clocked converters will be presented in a later chapter on Interconnect (Chapter 6). Clocked circuits tend to be more reliable, however.

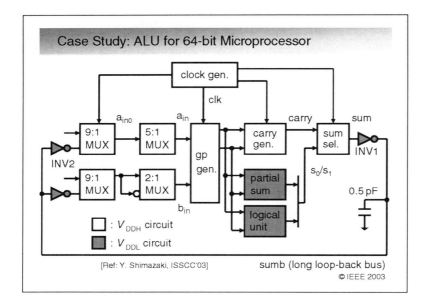

Slide 4.35

A real-life example of a high-performance Itanium-class (©Intel) data path helps to demonstrate the effective use of dual-V_{DD}. From the block diagram, it is apparent that the critical component from an energy perspective is the very large output capacitance of the ALU, which is due to its high fan-out. Hence, lowering the supply voltage on the output bus yields the largest potential for power reduction.

The shared-well technique was chosen for the implementation of this 64-bit ALU module, which is composed of the ALU, the loop-back bus driver, the input operand selectors, and the register files. For performance reasons, a domino circuit-style was adopted. As the carry generation is the most critical operation, circuits in the carry tree are assigned to the V_{DDH} domain. On the other hand, the partial-sum generator and the logical unit are assigned to the V_{DDL} domain. In addition, the bus driver, as the gate with the largest load, is also supplied from V_{DDL}. The level conversion from the V_{DDL} signal to the V_{DDH} signal is performed by the sum selector and the 9:1 multiplexer.

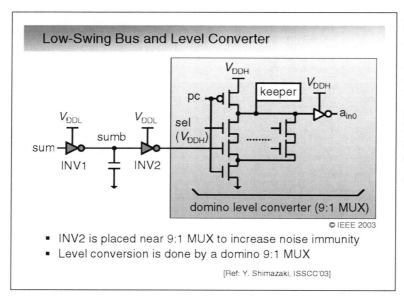

Slide 4.36

This schematic shows the low-swing loop-back bus and the domino-style level converter. Since the loop-back bus *sumb* has a large capacitive load, low-voltage implementation is quite attractive. Some issues deserve special attention:

- One of the concerns of the shared-well approach is the reverse biasing on the PMOS transistor. As *sum* is a monotonically rising signal (output of a domino stage), this does not impact the performance of the important gate INV1.

Optimizing Power @ Design Time – Circuit-Level Techniques

- In dynamic-logic designs, noise is one of the critical issues. To eliminate the effects of disturbances on the loop-back bus, the receiver INV2 is placed near the 9:1 multiplexer to increase noise immunity.
- The output of INV2, which is a V_{DDL} signal, is converted V_{DDH} by the 9:1 multiplexer. The level conversion is fast, as the precharge levels are independent of the level of the input signal.

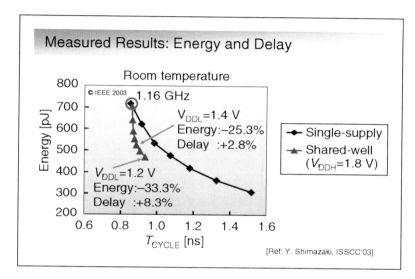

Slide 4.37

This figure plots the familiar energy–delay plots of the ALU (as measured). The energy–delay curve for single-supply operation is drawn as a reference. At the nominal supply voltage of 1.8 V (for a 180 nm CMOS technology), the chip operates at 1.16 GHz. Introducing a second supply yields an energy saving of 33% at the small cost of 8% in delay increase. This example demonstrates that the theoretical results derived in the earlier slides of this chapter are actually for real.

Slide 4.38

Transistor sizing is the other high-impact design parameter we have explored at the circuit level so far. The theoretical analysis assumes a continuous sizing model, which is only a possibility in purely custom design. In ASIC design flows, transistor sizes are predetermined in the cell library. In the early days of application-specific integrated circuit (ASIC) design and automated synthesis, libraries used to be quite small, counting between 50 and 100 cells. Energy considerations have changed the picture substantially. With the need for various sizing options for each logical cell, industrial libraries now count close to 1000 cells. As with supply voltages, it is necessary to move from a continuous model to a discrete one. Similarly, the overall impact on energy efficiency of doing so can be quite small.

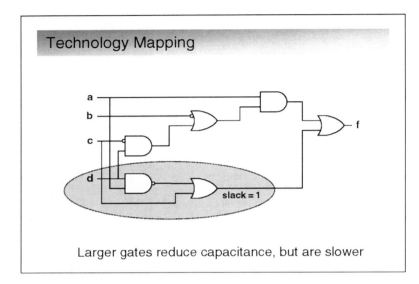

Slide 4.39

In the ASIC design flow, it is in the "technology mapping" phase that the actual library cells are selected for the implementation of a given logical function. The logic network, resulting from "technology-independent" optimizations, is mapped onto the library cells such that performance constraints are met and energy is minimized. Hence, this is where the transistor (gate) sizing actually happens. Beyond choosing between identical cells with different sizes, technology mapping also gets to choose between different gate mappings: simple cells with small fan-in, or more complex cells with large fan-in. Over the last decade(s), it has been common understanding that simple gates are good from a performance perspective – delay is a quadratic function of fan-in. From an energy perspective, complex gates are more attractive, as the intrinsic capacitance of these is substantially smaller than the inter-gate routing capacitances of a network of simple gates. Hence, it makes sense for complex gates to be preferentially used on non-critical paths.

Technology Mapping

Example: four-input AND
- (a) Implemented using four-input NAND + INV
- (b) Implemented using two-input NAND + two-input NOR

Gate type	Area (cell unit)	Input cap. (fF)	Library 1: High-Speed Average delay (ps)	Library 2: Low-Power Average delay (ps)
INV	3	1.8	$7.0 + 3.8C_L$	$12.0 + 6.0C_L$
NAND2	4	2.0	$10.3 + 5.3C_L$	$16.3 + 8.8C_L$
NAND4	5	2.0	$13.6 + 5.8C_L$	$22.7 + 10.2C_L$
NOR2	3	2.2	$10.7 + 5.4C_L$	$16.7 + 8.9C_L$

(delay formula: C_L in fF)
(numbers calibrated for 90 nm)

Slide 4.40
This argument is illustrated with an example. In this slide, we have summarized the area, delay, and energy properties of four cells (INV, NAND2, NOR2, NAND4) implemented in a 90 nm CMOS technology. Two different libraries are considered: a low-power and a high-performance version.

Technology Mapping – Example

four-input AND	(a) NAND4 + INV	(b) NAND2 + NOR2
Area	8	11
HS: Delay (ps)	$31.0 + 3.8C_L$	$32.7 + 5.4C_L$
LP: Delay (ps)	$53.1 + 6.0C_L$	$52.4 + 8.9C_L$
Sw Energy (fF)	$0.1 + 0.06C_L$	$0.83 + 0.06C_L$

- Area
  - Four-input more compact than two-input (two gates vs three gates)
- Timing
  - Both implementations are two-stage realizations
  - Second-stage INV (a) is better driver than NOR2 (b)
  - For more complex blocks, simpler gates will show better performance
- Energy
  - Internal switching increases energy in the two-input case
  - Low-power library has worse delay, but lower leakage (see later)

Slide 4.41
These libraries are used to map the same function, an AND4, using either two-input or four-input gates (NAND4 + INV or NAND2 + NOR2). The resulting metrics show that the complex gate implementation yields a substantial reduction in energy and also reduces area. For this simple example, the complex-gate version is just as fast, if not faster. However this is due to the somewhat simplistic nature of the example. The situation becomes even more pronounced if the library would contain very complex gates (e.g., fan-in of 5 or 6).

Slide 4.42
Technology mapping has brought us almost seamlessly to the next abstraction level in the design process – the logic level. Transistor sizes, voltage levels, and circuit style are the main optimization knobs at the circuit level. At the logic level, the gate–network topology to implement a given

Gate-Level Trade-offs for Power

- **Technology mapping**
  - Gate selection
  - Sizing
  - Pin assignment
- **Logical Optimizations**
  - Factoring
  - Restructuring
  - Buffer insertion/deletion
  - Don't-care optimization

function is chosen and fine-tuned. The link between the two is the already discussed technology-mapping process. Beyond gate selection and transistor sizing, technology mapping also performs pin assignment. It is well known that, from a performance perspective, it is a good idea to connect the most critical signal to the input pin "closest" to the output node. For a CMOS NAND gate, for instance, this would be the top transistor of the NMOS pull-down chain. From a power reduction point of view, on the other hand, it is wise to connect the most active signal to that node, as this minimizes the switching capacitance.

The technology-independent part of the logic-synthesis process consists of a sequence of optimizations that manipulate the network topology to minimize delay, power, or area. As we have become used to, each such optimization represents a careful trade-off, not only between power and delay, but sometimes also between the different components of power such as activity and capacitance. This is illustrated with a couple of examples in the following slid

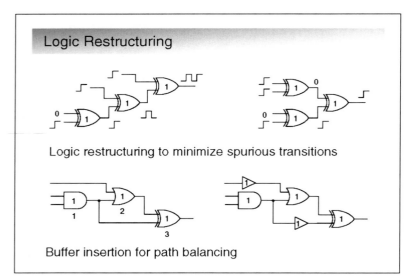

Slide 4.43
In Chapter 3, we have established that the occurrence of dynamic hazards in a logic network is minimized when the network is *balanced* from a timing perspective – that is, most timing paths are of similar lengths. Paths of unequal length can always be equalized with respect to time in a number of ways: (1) through the restructuring of the network, such that an equivalent network with balanced paths is obtained; (2) through the introduction of non-inverting buffers on the fastest paths. The attentive reader realizes that although the latter helps to minimize glitching, the buffers themselves add extra switching capacitance. Hence, as always, buffer insertion is a careful trade-off process. Analysis of circuits

generated by state-of-the-art synthesis tools have shown that simple buffers are responsible for a considerable part of the overall power budget of the combinatorial modules.

Slide 4.44

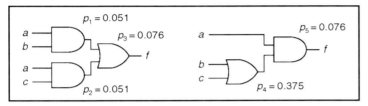

Factoring is another transformation that may introduce unintended consequences. From a capacitance perspective, it seems obvious that a simpler logical expression would require less power as well. For instance, translating the function $f = a \cdot b + a \cdot c$ into its equivalent $f = a \cdot (b + c)$ seems a no-brainer, as it requires one less gate. However, it may also introduce an internal node with substantially higher transition probabilities, as annotated on the slide. This may actually increase the net power. The lesson to be drawn is that power-aware logical synthesis must not only be aware of network topology and timing, but should – to the best possible extent – incorporate parameters such as capacitance, activity, and glitching. In the end, the goal is again to derive the pareto-optimal energy–delay curves, which we are now so familiar with, or to reformulate the synthesis process along the following lines: choose the network that minimizes power for a given maximum delay or minimizes the delay for a maximum power.

Slide 4.45

Lessons from Circuit Optimization

- Joint optimization over multiple design parameters possible using sensitivity-based optimization framework
  - Equal marginal costs ⇔ Energy-efficient design
- Peak performance is VERY power inefficient
  - About 70% energy reduction for 20% delay penalty
  - Additional variables for higher energy-efficiency
- Two supply voltages in general sufficient; three or more supply voltages only offer small advantage
- Choice between sizing and supply voltage parameters depends upon circuit topology
- But ... leakage not considered so far

Based on the preceding discussions, we can now draw a clear set of guidelines for energy–delay optimization at the circuit and logical levels. An attempt of doing so is presented in this slide.

Yet, so far we have only addressed dynamic power. In the rest of the chapter we tackle the other important contributor of power in contemporary networks: leakage.

Slide 4.46

Considering Leakage at Design Time

- Considering leakage as well as dynamic power is essential in sub-100 nm technologies
- Leakage is not essentially a bad thing
  - Increased leakage leads to improved performance, allowing for lower supply voltages
  - Again a trade-off issue ...

Leakage has so far been presented as an evil side effect of nanometer-size technology scaling, something that should be avoided by all cost. However, given an actual technology node, this may not necessarily be the case. For instance, a lower threshold (and increased leakage) allows for a lower supply voltage for the same delay — effectively trading off dynamic power for static power. This was already illustrated graphically in Slide 3.41, where power and delay of a logical function were plotted as a function of supply and threshold voltages. Once one realizes that allowing for an amount of static power may actually be a good thing, the next question inevitably arises: is there an optimal balance between dynamic and static power, and if so, what is the "golden" ratio?

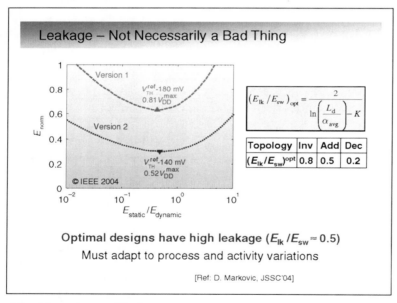

Slide 4.47

The answer is an unequivocal yes. This is best illustrated by the graph in this slide, which plots the normalized minimum energy per operation for a given function and a given delay as a function of the ratio between static and dynamic power. The same curve is also plotted for a modified version of the same function.

A number of interesting observations can be drawn from this set of graphs:

- The most energy-efficient designs have a considerable amount of leakage energy.
- For both the designs, the static energy is approximately 50% of the dynamic energy (or one-third of the total energy), and does not vary very much between the different circuit topologies.
- The curves are fairly flat around the minimum, making the minimum energy somewhat insensitive to the precise ratio.

This ratio does not change much for different topologies except if activity changes by orders of magnitude, as the optimal ratio is a logarithmic function of activity and logic depth. Still, looking into significantly different circuit topologies in the last few slides, we found that optimal ratio of the leakage-to-switching energy did not change much. Moreover, in the range defined by these extreme cases, energy of adder-based implementations is still very close to minimum, from 0.2 to 0.8 leakage-to-switching ratio, as shown in this graph. A similar situation occurs if we analyze inverter chain and memory decoder circuits assuming an optimal leakage-to-switching ratio of 0.5.

From this analysis, we can derive a very simple general result: *energy is minimized when the leakage-to-switching ratio is about 0.5*, regardless of logic topology or function. This is an important practical result. We can use this knowledge to determine the optimal V_{DD} and V_{TH} in a broad range of designs.

Refining the Optimization Model

- Switching energy

$$E_{dyn} = \alpha_{0 \to 1} K_e S(\gamma + f) V_{DD}^2$$

- Leakage energy

$$E_{stat} = S I_0(\Psi) e^{\frac{-V_{TH} + \lambda_d V_{DD}}{kT/q}} V_{DD} T_{cycle}$$

with:

$I_0(\Psi)$: normalized leakage current with inputs in state Ψ

Slide 4.48
The effect of leakage is easily introduced in our earlier-defined optimization framework. Remember that the leakage current of a module is a function of the state of its inputs. However, it is often acceptable to use the average leakage over the different states. Another observation is that the ratio between dynamic and static energy is a function of the cycle time and the average activity per cycle.

Reducing Leakage @ Design Time

- Using longer transistors
  - Limited benefit
  - Increase in active current
- Using higher thresholds
  - Channel doping
  - Stacked devices
  - Body biasing
- Reducing the voltage!!

Slide 4.49
When trying to manipulate the leakage current, the designer has a number of knobs at her disposition – In fact, they are quite similar to the ones we used for optimizing the dynamic power: transistor sizes, and threshold and supply voltages. How they influence leakage current is substantially different though. The choice of the threshold voltage is especially important.

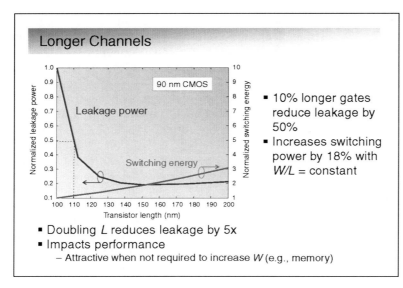

Slide 4.50
While wider transistors obviously leak more, the chosen transistor length has an impact as well. As already shown in Slide 2.15, very short transistors suffer from a sharp reduction in threshold voltage, and hence an exponential increase in leakage current. In leakage-critical designs such as memory cells, for instance, it makes sense to consider the use of transistors with longer channel lengths rather than the ones prescribed by the nominal process parameters. This comes at a penalty in dynamic power though, but that increase is relatively small. For a 90 nm CMOS technology, it was shown that increasing the channel length by 10% reduces the leakage current by 50%, while raising the dynamic power by 18%. It may seem strange to deliberately forgo one of the key benefits of technology scaling – that is, smaller transistors – yet sometimes the penalty in area and performance is inconsequential, whereas the gain in overall power consumption is substantial.

Slide 4.51
Using multiple threshold voltages is an effective tool in the static-power optimization portfolio. In contrast to the usage of multiple supply voltages, introducing multiple thresholds has relatively little impact on the design flow. No level converters are needed, and no special layout strategies are required. The real burden is the added cost to the manufacturing process. From a design perspective, the challenge is on the technology mapping process, which is where the choice between cells with different thresholds is really made.

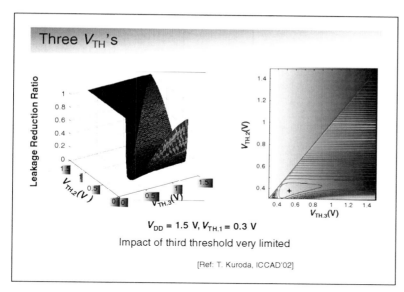

Slide 4.52

The immediate question is how many threshold voltages are truly desirable. As with supply voltages, the addition of more levels comes at a substantial cost, and most likely yields a diminishing return. A number of studies have shown that although there is still some benefit in having three discrete threshold voltages for both NMOS and PMOS transistors, it is quite marginal. Hence, two thresholds for both devices have become the norm in the sub-100 nm technologies.

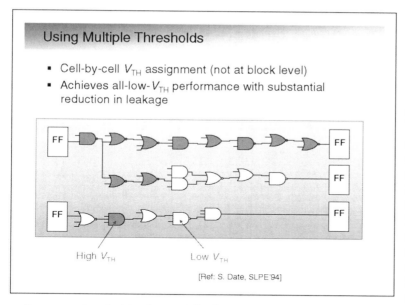

Slide 4.53

As was the case with dynamic power reduction, the strategy is to increase the threshold voltages in timing paths that are not critical, leading to static leakage power reduction at no performance and dynamic power costs. The appealing factor is that high-threshold cells can be introduced anywhere in the logic structure without major side effects. The burden is clearly on the tools, as timing slack can be used in a number of ways: reducing transistor sizes, supply voltages, or threshold voltages. The former two reduce both dynamic and static power, whereas the latter only influences the static component. Remember however that an optimal design carefully balances both components.

Slide 4.54

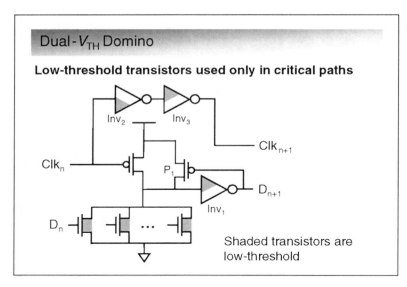

Most of the discussion on leakage so far has concentrated on static logic. I reckon that dynamic-circuit designers are even more worried: for them, leakage means not only power dissipation but also a serious degradation in noise margin. Again, a careful selection between low- and high-threshold devices can go a long way. Low-threshold transistors are used in the timing-critical paths, such as the pull-down logic module. Yet even with these options, it is becoming increasingly apparent that dynamic logic is facing serious challenges in the extreme-scaling regimes.

Slide 4.55

Multiple Thresholds and Design Methodology

- Easily introduced in standard-cell design methodology by extending cell libraries with cells with different thresholds
  - Selection of cells during technology mapping
  - No impact on dynamic power
  - No interface issues (as was the case with multiple V_{DD}s)

- Impact: Can reduce leakage power substantially

Repeating what was stated earlier, the concept of multiple thresholds is introduced quite easily in the existing commercial design flows. In hindsight, this is clearly a no-brainer. The major impact is that the size of the cell library doubles (at least), which increases the cost of the characterization process. This, combined with the introduction of a range of size options for each cell, has led to an explosion in the size of a typical library. Libraries with more than 1000 cells are not an exception.

Slide 4.56

Dual-V_{TH} for High-Performance Design

	High-V_{TH} Only	Low-V_{TH} Only	Dual-V_{TH}
Total Slack	−53 ps	0 ps	0 ps
Dynamic Power	3.2 mW	3.3 mW	3.2 mW
Static Power	914 nW	3873 nW	1519 nW

All designs synthesized automatically using Synopsys Flows

[Courtesy: Synopsys, Toshiba, 2004]

In this experiment, performed jointly by Toshiba and Synopsys, the impact of the introduction of cells with multiple thresholds in a high-performance design is analyzed. The dual-threshold strategy leaves timing and dynamic power unchanged, while reducing the leakage power by half.

Slide 4.57

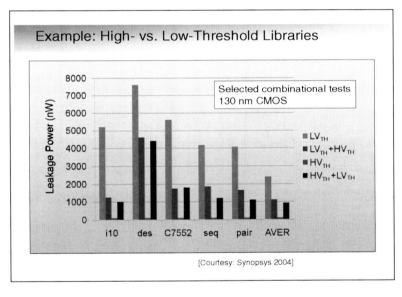

[Courtesy: Synopsys 2004]

A more detailed analysis is shown in this slide, which also illustrates the impact of the chosen design flow over a set of six benchmarks with varying complexity. It compares the high-V_{TH} and low-V_{TH} designs (the extremes) with a design starting from low-V_{TH} transistors only followed by a gradual introduction of high-V_{TH} devices, and vice-versa. It shows that the latter strategy – that is, starting exclusively with high-V_{TH} transistors and introducing low-V_{TH} transistors only in the critical paths to meet the timing constraints – yields better results from a leakage perspective.

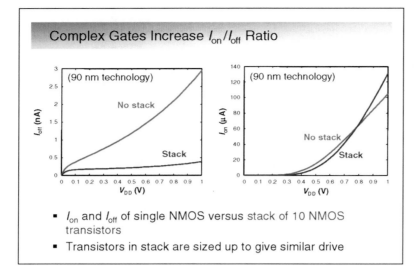

Slide 4.58

In earlier chapters, we have already introduced the notion that stacking transistors reduces the leakage current super-linearly primarily due to the DIBL effect. The stacking effect is an effective means of managing leakage current at design time. As illustrated in the graphs, the combination of stacking and transistor sizing allows us to maintain the on-current, while keeping the off-current in check, even for higher supply voltages.

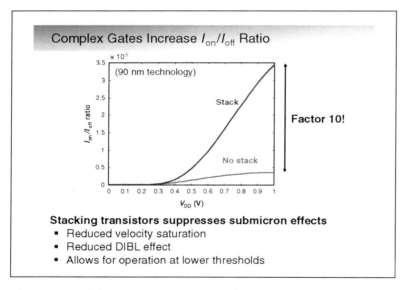

Slide 4.59

This combined effect is put in a clear perspective in this graph, which plots the I_{on}/I_{off} ratio of a transistor stack of 10 versus a single transistor as a function of V_{DD}. For a supply voltage of 1 V, the stacked transistor chain features an on-versus-off current ratio that is 10 times higher. This enables us to lower thresholds to values that would be prohibitive in simple gates. Overall, it also indicates that the usage of complex gates, already beneficial in the reduction of dynamic power, helps to reduce static power as well. From a power perspective, this is a win–win situation.

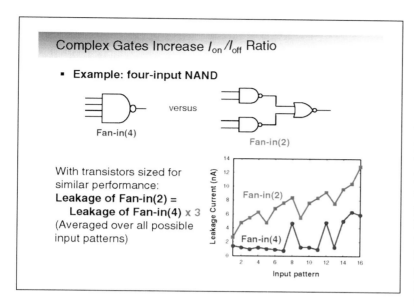

Slide 4.60

The advantage of using complex gates is illustrated with a simple example: a fan-in(4) NAND versus a fan-in(2) NAND/NOR implementation of the same function. The leakage current is analyzed over all 16 input combinations (remember that leakage is state-dependent). On the average, the complex-gate topology has a leakage current that is three times smaller than that of the implementation employing simple gates. One way of looking at this is that, for the same functionality, complex gates come with fewer leakage paths. However, they also carry a performance penalty. For high-performance designs, simple gates are a necessity in the critical-timing paths.

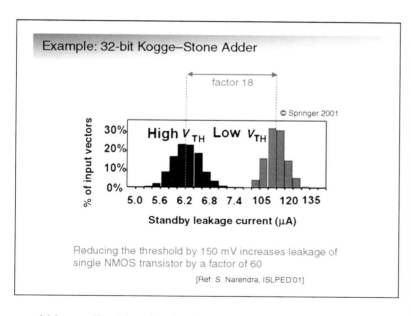

Slide 4.61

The complex-versus-simple gate trade-off is illustrated with the example of a complex Kogge–Stone adder (from [Narendra, ISLPED'01]). This is the same circuit we studied earlier in this chapter. The histogram of the leakage currents over a large range of random input signals is plotted. It can be observed that the average leakage current of the low-V_{TH} version is only 18 times larger than that of the high-V_{TH} version, which is substantially smaller than what would be predicted by the threshold ratios. For a single NMOS transistor, reducing the threshold by 150 mV would cause the leakage current to go up by a factor of 60 (for the slope factor $n = 1.4$).

Summary

- Circuit optimization can lead to substantial energy reduction at limited performance loss
- Energy–delay plots are the perfect mechanisms for analyzing energy–delay trade-offs
- Well-defined optimization problem over W, V_{DD} and V_{TH} parameters
- Increasingly better support by today's CAD flows
- Observe: leakage is not necessarily bad – if appropriately managed

Slide 4.62
In summary, the energy–delay trade-off challenge can be redefined into a perfectly manageable optimization problem. Transistor sizing, multiple supply and threshold voltages, and circuit topology are the main knobs available to a designer. Also worth remembering is that energy-efficient designs carefully balance the dynamic and static power components, subject to the predicted activity level of the modules. The burden is now on the EDA companies to translate these concepts into generally applicable tool flows.

References

Books:
- A. Bellaouar and M.I Elmasry, *Low-Power Digital VLSI Design Circuits and Systems*, Kluwer Academic Publishers, 1st ed, 1995.
- D. Chinnery and K. Keutzer, *Closing the Gap Between ASIC and Custom*, Springer, 2002.
- D. Chinnery and K. Keutzer, *Closing the Power Gap Between ASIC and Custom*, Springer, 2007.
- J. Rabaey, A. Chandrakasan and B. Nikolic, *Digital Integrated Circuits: A Design Perspective*, 2nd ed, Prentice Hall 2003.
- I. Sutherland, B. Sproul and D. Harris, *Logical Effort: Designing Fast CMOS Circuits*, Morgan-Kaufmann, 1st ed, 1999.

Articles:
- R.W. Brodersen, M.A. Horowitz, D. Markovic, B. Nikolic and V. Stojanovic, "Methods for True Power Minimization," Int. Conf. on Computer-Aided Design (ICCAD), pp. 35–42, Nov. 2002.
- S. Date, N. Shibata, S. Mutoh, and J. Yamada, "1-V 30-MHz Memory-Macrocell-Circuit Technology with a 0.5 gm Multi-Threshold CMOS," Proceedings of the 1994 Symposium on Low Power Electronics, San Diego, CA, pp. 90–91, Oct. 1994.
- M. Hamada, Y. Ootaguro and T. Kuroda, "Utilizing Surplus Timing for Power Reduction," IEEE Custom Integrated Circuits Conf., (CICC), pp. 89–92, Sept. 2001.
- F. Ishihara, F. Sheikh and B. Nikolic, "Level Conversion for Dual-Supply Systems," Int. Conf. Low Power Electronics and Design, (ISLPED), pp. 164–167, Aug. 2003.
- P.M. Kogge and H.S. Stone, "A Parallel Algorithm for the Efficient Solution of General Class of Recurrence Equations," *IEEE Trans. Comput.*, C-22(8), pp. 786–793, Aug 1973.
- T. Kuroda, "Optimization and control of V_{DD} and V_{TH} for Low-Power, High-Speed CMOS Design," Proceedings ICCAD 2002, San Jose, Nov. 2002.

Slide 4.63 and 4.64
Some references . . .

References

Articles (cont.):

- H.C. Lin and L.W. Linholm, "An optimized output stage for MOS integrated circuits," *IEEE Journal of Solid-State Circuits*, SC-102, pp. 106–109, Apr. 1975.
- S. Ma and P. Franzon, "Energy control and accurate delay estimation in the design of CMOS buffers," *IEEE Journal of Solid-State Circuits*, (299), pp. 1150–1153, Sep. 1994.
- D. Markovic, V. Stojanovic, B. Nikolic, M.A. Horowitz and R.W. Brodersen, "Methods for true energy-Performance Optimization," *IEEE Journal of Solid-State Circuits*, 39(8), pp. 1282–1293, Aug. 2004.
- MathWorks, http://www.mathworks.com
- S. Narendra, S. Borkar, V. De, D. Antoniadis and A. Chandrakasan, "Scaling of stack effect and its applications for leakage reduction," Int. Conf. Low Power Electronics and Design, (ISLPED), pp. 195–200, Aug. 2001.
- T. Sakurai and R. Newton, "Alpha-power law MOSFET model and its applications to CMOS inverter delay and other formulas," *IEEE Journal of Solid-State Circuits*, 25(2), pp. 584–594, Apr. 1990.
- Y. Shimazaki, R. Zlatanovici and B. Nikolic, "A shared-well dual-supply-voltage 64-bit ALU," Int. Conf. Solid-State Circuits, (ISSCC), pp. 104–105, Feb. 2003.
- V. Stojanovic, D. Markovic, B. Nikolic, M.A. Horowitz and R.W. Brodersen, "Energy-delay tradeoffs in combinational logic using gate sizing and supply voltage optimization," European Solid-State Circuits Conf., (ESSCIRC), pp. 211–214, Sep. 2002.
- M. Takahashi et al., "A 60mW MPEG video codec using clustered voltage scaling with variable supply-voltage scheme," IEEE Int. Solid-State Circuits Conf., (ISSCC), pp. 36–37, Feb. 1998.

Chapter 5
Optimizing Power @ Design Time – Architecture, Algorithms, and Systems

Slide 5.1

Optimizing Power @ DesignTime

Architectures, Algorithms, and Systems

Jan M. Rabaey
Dejan Marković

This chapter presents power–area–performance optimization at the higher levels of the design hierarchy – this includes joint optimization efforts at the circuit, architecture, and algorithm levels. The common goal in all these optimizations is to reach a global optimum in the power–area–performance space for a given design.

The complexity of global optimization involving variables from all layers of the design-abstraction chain can be quite high. Fortunately, it turns out that many of the variables can be independently tuned, so a designer can partition optimization routines into smaller tractable problems. This modular approach helps gain insight into individual variables and provides a way to navigate top-level optimization through inter-layer interactions.

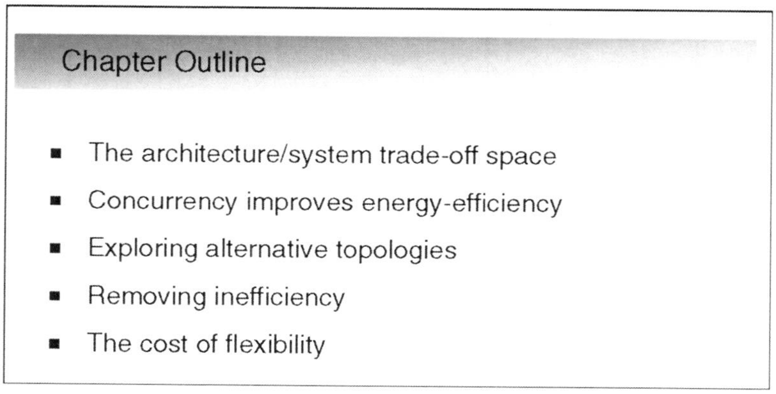

Slide 5.2

Chapter Outline

- The architecture/system trade-off space
- Concurrency improves energy-efficiency
- Exploring alternative topologies
- Removing inefficiency
- The cost of flexibility

The goal of system-level power (energy) optimizations is to transform the energy–delay space such that a broader range of options becomes available at the logic or circuit levels. In this chapter, we classify these transformations into a number of classes: the usage of concurrency, considering alternative topologies for the same function, and eliminating waste. The latter deserves some special attention. To reduce the non-recurring expenses and to encourage re-use, programmable architectures are becoming the implementation platform of choice. Yet, this comes at a huge expense in energy efficiency. The exploration of architectures that combine flexibility and efficiency is the topic of the last part of this chapter.

Slide 5.3

The main challenge in hierarchical optimization is the interaction between the layers. One way to look at this is that optimizations at the higher abstraction layers enlarge the optimization space, and allow circuit-level techniques such as supply voltage or sizing to be more effective. Other optimizations may help to increase the computational efficiency for a given function.

Motivation

- Optimizations at the architecture or system level can enable more effective power minimization at the circuit level (while maintaining performance), such as
  - Enabling a reduction in supply voltage
  - Reducing the effective switching capacitance for a given function (physical capacitance, activity)
  - Reducing the switching rates
  - Reducing leakage
- Optimizations at higher abstraction levels tend to have greater potential impact
  - While circuit techniques may yield improvements in the 10–50% range, architecture and algorithm optimizations have reported power reduction by orders of magnitude

e 5.4

Consider the energy–delay design space exploration exploiting size as well as supply and threshold voltages as parameters, as discussed in Chapter 4. For a 64-bit tree adder and a given technology, a pareto-optimal energy–delay curve is obtained showing some nice energy or delay improvements over the reference design. Yet the overall optimization space is restricted by the topology of the adder. Larger energy savings could be obtained by choosing a different adder topology such as a ripple adder. To accomplish these larger gains (both in delay and energy), accompanying transformations at the micro-architecture or system architecture level are needed. Over the past decades, it has been shown that this can lead to orders of magnitude in energy-efficiency improvement – quite impressive compared to the 30% range that is typically obtained at the circuit level. In this chapter, we present a methodological approach to extend the techniques introduced so far to the higher abstraction layers.

Lessons Learned from Circuit Optimization

Case study: Tree adder
Result of joint (V_{DD}, V_{TH}, W) optimization:
- 65% of energy saved without delay penalty
- 25% smaller delay without energy cost

Ref: min delay at nominal V_{DD}, V_{TH}

Circuit Optimization Limited in Range

Need higher-level optimizations for larger gain

[Ref: D. Markovic, JSSC'04]

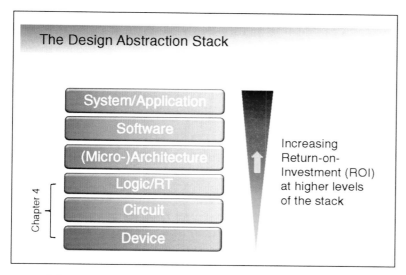

Slide 5.5

The digital design abstraction stack is pictured in this slide. So far, we have mostly covered the device, circuit, and logic levels. Now, we will explore the micro-architecture and architecture levels. While software- and system-level optimizations may have a huge impact as well, they are somewhat out of the scope of this text, and are discussed only in passing.

To make higher-level exploration effective however, it is essential that information from the lower-level layers percolates upward and is available as information to the architecture or system designer. For example, the energy–delay curve of a given adder in a given technology determines the design space that can be offered by that adder (overall, its design parameters). Here the information propagates "bottom-up". At the same time, the application at hand imposes constraints on the implementation (such as, for instance, the minimum performance or maximum energy). These constraints propagate in the "top-down" fashion. Exploration at a given level of design abstraction hence becomes an exercise in marrying the top-down constraints to the bottom-up information. This process is called "meet-in-the-middle".

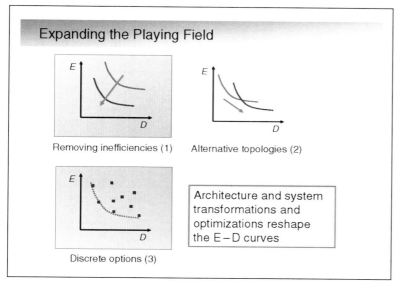

Slide 5.6

The design parameters at the circuit level were mostly continuous (sizing, choice of threshold and/or supply voltage). At the higher abstraction levels, the choices are rather more discrete: which adder topology to use, how much pipelining to introduce, etc. From an exploration perspective, these discrete choices help to expand the energy–delay space. For instance, when two adder topologies are available, each of them comes with its own optimal energy–delay curve. The design space is now the combination of the two, and a new optimal E–D curve emerges as shown in trade-off plot 2. In some cases, one version of a function is always superior to another, which makes the selection process very simple (trade-off plot 1). This may not be obvious at a first glance, and may only be revealed after a rigorous analytical inspection. Using an informed design-space exploration, we will demonstrate in this chapter that some adder topologies are inferior under all circumstances and should never be used (at least, not in modern CMOS processes). A third scenario is where the exploration space consists of many discrete options (such as the sizes and the number of register files). In this case, we can derive an optimal composite E–D curve by selecting the best available option for each performance level (trade-off plot 3).

Although the E–D space represents one very interesting projection of the overall design space, the designer should be aware that the overall design space is far more complex, and that other metrics such as area and design cost are relevant as well. On the first two scenarios, the gray arrows points towards implementations with smaller area. In most cases, implementations with lower energy also are smaller in size, but that is not necessarily always the case. It holds however for most instances of the first two exploration scenarios. This is indicated on the graphs by the gray arrows, which point toward smaller implementations.

Complex systems are built through by composing a number of simpler modules in a hierarchical fashion. Deriving the E–D curves of the composition can be quite involved. The energy component is purely additive, and hence quite straightforward. Delay analysis may be more complex, yet is very well understood.

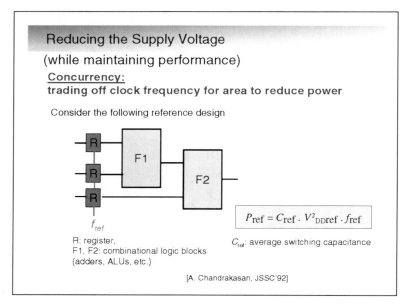

Slide 5.7
A first (and famous) example of architectural energy–delay trade-off is the exploitation of *concurrency* to enable aggressive supply voltage scaling [Chandrakasan, JSSC'92], or equivalently, to provide improved performance at a fixed energy per operation (EOP). To demonstrate the concept, we start from a simple reference design that operates at a nominal (reference) supply voltage $V_{DD\,ref}$ and a frequency f_{ref}. The average switched capacitance of this design is C_{ref}.

Slide 5.8
A parallel implementation of the same design essentially replicates the design such that parallel branches process interleaved input samples. Therefore, the inputs coming into each parallel branch are effectively down-sampled. An output multiplexer is needed to recombine the outputs, and produce a single data stream.

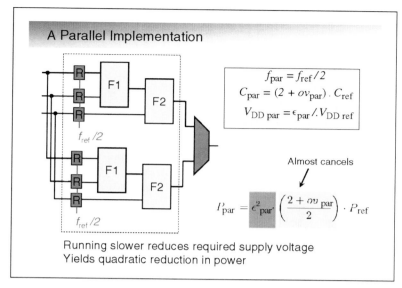

Owing to the *parallelism*, branches now can operate at half the speed, hence $f_{par} = f_{ref}/2$. This reduced delay requirement enables a reduction of the supply voltage by a factor ϵ_{par}. It is the squared effect of that reduction that makes this technique so effective. The multiplexing overhead is typically small, especially when parallelism is applied to large blocks. Notice that though the overhead in switching capacitance is minimal, the area overhead is substantial (effectively larger than the amount of concurrency introduced).

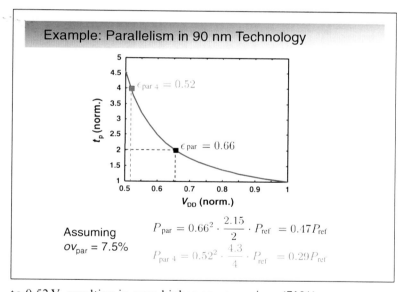

Slide 5.9
The impact of introducing concurrency to reduce EOP for a fixed performance hinges on the delay–supply voltage relationship. For a 90 nm technology, increasing the delay by a factor of 2 is equivalent to reducing the supply voltage by a factor of 0.66. Squaring this translates into an energy reduction by a factor of more than one half (including the overhead). Increasing the parallelism by another factor of 2 reduces the required supply voltage to 0.52 V, resulting in even higher power savings (71%).

Slide 5.10
Other forms of introducing concurrency can be equally effective in reducing the supply voltage, and hence reducing power dissipation. An example of such is *pipelining*, which improves throughput at the cost of latency by inserting extra registers between logic gates. The area overhead of pipelining is much smaller than that of parallelism – the only cost being the extra registers, compared to replicating the design and adding multiplexers. However, owing to the

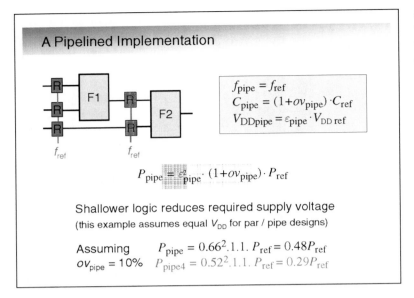

extra switching capacitance introduced by the registers (and the extra clock load), pipelined implementations typically come with a higher switched capacitance than parallel designs. Assuming a 10% pipelining overhead, power savings are similar to those obtained with parallelism. The area cost is substantially lower, though.

Slide 5.11

As we have learned from our earlier discussions on circuit-level optimization, the effect of reducing the supply voltage quickly saturates – especially when the V_{DD}/V_{TH} ratio gets small. Under those conditions, a small incremental reduction in V_{DD} translates into a large increase in delay, which must be compensated by even more concurrency. As shown for a typical 90 nm technology, concurrency levels higher than eight do little to further improve the power dissipation.

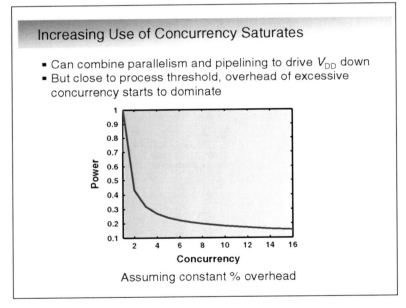

Slide 5.12

The reality is even worse. The introduction of concurrency comes with an overhead. At low voltage levels (and hence high levels of concurrency), that overhead starts to dominate the gain made by the further reduction in supply voltage, and the power dissipation actually increases anew. Leakage also has a negative impact, as parallelism decreases the activity factor. The

Optimizing Power @ Design Time – Architecture, Algorithms, and Systems

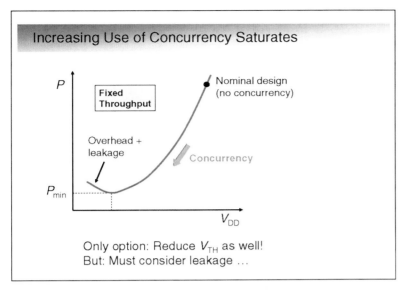

presence of a large number of gates with long delays tends to emphasize static over dynamic power.

The only way to keep increasing concurrency levels (and hence the EOP) is to reduce the threshold as well allowing for a further reduction in voltage without a major performance penalty. However, this requires a careful management of the leakage currents, which is non-trivial (as you must be convinced about by now).

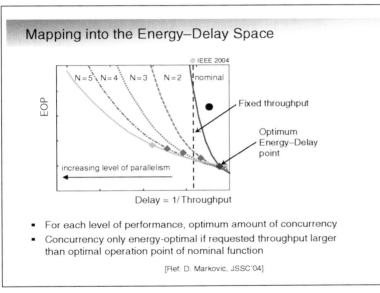

Slide 5.13

The overall impact of the introduction of concurrency and its potential benefits are best understood in our familiar energy–delay space. In this plot, we have plotted the optimal energy–delay curves for an ALU design implemented with varying degrees of concurrency (for each implementation, the pareto-optimal curve is obtained using the techniques described in Chapter 4). Again, the curves can be combined to yield a single optimal E–D curve.

Two different optimization scenarios using concurrency can be considered:

- Fixed Performance: Adding concurrency reduces the EOP until a given point at which the overhead starts to dominate. Hence, for every performance level there *exists an optimum level of concurrency that minimizes the energy*.
- Fixed EOP: Introducing concurrency helps to *improve performance at no EOP cost*. This is in contrast to the traditional approach, where an increase in performance is equivalent to higher clock frequencies, and hence larger dynamic power.

It is interesting to observe that each design instance is optimal over a limited delay range. For instance, if the requested throughput is small, using a high level of parallelism is an inferior option, as the overhead dominates. The attraction of the energy–delay curve representation is that it allows the designer to make architectural decisions in an informed way.

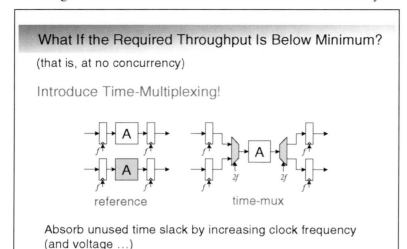

Slide 5.14
The question now arises as to what to do when the requested throughput is *really low* (for instance, in the case of the microwatt nodes described in Chapter 1). This is especially valid with scaling of technology, where the speed of the transistors may be more than what is needed for a given application, and the nominal – that is, no concurrency – implementation is still too fast. In such cases, the solution is to introduce the inverse of concurrency, which is *time-multiplexing*, to trade off the excess speed for reduced area.

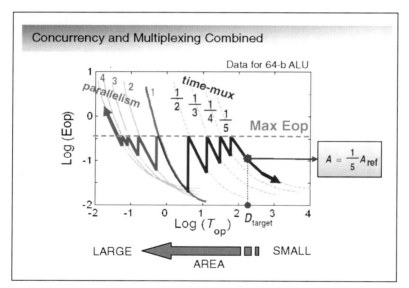

Slide 5.15
Reverting to energy–delay space, we observe that with parallelism and time-multiplexing we can span a very wide range on the performance axis. Relaxed-delay (low-throughput) targets prefer time-multiplexed solutions, whereas increased concurrency is the right option when high throughput is needed. One additional factor to bring to the game is the area cost, which we would like to minimize for a given set of design constraints. Let us consider different scenarios:

- For a given maximum delay (D_{target} in the figure): if the goal is to minimize the EOP, then there exists an optimum amount of concurrency ($= 1/2$); on the other hand, if the goal is to minimize the area for a given EOP, a smaller amount of concurrency is desirable ($= 1/5$).
- For a given maximum of EOP, we choose the amount of concurrency that meets the minimum performance and minimizes the area, as indicated by the red and blue curves on

the plot. In this scenario, concurrency and time-multiplexing provide an efficient way to trade off throughput for area.

Some Energy-Inspired Design Guidelines

For maximum performance
- Maximize use of concurrency at the cost of area

For given performance
- Optimal amount of concurrency for minimum energy

For given energy
- Least amount of concurrency that meets performance goals

For minimum energy
- Solution with minimum overhead (that is – direct mapping between function and architecture)

Slide 5.16
To summarize, maximum-performance designs request the maximum possible concurrency at the expense of increased area. For a given performance, however, one should optimize the amount of concurrency to minimize energy.

Equivalently, for a given energy budget, the least amount of concurrency that meets the performance goals should be used. For the absolute minimum energy, a direct mapping architecture (concurrency = 1) should be used because it has no overhead in switched capacitance, provided of course that this architecture meets the design constraints.

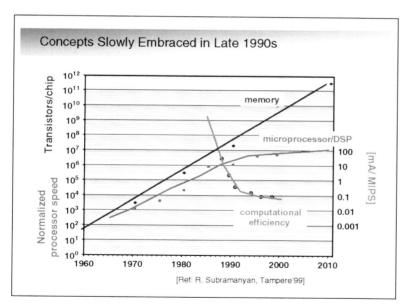

Slide 5.17
The ideas that were put forward in the previous slides originated in the early 1990s. Yet it took some time for them to be fully embraced by the computing community. That happened only when people started to realize that the traditional performance improvement strategies for processors – that is, technology scaling combined with increasing the clock frequency – started to yield less, as is shown in this graph. It was mostly power constraints that slowed down the increases in clock frequency, and ultimately conspired to halt it altogether. Remember how clock frequency was the main differentiator in the advertisements of new processors in the 1980s and 1990s? With stalling clock frequencies, the only way to maintain the scaling of the performance of the single processor was to increase the instructions per cycle (IPC) through the introduction of extra architectural performance-enhancing techniques such as multi-threading and

speculation. All of these add to the complexity of the processor, and come at the expense of energy efficiency. (Caveat: this is an in-a-nutshell summary – the real equation is a lot more complex).

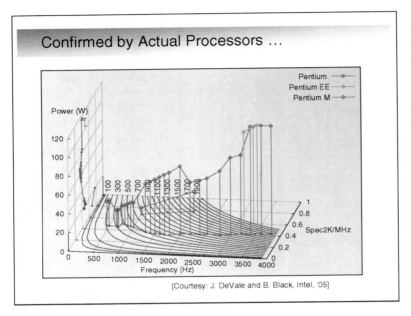

Slide 5.18

The reality of these concerns is quite clearly illustrated in this three-dimensional chart, which plots three Intel processor families in the power–clock frequency–Spec2K/MHz space. The latter metric measures the effective performance of a processor, independent of the clock frequency. It shows that ultimately the effective performance of the processors within a single family increased little over time, and that clock frequency was the primary tuning knob. Unfortunately this translated directly into massive increases in power dissipation.

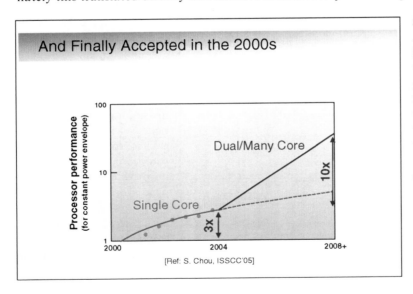

Slide 5.19

After a slow adoption, the idea of using concurrency as the way out of the quagmire gathered full steam in the 2000s, when all major microprocessor vendors agreed that the only way to improve performance within a given power envelope is to adopt concurrency, leading to the multitude of multi-core architectures we are seeing today.

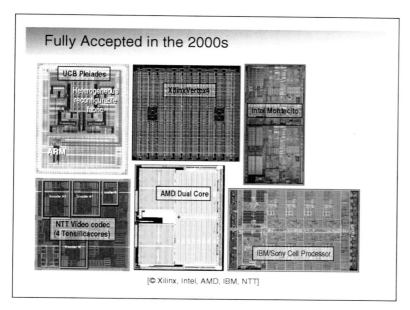

Slide 5.20

This slide just shows a sample of the many multi-core architectures that were introduced starting 2005. Initially adopted in application-specific processors (telecommunications, media processing, graphics, gaming), the multi-core idea spread to general-purpose processing starting with the dual-core architecture, expanding rapidly to four and more cores on a die thereafter.

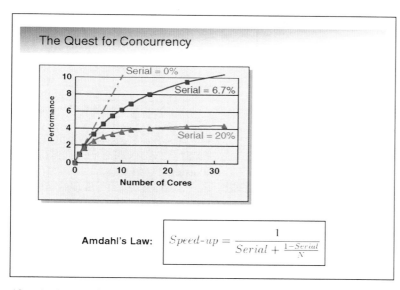

Slide 5.21

There is one important caveat though. Conceptually, we can keep on improving the performance (for a given technology) at a fixed EOP cost by providing more concurrency. This requires however that the concurrency is available in the application(s) at hand. From the (in)famous Amdahl's law, we know that the amount of speed-up attainable through concurrency is limited by the amount of serial (non-parallelizable) code. Even if only 20% of the code is sequential, the maximum amount of performance gain is limited to just a little more than 4.

Slide 5.22

The impact of this is clearly illustrated in the following case study performed at Intel. A set of benchmarks is mapped on three different multi-core architectures with different granularity and different amount of concurrency (12 large, 48 medium, or 144 small processors). Each realization is such that it occupies the same area (13 mm on the side), and dissipates the same maximum power (100 W) in a speculative 22 nm CMOS technology. The large processors are more powerful from a computational throughput perspective, but operate at a higher EOP cost.

When comparing the overall performance of the three alternatives over a set of benchmarks with different levels of concurrency, it shows that large-processor version outperforms the others when little parallelism is available. However, when plenty of concurrency is present, the "many small cores" option outperforms the others. This is especially the case for the TPT (totally parallel) benchmark.

In summary, it becomes clear that "massively parallel" architectures only pay off for applications where there is sufficient concurrency. There is no clear answer about the right granularity of the computational elements. It is fair to assume that architectures of the future will combine a variety of processing elements with different granularities (such as, for instance, the IBM cell processor^{TM}, or the Xilinx Vertex^{TM}).

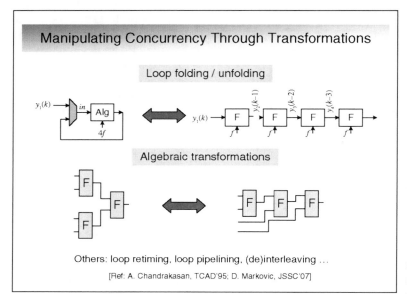

Slide 5.23

One option to improve the odds is to make the applications more parallel with the aid of optimizing transformations. Loop transformations such as loop unrolling and loop retiming/pipelining are the most effective. Algebraic transformations (commutativity, associativity, distributivity) come in very handy as well. Automating this process is not easy though. The (dormant, but reviving) field of high-level synthesis actually created some real breakthroughs in this area, especially in the domain of signal processing. In the latter, an infinite loop (i.e., time) is always present, allowing for the creation of almost limitless amounts of concurrency [Chandrakasan, TCAD'95].

For example, the loop-unfolding transformation translates a sequential execution (as captured in a recursive loop) into a concurrent one by unrolling the loop a number of times and introducing pipelining. The reverse is obviously also possible.

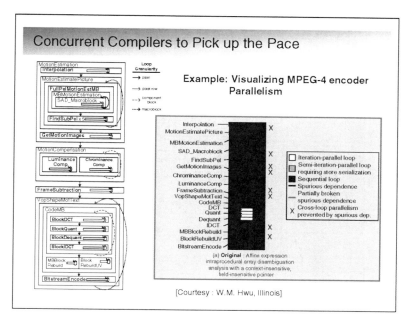

Slide 5.24
The effectiveness of today's optimizing compilers in exposing concurrency is illustrated with the well-known example of MPEG-4 video encoder. The hierarchical composition of the computational core of MPEG-4, that is the motion compensation, is shown on the left. On the right, the amount of available parallelism in the reference design is visualized. White represents fully concurrent loops, whereas black stands for fully sequential ones. Gray represents a loop that is only partially concurrent. Also important are the "spurious dependencies", which prevent the code from being concurrent, but only occur intermittently, or even may be false. It is quite obvious that the reference design, as it is, is not very friendly for concurrent architectures.

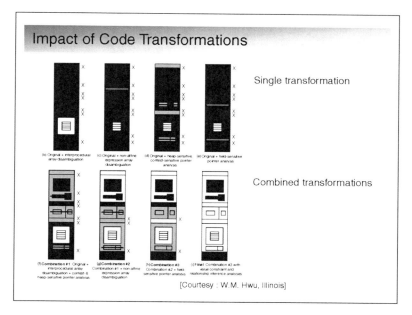

Slide 5.25
The application of a single transformation (such as pointer analysis or disambiguation) can make some major inroads, but it is only the combined effect of multiple transformations executed in concert that can make the code almost fully concurrent (which is an amazing accomplishment in itself). The reader should realize that the latter cannot be performed automatically right now, but needs the manual intervention of the software developer. It is not clear when, if ever, fully automatic parallelization of sequential code will become feasible. Very often, a high-level perspective is needed, which is hard to accomplish through localized transformations. User intervention guided by the appropriate visualization may be a necessity. Even better would be to train software engineers to write

concurrent code from scratch. This is best accomplished through the use of programming environments that make concurrency explicit (such as the Mathworks Simulink™ environment).

In summary, concurrency is a great tool to keep energy in check, or to even reduce it. It requires a rethinking however on how complex design is done.

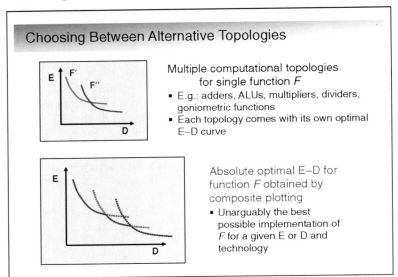

Slide 5.26
Beyond the introduction of concurrency, other architectural strategies can be explored, such as considering alternative implementation topologies for a given function. For each topology, an energy–delay curve can be obtained in a bottom-up fashion. The architectural exploration process then selects the most appropriate implementation for a given set of performance or energy constraints.

The E–D curves of the alternative topologies can be combined to define a composite, globally optimum, trade-off curve for the function (bottom graph). The boundary line is optimal, because all other points consume more energy for the same delay, or have longer delay for the same energy.

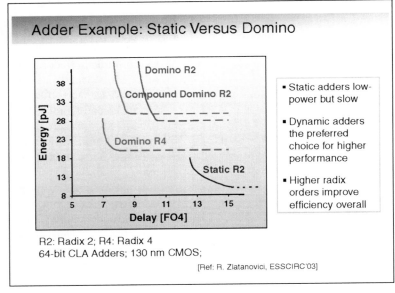

Slide 5.27
Consider, for instance, the case of the adder module, which is often the most performance- and energy-critical component in a design. The quest for the ultimate adder topology has filled many conference rooms, journal articles, and book chapters. The choices are between ripple, carry-select, carry-bypass, and various styles of look-ahead adders (amongst others). Within each of these categories, an extra number of design choices must be made, such as the exact topology (e.g., the radix in a look-ahead adder) or the circuit style to be used. Although it seems that the number of options is overwhelming, using some hierarchy in

the decision-making process makes the selection process a lot more amenable, and may even lead to the establishment of some fundamental selection rules and some definitive truths.

Let us consider the case of a 64-bit carry look-ahead adder. Design parameters include the radix number (fan-out in the tree) as well as the circuit style. Optimizing transistor sizes and supply voltages (using the methodology introduced in Chapter 4), optimal energy–delay curves can be provided for each option, as shown in the slide. As a result, the decision process is substantially simplified: For high performance, use higher-radix solutions and dynamic logic, whereas static logic and lower-radix numbers are preferred for low-energy solutions. The beauty of this E–D exploration is that designers can rely on objective comparisons to make their decisions.

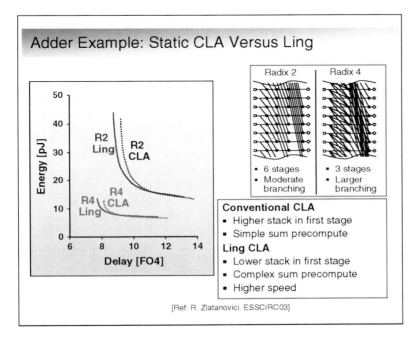

Slide 5.28

The same approach equally applies to higher levels of the decision hierarchy. Even with the carry look-ahead topology, many options exist, an example of which is the Ling adder that uses some logic manipulations to get even higher performance. Again, some ground truths can be established. Radix-2 CLA adders rarely seem to be attractive. If performance is the primary goal, then radix-4 Ling adder is the preferred choice.

The bottom line of this discussion is that energy–delay trade-off can be turned into an engineering science, and should not be a black art.

Slide 5.29

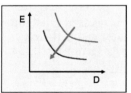

Improving Computational Efficiency

Implementations for a given function maybe inefficient and can often be replaced with more efficient versions **without penalty in energy or delay**

Inefficiencies arise from:
- Over-dimensioning or over-design
- Generality of function
- Design methodologies
- Limited design time
- Need for flexibility, re-use, and programmability

The plots in the previous slides already indicate that some architectural options are truly inferior in all aspects, and hence should be discarded from the designer's consideration. Such inefficiencies may arise from a number of reasons including legacy or historical issues, over-design, use of inferior logic style, inferior design methodology, etc. Although this may seem obvious, it is surprising that inefficient options are still in frequent use, often because of lack of knowledge or lack of exploration capabilities and most often because of adherence to generally accepted design paradigms or methodologies.

Slide 5.30

Improving Computational Efficiency

Some simple guidelines:
- Match computation and architecture
  - Dedicated solutions superior by far
- Preserve locality present in algorithm
  - Getting data from far away is expensive
- Exploit signal statistics
  - Correlated data contains less transitions than random data
- Energy on demand
  - Only spend energy when truly needed

Some simple and general guidelines are valid when it comes to improving computational efficiency. Whereas the slide enumerates a number of concepts that are worked out in more detail in subsequent pages, it is worth condensing this to an even smaller number of ground truths.

- Generality comes with a major penalty in efficiency.
- It pays to have the architecture match the intent of the computation.
- Never have anything consume power when it is not in use.

Slide 5.31

Consider first the issue of matching computation to architecture. To illustrate this, let us consider a simple example of computing a second-order polynomial. In a traditional Von–Neumann style processor, the computation is taken apart into a set of sequential instructions, multiplexed on a generic ALU. The architecture of the computational engine and the topology of the algorithm are

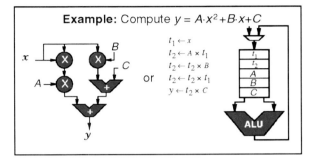

totally disjoint and have little in common. Mapping the two onto each other leads to substantial overhead and inefficiency. Another option is to have algorithm and architecture match each other directly (as shown on the left). The advantage of this is that every operation and every communication is performed without any overhead. This style of programmable architecture is called "spatial programming", and is best suited for reconfigurable hardware platforms. The former approach is also known as "temporal programming". As we will show later, the difference in energy efficiency between the two is huge.

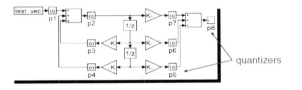

Slide 5.32
Another example of the matching between algorithm and architecture is the choice of the word length. Most programmable processors come with a fixed word length (16, 32, or 64 bit), although the actual computation may need far less. During execution, this leads to a sizable amount of switching energy (as well as leakage) being wasted. This is avoided if the word length of the computational engine can either be matched or adjusted to the algorithmic needs.

Slide 5.33
To illustrate the impact of the architectural optimizations discussed so far and their interaction, we use a case study of a singular-value decomposition (SVD) processor for multiple-input and multiple-output (MIMO) communications [D. Marković, JSSC'07]. Multi-antenna techniques are used to improve robustness or increase capacity of a wireless link. Link robustness is improved by averaging the signal over multiple propagation paths as shown in the illustration. The number of averaging paths can be artificially increased by sending the same signal over multiple antennas. Even more aggressively, the

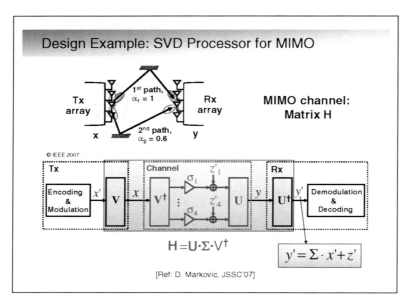

capacity of the link can be further increased by sending independent and carefully tailored data streams over the transmit antennas. This is called spatial multiplexing.

In a MIMO system, the channel is a complex matrix H formed of transfer functions between individual antenna pairs. x and y are vectors of Tx and Rx symbols. Given x and y, the question is how to estimate gains of these spatial sub-channels. An optimal way to extract the spatial multiplexing gains is to use singular-value decomposition.

This algorithm is however quite complex, and involves hundreds of additions and multiplications, as well as divisions and square roots, all of which have to be executed real-time at the data rate (which is in the range of 100s of MHz). This far exceeds the complexity of standard communication blocks such as FFT or Viterbi en(de)coding. The challenge is to come up with an architecture that is both energy- and area-efficient. In this particular case, we study a multi-antenna algorithm that can achieve around 250 Mbps over 16 frequency sub-channels using a 4×4 antenna system.

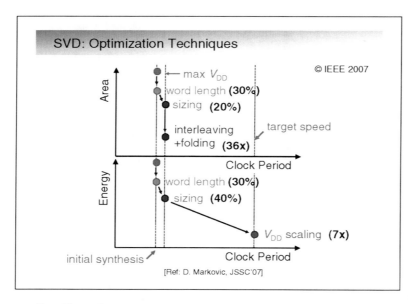

Slide 5.34

This slide illustrates how various optimization techniques are used to reach the target speed with minimal power and area. The process starts with a fully parallel implementation, which is both very large and too fast. The excess performance is traded for area and energy reduction. Qualitatively, word-length optimization reduces both area and energy; interleaving and folding mainly impact area and have a small impact on energy (neglected in this simplified diagram); gate sizing primarily affects the energy (small change in area of standard-cell based design); and, finally, voltage scaling has a major impact on energy.

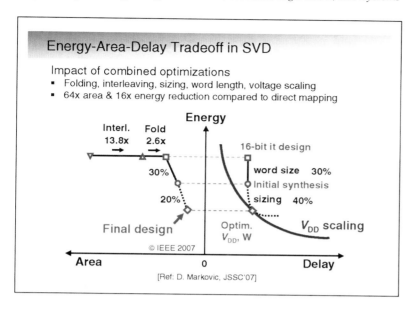

Slide 5.35

The Energy–Delay–Area diagram is a convenient way to look at the combined effect of all optimization steps. As the slide shows, the major impact on energy comes from supply voltage scaling and gate sizing, whereas area is primarily reduced by interleaving and folding.

The process proceeds as follows: Starting from a 16-bit realization of the algorithm, word-length optimization yields a 30% reduction in energy and area. The next step is logic synthesis, which includes gate sizing and supply voltage optimizations. From prior discussions, we know that sizing is most effective at small incremental delays compared to the minimum delay, so we synthesize the design with 20% slack and perform incremental compilation to utilize benefits of sizing for a 40% reduction in energy and a 20% reduction in area. Standard cells are characterized for 1 V supply, so we translate timing specifications to that voltage. At the optimal V_{DD} and W, energy–delay curves of sizing and V_{DD} are tangential, which means that the sensitivities are equal. The final design is 64 times smaller and consumes 16 times less energy than the original 16-bit direct-mapped parallel realization.

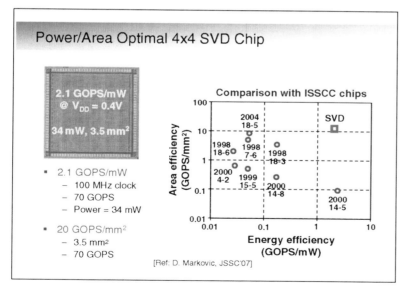

Slide 5.36

The measured performance data of a chip implementation of the SVD algorithm are shown in this slide. Implemented in a 90 nm CMOS technology, the optimal supply voltage is 0.4 V for a 100 MHz clock. The chip is actually functional all the way down to 255 mV, running with a 10 MHz clock. The leakage power is 12% of the total power in the worst case, and clocking power is 14 mW, including leakage.

A comparison against a number of custom chips

from the multimedia and wireless tracks of the ISSCC conference shows how this combined set of optimizations leads to a design that simultaneously excels in area and energy efficiency. Publication year and paper number are indicated; figures are normalized to a 90 nm, 1 V process.

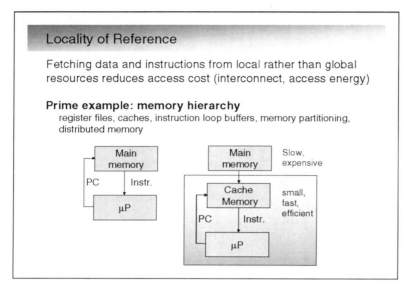

Slide 5.37
Maintaining *locality of reference* is another mechanism to increase the efficiency of an architecture. This in fact not only is true for power, but also helps performance and area (in other words, a win–win). Anytime a piece of data or an instruction has to be fetched from a long distance, it comes at a cost in energy and delay. Hence, keeping relevant or often-used data and instructions close to the location where they are processed is a good idea. This is, for instance, the main motivation behind the construction of memory hierarchies, such as multi-level caches.

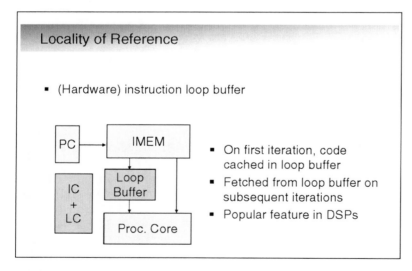

Slide 5.38
The instruction loop buffer (ILB) is an example of how locality of reference is effectively used in Digital Signal Processors (DSPs). Many DSP algorithms such as FIR filters, correlators, and FFTs can be described as short loops with only a few instructions. Rather than fetching the instructions from a large instruction memory or cache, it is far more energy-efficient to load these few instructions into a small buffer memory on the first execution of the loop, and fetch them from there on the subsequent iterations.

Slide 5.39
Similar considerations are true for data locality. Sometimes a careful reorganization of the algorithm or code is sufficient to realize major benefits – without needing any extra hardware

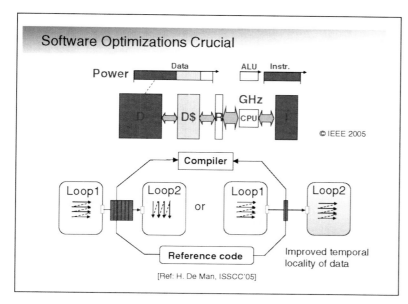

support. Consider, for example, the case of image or video processing. Many algorithms require that a video frame is traversed first in the horizontal, next in the vertical direction. This requires the intermediate result to be stored in a frame buffer (which can be quite large), translating into a lot of memory accesses into a large SRAM memory. If the code could be reorganized (either manually or by the compiler) to traverse the data in the horizontal direction twice, the intermediate storage requirements are reduced substantially (that is, to a single line), leading to both energy reduction and performance improvement.

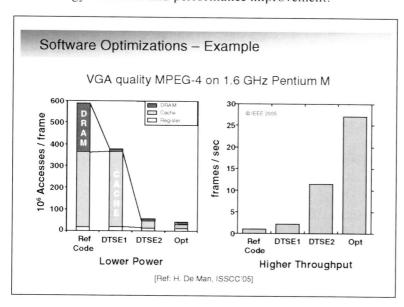

Slide 5.40
Even on generic architectures such as the Pentium, the impact of optimizations for data locality can be huge. Starting from generic reference code for MPEG-4, memory accesses can be reduced by a factor of 12 through a sequence of software transformations. This translates almost directly into energy savings, simultaneously improving the performance by a factor of almost 30.

Slide 5.41
Architectural optimizations can also be used to minimize activity – an important component of the dynamic power component. In fact, when taking a close look at many integrated circuits, we can discover a large amount of spurious activity which little or even zero computational meaning, and some of it being a direct consequence of architectural choices.

A simple example can help to illustrate this. Many data streams exhibit temporal correlations. Examples are speech, video, images, sensor data, etc. Under such conditions, the probability of a

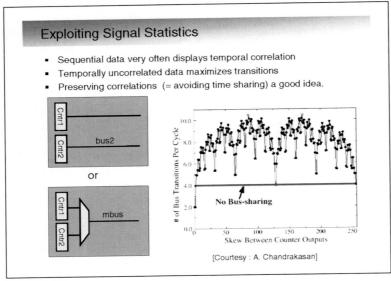

[Courtesy : A. Chandrakasan]

bit to undergo a transition from sample to sample is substantially smaller than when the data is purely random. In the latter case, the transition probability per bit would be exactly 1/2. Consider now the case of an N-bit counter. The average transition probability (per bit) equals $2/N$ (for large N), which is substantially lower than the random case for $N>4$.

Time multiplexing two or more unrelated streams over the same bus destroys these correlations, and turns every signal into a random one, hence maximizing the activity. When data is very strongly correlated (and the load capacitance is large), it is often advisable to avoid multiplexing.

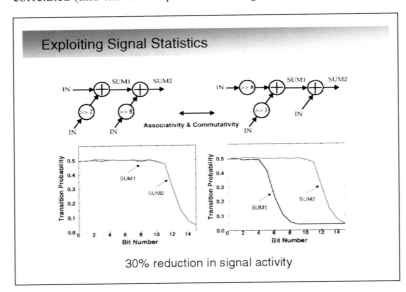

30% reduction in signal activity

Slide 5.42

Another example on how data correlations can help to reduce activity is shown in this slide. In signal processing applications, multiplication with a fixed number is often replaced by a sequence of adds and shifts. This avoids the use of expensive multipliers, is faster, and saves energy. The order in which we add those numbers (which is totally arbitrary owing to the associativity of the add function) impacts the activity in the network. In general, it is advisable to combine closely correlated signals first (such as $x>>8$ and $x>>7$), as this preserves the correlation and minimizes spurious activity.

Slide 5.43

One dominant source of energy inefficiency in contemporary integrated circuits and systems is the provision for "flexibility" or "programmability". Although programmability is a very attractive proposition from an economic and business perspective, it comes at a huge efficiency cost. The design challenge is to derive effective architectures that combine flexibility and programmability with computational efficiency.

The Cost of Flexibility

- Programmable solutions very attractive
  - Shorter time to market
  - High re-use
  - Field updates (reprogramming)
- But come at a large efficiency cost
  - Energy per function and throughput-latency per function substantially higher than in dedicated implementation
- How to combine flexibility and efficiency?
  - Simple versus complex processors
  - Stepping away from "completely flexible" to "somewhat dedicated"
  - Concurrency versus clock frequency
  - Novel architectural solutions such as reconfiguration

Slide 5.44

Quantifying flexibility is not simple. The only real way to evaluate the flexibility of a given architecture is to analyze its performance metrics over a set of representative applications, or benchmark sets. A fitting framework is our energy–delay graph, extended with an extra "application axis". A dedicated or custom architecture only executes a single application, and yields a single energy–delay curve. For a more flexible architecture, energy–delay curves can be generated for each member of the test bench. One (preferred) option is then to have the average of these curves represent the architecture. Another option is to use the circumference of the E–D plots over all individual applications.

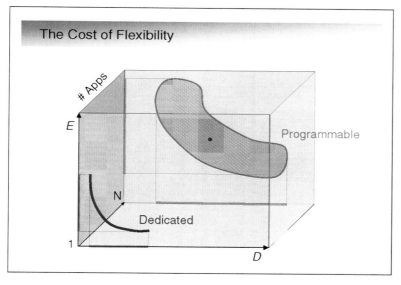

Slide 5.45

Whatever representation is chosen, it is important to realize that the energy–delay curves of the same application implemented in custom style and on a programmable processor are generally quite far apart. It has been reported a number of times that for the same performance the EOP (measured in MIPS/mW or billion operations/Joule) for different implementation styles can be as much as *three orders of magnitude* apart, with custom implementation and general-purpose

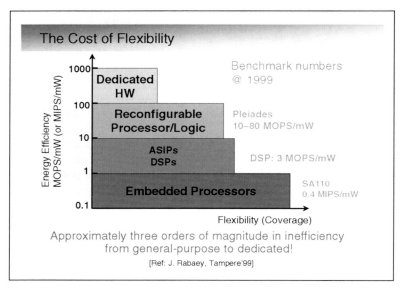

programmable logic at the extreme ends. To fill in the gap between the two, designers have come up with some intermediate approaches such as application-specific instruction processors (ASIPs) and DSPs, which trade generality for energy efficiency. For instance, by adding dedicated hardware and instructions, a processor can become more efficient for a specific class of applications at the expense of others. Even closer to dedicated hardware are the configurable spatial programming approaches, in which dedicated functional units are reconfigured to perform a given function or task.

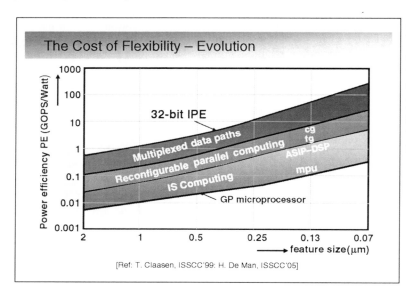

Slide 5.46
T. Claasen and H. De Man observed similar ratios in their ISSCC keynotes. Observe that these graphs project that the extremes grow even further apart with the scaling of technology.

Slide 5.47
The trade-off between flexibility and energy efficiency is beautifully illustrated in this example, which explores a variety of implementations of correlators for CDMA (Code-division multiple access), as used in 3G cellular telephony. All implementations are normalized to the same technology node. Implementing this compute-intensive function as a custom module reduces the energy per execution by a factor of approximately 150 over a software implementation on a DSP. Observe that the area cost is reduced substantially as well. A reconfigurable solution gets closer to the custom one, but still needs six times more energy.

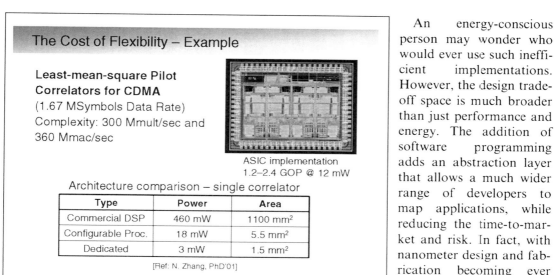

An energy-conscious person may wonder who would ever use such inefficient implementations. However, the design trade-off space is much broader than just performance and energy. The addition of software programming adds an abstraction layer that allows a much wider range of developers to map applications, while reducing the time-to-market and risk. In fact, with nanometer design and fabrication becoming ever more complex and challenging, the trend is clearly toward ever more flexibility and programmability. The challenge for the micro-architecture and system-on-a-chip designer now is how to effectively combine the two. We have already mentioned the use of concurrent architectures. However, the range of options and choices is much broader than that. Without a structured exploration strategy, solutions are mostly chosen in an ad hoc fashion, and decisions are made based on rules of thumb or pure intuition.

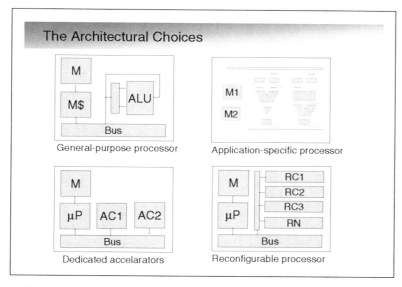

Slide 5.48

To illustrate the process, let us first consider the case of a single-instruction processor to be used in a system-on-a-chip. The first option is to use a generic processor, which can be obtained either for free, or from an intellectual-property (IP) company. Even within that constrained framework, a range of options exist, such as choosing the width of the data path or the structure of the memory architecture. If energy efficiency is crucial, further options should be considered as well, some of which are indicated on this slide and are elaborated on in the subsequent ones.

Given the potential benefits, industry has been intensely exploring all these options (and many others). Companies in the networking, communications, and signal processing domains all have their own (and different) processor recipes. Numerous start-ups have had high hopes that their

138   Chapter #5

unique concept presents the perfect solution (and the road to richness). Only a few have succeeded, though.

Simple Versus Complex Processors?

- Best explored using Energy–Delay curves

- For each proposed architecture and parameter set, determine average energy–delay over a library of benchmark examples

- Modern computer-aided design tools allow for quick synthesis and analysis
  – Leads to fair comparison

- Example: Subliminal Project – University of Michigan
  – Explores processor architecture over the following parameters: Depth and number of pipeline stages; Memory: Von Neumann or Harvard; ALU Width(8/16/32); With or without explicit register file

Slide 5.49
Consider first the case of a single generic processor designed to cover a set of representative benchmarks. To determine whether a simple or a more complex processor is the best choice for a given performance or EOP specification, a structured-exploration approach can be used. Given the relevant design parameters, a set of meaningful instances are generated, synthesized, and extracted (courtesy of David Blaauw). The energy–delay metrics are found by simulating the obtained instances over the benchmark set, and averaging the results.

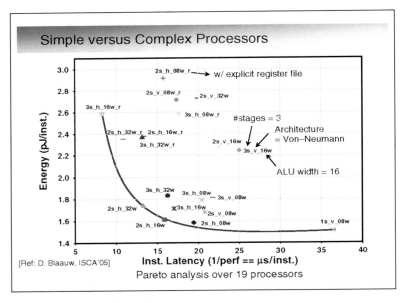

Slide 5.50
Each of the processor instances is now a single point in the architecture's energy–delay space. With sufficient representative points, we see the common hockey-stick pareto-optimal curve emerge. A couple of quick observations: for the absolute lowest energy, choose the simplest processor; for the highest performance, go for the complex one. Also, a large number of processor options are inferior from any perspective, and should be rejected right off the bat.

The results shown in this slide suggest again that the energy–delay trade-off game takes on identical formats at all levels of the design hierarchy.

Application-Specific Processors

- Tailored processor to be efficient for a sub-set of applications
  - Memory architecture, interconnect structure, computational units, instructions
- Digital signal processors best known example
  - Special memory architecture provides locality
  - Datapath optimized for vector-multiplication (originally)
- Examples now available in many other areas (graphics, security, control, etc.)

Slide 5.51
To further improve energy and extend the energy–delay space, the processor can be fine-tuned to better fit the scope of a sub-set of applications (such as communications, graphics, multimedia, or networking). Often, recurring operations or functions are captured in dedicated hardware in the data path, and special instructions are introduced. In addition, register files, interconnect structure, and memory architecture can be modified. This leads to the so-called ASIP.

Example 1: DSPs

- The first type of application-specific processor to become popular
- Initially mostly for performance, but energy benefit also recognized now
- Key properties: dedicated memory architecture (multiple data memories), data path specialized for specific functions such as vector multiplies and FFTs
- Over time: introduction of more and more concurrency (VLIW)

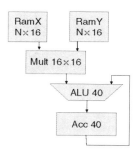

Slide 5.52
The premier example of an ASIP is the DSP. Introduced first by Intel and AT&T in the late 1970s, the concept was ultimately popularized by Texas Instruments. Whereas performance was the initial driving force, it was soon realized that focusing on a specific application domain (such as FIR filtering, FFTs, modems, and cellular processors) ultimately leads to energy efficiency as well.

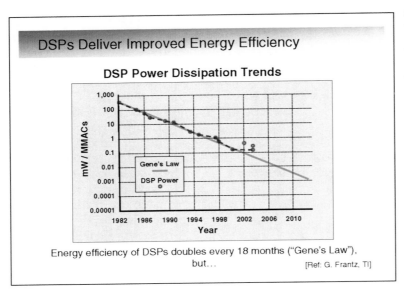

Slide 5.53
It was Gene Frantz of TI fame who was the first one to observe that the energy efficiency of DSPs doubles every 18 months as a result of both technology scaling and architectural improvements, pretty much along the same lines as performance in general-purpose processors (GPP). It is interesting to observe that the efficiency improvements started to saturate in the early 2000s, in close resemblance to the performance in GPPs.

Slide 5.54
Extrapolating from the past predicts that DSPs may require only 1 μW per MIPS by 2012. Given the technological trends, this is highly unlikely. In fact, in recent years the DSP concept has been severely challenged, and other ideas such as hardware accelerators and co-processors have firmly gained ground.

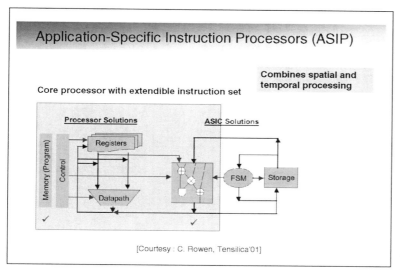

Slide 5.55
With the growing importance of energy efficiency, the ASIP concept has gained considerable ground. One option that is very attractive is to start from a generic processor core, and to extend the instruction set by adding dedicated hardware units to the data path – based on the application domain at hand. The advantage of this approach is that it is incremental. The compilers and software tools can be automatically updated to embrace the extended architecture. In a sense, this approach combines temporal (through the generic processor core) and spatial processing (in the dedicated concurrent hardware extensions).

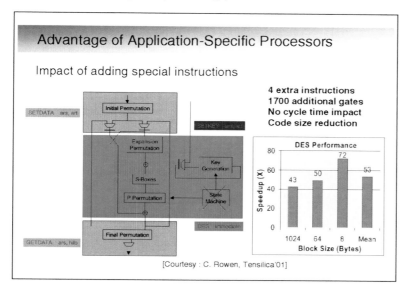

Slide 5.56
Consider, for example, a processor for security applications, in which the DES encryption algorithm prominently features. The basic processor core is extended by adding a dedicated hardware module, which efficiently executes the permutations and the S-boxes that form the core of the DES algorithm. The extra hardware only takes 1700 additional gates. From a software perspective, these additions are translated into just 4 extra instructions. The impact on performance and energy efficiency is huge though.

Slide 5.57

The second and more recent example of the attractiveness of the extensible-processor approach is in the domain of video processing, particularly in realizing a decoder for the popular H.264 standard. One interesting option of that standard is the highly effective, but compute-intensive, CABAC coder, which is based on arithmetic coding. Arithmetic coding requires a large number of bit-level operations, which are not well-supported in a general-purpose core. Adding some dedicated instructions improves the CABAC performance by a factor of over 50, and the energy

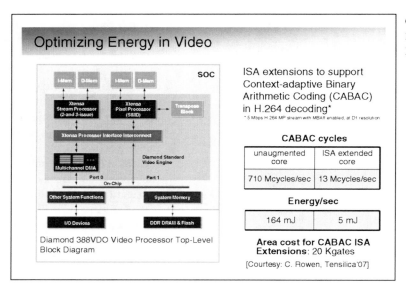

efficiency by a factor of 30, for the extra cost of just 20K extra gates

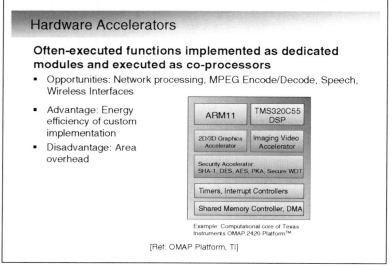

Slide 5.58

One step that goes a bit further is to spawn off complete functions as co-processors (often called accelerators). The difference with the extended ISA is that the co-processors not only perform data operations but also feature their own sequencers, and hence operate independently from the core processor. Once started, the co-processor operates until completion upon which control is passed back to the main core. Most often, these accelerators embody small loops with large iteration counts. The advantage is even better energy efficiency at the expense of area overhead. Typical examples of co-processors are correlators for wideband CDMA and FFT units for wireless OFDM (as used in WiFi implementations).

As an example, we show the computational core (all periphery omitted) of the Texas Instruments OMAP 2420 platform™, targeting mobile wireless applications. In addition to an ARM general-purpose core and a TI C55 DSP processor, the system-on-a-chip contains a number of accelerator processors for graphics, video, and security functions.

Optimizing Power @ Design Time – Architecture, Algorithms, and Systems   143

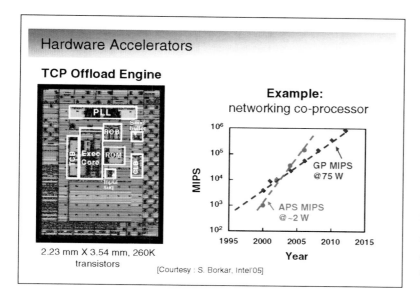

Slide 5.59

The effectiveness of the accelerator approach is demonstrated by this example (provided by Intel). It features an accelerator for the prevalent TCP networking stack. Given a power budget of 2 W, the TCP accelerator processor outperforms a general-purpose processor (with a 75 W budget). In fact, the gap between the two is shown to be increasing over time (as we have already observed in Slide 5.46).

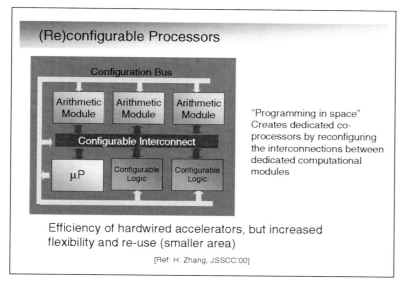

Slide 5.60

One of the main negatives of the accelerator approach is the hardware overhead – Though energy-efficient, the dedicated accelerators are only used a fraction of the time, and their area efficiency is quite low. Reconfigurable spatial programming presents a means to provide simultaneously energy and area efficiencies. The idea is to create accelerators on a temporary basis by assembling a number of functional units into a dedicated computational engine. When the task at hand is finished, the structure is de-assembled, and the computational units can be re-used in other functions. The underlying tenet is that reconfiguration is performed on a per-task basis. Reconfiguration events hence are rare enough that their overhead is small compared to the energy benefits of spatial computing and the area benefits of re-use.

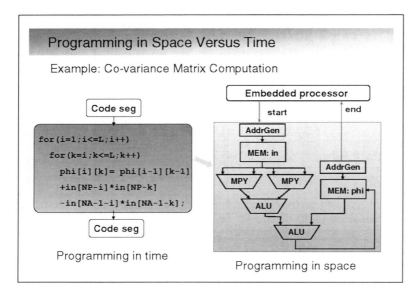

Slide 5.61

An example again goes a long way in illustrating the concept. A number of signal-processing applications, such as speech compression, require the computation of a co-variance matrix. The sequential "C" language code implementing this function is shown on the left. However, rather than implementing it on a sequential instruction set processor, the same function can be more efficiently implemented in a spatial fashion by connecting a number of functional units such as ALUs, multipliers, memories, and most importantly address generators. The latter replace, in a sense, the indexes of the nested loops. Observe that all the units are quite generic. Creating a dedicated function requires two actions: setting the parameters of the computational modules and programming the interconnections.

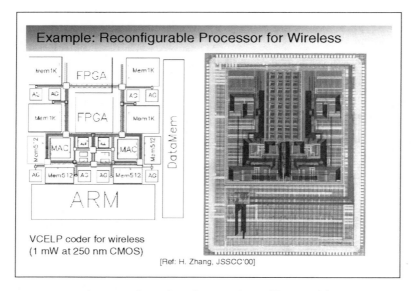

Slide 5.62

There are many ways of translating this idea into silicon. This slide shows one particular incarnation. The intended application area was speech processing for cellular telephony (by now, this application only represents a small fraction of the computational needs of a mobile communicator, but in the late 1990s it was a big thing). The system-on-a-chip consists of an embedded core (ARM) and an array of computational units, address generators, and memories of various stripes. To provide even more spatial programmability for small-granularity functions, two FPGA modules are included as well. The key element linking all these modules is the reconfigurable network.

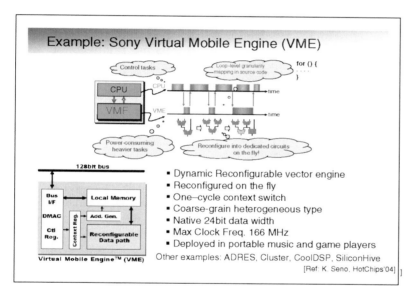

Slide 5.63
When mapping a VCELP coder on this architecture, it turns out that approximately 80% of the computational cycles can be performed on the reconfigurable fabric, whereas 20% remains on the ARM core (remember Amdahl's law mentioned earlier in the chapter). The net effect is a gain of almost 20 over equivalent implementations on ASIP processors.

Slide 5.64
Reconfigurable accelerators are now used in a wide range of applications, including high-volume components such as CD, DVD, and MP3 players. For example, Sony has been using an architecture called the Virtual Mobile Engine (VME) in quite a number of its consumer applications. The VME architecture bears close resemblance to the reconfigurable processor structure presented in the previous slides. A quick scan of today's system-on-a-chip architectures reveals many cases that closely fit this model.

Slide 5.65
Yet, the effectiveness of the ASIP and accelerator ideas is limited by just the same factor that hampers the multi-core idea, presented earlier in the chapter: Amdahl's law. In other words, the potential gains in energy efficiency are bounded by the fraction of the application or algorithm that is purely sequential. Augmenting concurrency through input languages with explicit concurrency semantics, automated transformations, and/or algorithmic innovations is absolutely essential.

Remember: Amdahl's Law Still Holds

- Effectiveness of alternative architectures (ASIP, Accelerator, Reconfigurable) determined by the amount of code spawned from GP

- Mostly effective for repetitive kernels

- 80%–20% rule typically seems to apply

- Transformations can help to improve effectiveness

- Most important: code development and algorithm selection that encourage concurrency

Although this is not a new idea at all, it has gained substantially more urgency over the past years. We recall a quote by John Tukey, one of the co-inventors of the popular FFT algorithm in the 1980s: "*In the past we had to concentrate on minimizing the number of computations, now it is more important for an algorithm to be parallel and regular.*"

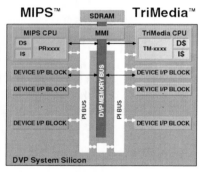

Slide 5.66

To bring all the aforementioned concepts together, it is worth analyzing some of the integrated solutions that industry is using today for embedded applications such as multimedia and communications, which are energy- and cost-constrained. To reduce design time and time-to-market, the industry has embraced the so-called *platform-based* design strategy [K. Keutzer, TCAD'00]. A platform is a structured approach to programmable architectures. Built around one or more general-purpose processors and/or DSPs and a fixed interconnect structure, one can choose to add a variety of special-purpose modules and accelerators to target the component for a particular product line and to ensure that the necessary performance and energy efficiency is obtained. Re-use dramatically reduces the cost and the time for a new design.

An example is the NXP Nexperia™ platform, targeting multimedia applications. The core of the platform is an interconnect structure and two cores (either of which can be omitted): a MIPS GPP and a Trimedia™ DSP. Nexperia provides a large library of I/O modules, memory structures, and fixed and reconfigurable accelerators. Various versions of the platform are now in use in HDTVs, DVD players, portable video players, etc.

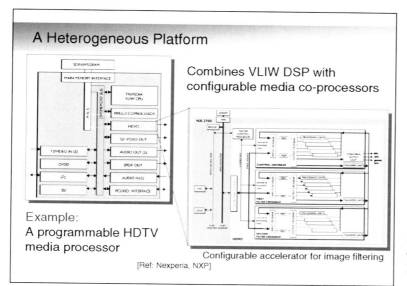

Slide 5.67
One instance of the Nexperia platform, a media processor for HDTV, is shown here. This incarnation only contains a DSP, no GPP. Most interesting are the wide range of input–output processing modules and the MPEG-2 and HDVO accelerator units. The latter is a reconfigurable co-processor for image filtering, combining flexibility, performance, and energy efficiency at a low area overhead.

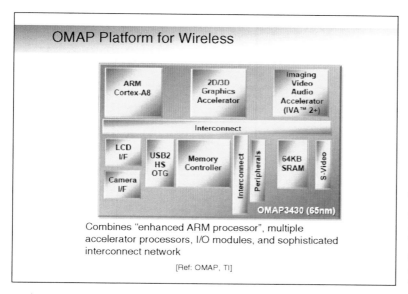

Slide 5.68
Another example of the platform approach to energy-efficient programmable systems-on-a-chip is the already mentioned OMAP platform™ of Texas Instruments. The application target of this platform is the wireless-communications arena. With wireless portable devices embracing a wider range of functionality (ranging from MP3 player over TV and video playback to gaming), as well as a broad spectrum of interfaces (3G cellular, WiFi, Bluetooth, WiMAX, etc.), programmability and flexibility are essential. At the same time, the form factor limits the amount of energy available. Power is capped at 3 W, as we discussed in Chapter 1. This slide shows one instance of the OMAP platform (OMAP3430), focused in particular on providing graphics and multimedia functionality.

Summary and Perspectives

- Architectural and algorithmic optimization can lead to drastic improvements in energy efficiency

- Concurrency is an effective means to improve throughput at fixed energy or reduce energy for fixed throughput

- Energy-efficient architectures specialize the implementation of often-recurring instructions or functions

Slide 5.69

In summary, this chapter demonstrates that architectural and algorithmic innovation is one of the most effective ways in managing and reducing energy dissipation. In addition, for this innovation to become an integral part of the everyday design process, a structured exploration strategy is essential (instead of the ad hoc process mostly used today). Understanding the trade-offs in the energy–delay–(area, flexibility) space goes a long way in establishing such a methodology.

References

Theses:
- M. Potkonjak, "Algorithms for high level synthesis: resource utilization based approach," PhD thesis, UC Berkeley, 1991.
- N. Zhang, "Algorithm/Architecture Co-Design for Wireless Communication Systems," PhD thesis, UC Berkeley, 2001.

Articles:
- D. Blaauw and B. Zhai, "Energy efficient design for subthreshold supply voltage operation," *IEEE International Symposium on Circuits and Systems (ISCAS)*, April, 2006
- S. Borkar, "Design challenges of technology scaling," *IEEE Micro*, 19(4), pp. 23–29, July–Aug. 1999.
- A.P. Chandrakasan, S. Sheng and R.W. Brodersen, "Low-power CMOS digital design," *IEEE Journal of Solid-State Circuits*, 27(4), pp. 473–84, April 1992.
- A. Chandrakasan, M. Potkonjak, J. Rabaey and R. Brodersen, "Optimizing power using transformations", *IEEE Transactions on Computer Aided Design*, 14(1), pp. 12–31, Jan. 1995.
- S. Chou, "Integration and innovation in the nanoelectronics era," Keynote presentation, Digest of Technical Papers, International Solid-State Circuits Conference (ISSCC05), pp. 36–41, Feb. 2005.
- T. Claasen, "High speed: not the only way to exploit the intrinsic computational power of silicon," Keynote presentation, Digest of Technical Papers, International Solid-State Circuits Conference (ISSCC99), pp. 22–25, Feb. 1999.
- H. De Man, "Ambient intelligence: gigascale dreams and nanoscale realities," Keynote presentation, Digest of Technical Papers, International Solid-State Circuits Conference (ISSCC '05), pp. 29–35, Feb. 2005.
- G. Frantz, http://blogs.ti.com/2006/06/23/what-moore-didn%e2%80%99t-tell-us-about-ics/
- K. Keutzer, S. Malik, R. Newton, J. Rabaey and A. Sangiovanni-Vincentelli, "System level design: orthogonalization of concerns and platform-based design," *IEEE Transactions on Computer-Aided Design of Integrated Circuits & Systems*, 19(12), pp.1523–1543, Dec. 2000.

Slides 5.70 and 5.71

Some references...

References

Articles (contd.)

- T. Kuroda and T. Sakurai, "Overview of low-power ULSI circuit techniques," *IEICE Trans. on Electronics*, E78-C(4), pp. 334–344, April 1995.
- D. Markovic, V. Stojanovic, B. Nikolic, M.A. Horowitz and R.W. Brodersen, "Methods for true energy-performance optimization," *IEEE Journal of Solid-State Circuits*, 39(8), pp. 1282–1293, Aug. 2004.
- D. Markovic, B. Nikolic and R.W. Brodersen, "Power and area minimization for multidimensional signal processing," *IEEE Journal of Solid-State Circuits*, 42(4), pp. 922–934, April 2007.
- Nexperia, NXP Semiconductors, http://www.nxp.com/products/**nexperia**/about/index.html
- OMAP, Texas Instruments, http://focus.ti.com/general/docs/wtbu/wtbugencontent.tsp?templateId=6123&navigationId=11988&contentId=4638
- J. Rabaey, "System-on-a-Chip – A Case for Heterogeneous Architectures", Invited Presentation, Wireless Technology Seminar, Tampere, May 1999. Also in HotChips' 2000.
- K. Seno, "A 90nm embedded DRAM single chip LSI with a 3D graphics, H.264 codec engine, and a reconfigurable processor", HotChips 2004.
- R. Subramanyan, "Reconfigurable Digital Communications Systems on a Chip", Invited Presentation, Wireless Technology Seminar, Tampere, May 1999.
- H. Zhang, V. Prabhu, V. George, M. Wan, M. Benes, A. Abnous and J. Rabaey, "A 1V heterogeneous reconfigurable processor IC for baseband wireless applications," *IEEE Journal of Solid-State Circuits*, 35(11), pp. 1697–1704, Nov. 2000 (also ISSCC 2000).
- R. Zlatanovici and B. Nikolic, "Power-Performance Optimal 64-bit Carry-Lookahead Adders," in Proc. European Solid-State Circuits Conf. (ESSCIRC), pp. 321–324, Sept. 2003.

Chapter 6
Optimizing Power @ Design Time – Interconnect and Clocks

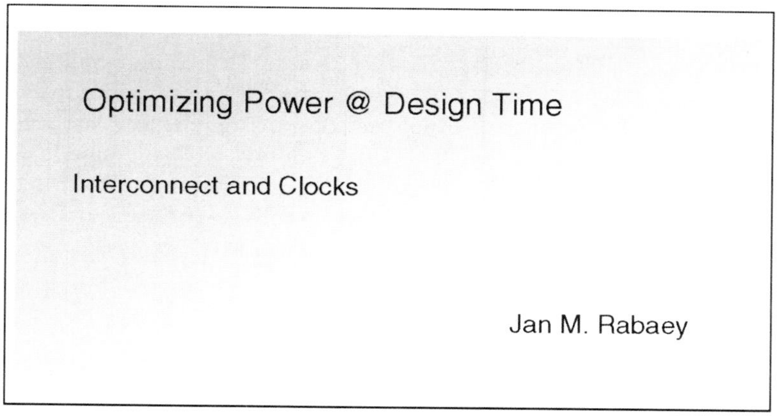

Slide 6.1

Optimizing Power @ Design Time

Interconnect and Clocks

Jan M. Rabaey

So far we have focused our discussion mostly on the energy efficiency of logic. However, interconnect and communication constitute a major component of the overall power budget, as we will demonstrate. They hence deserve some special attention, especially in light of the fact that the physics of interconnect scale somewhat differently than those of logic. As with logic, power optimization can again be considered at multiple levels of the design hierarchy.

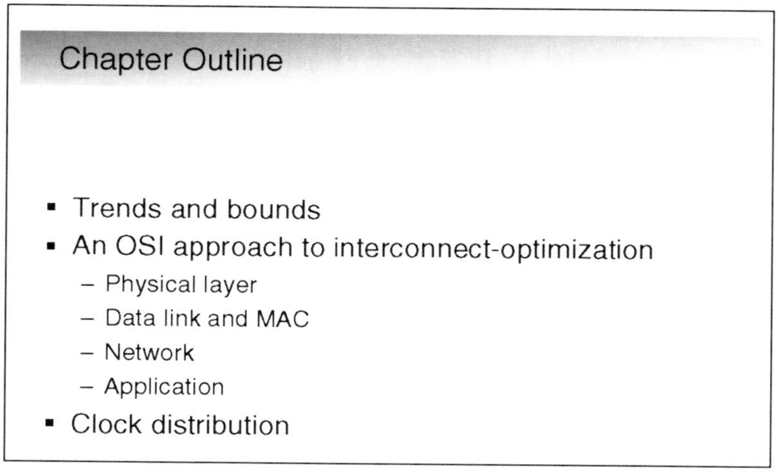

Slide 6.2

Chapter Outline

- Trends and bounds
- An OSI approach to interconnect-optimization
  - Physical layer
  - Data link and MAC
  - Network
  - Application
- Clock distribution

The chapter commences with an analysis of the scaling behavior of interconnect wires. Some fundamental bounds on the energy dissipation of interconnect are established. One particular aspect of this chapter is that it treats on-chip communication as a generic networking problem, and hence classifies the low-energy design techniques along the lines of the standard OSI layering (just as we would do for large-scale networking). The chapter concludes with a discussion of one class of wires that need special attention: the clock distribution network.

Slide 6.3

ITRS Projections			
Calendar Year	2012	2018	2020
Interconnect One Half Pitch	35 nm	18 nm	14 nm
MOSFET Physical Gate Length	14 nm	7 nm	6 nm
Number of Interconnect Levels	12–16	14–18	14–18
On-Chip Local Clock	20 GHz	53 GHz	73 GHz
Chip-to-Board Clock	15 GHz	56 GHz	89 GHz
# of Hi Perf. ASIC Signal I/O Pads	2500	3100	3100
# of Hi Perf. ASIC Power/Ground Pads	2500	3100	3100
Supply Voltage	0.7–0.9 V	0.5–0.7 V	0.5–0.7 V
Supply Current	283–220 A	396–283 A	396–283 A

If we consult the ITRS predictions on how interconnect will evolve in the coming decade, we observe that scaling is projected to go forward at the same pace as it does today. This leads to some staggering numbers. By 2020, we may have 14–18 (!) layers of interconnect with the lowest levels of the interconnect stack at a half pitch of only 14 nm. Clocks speeds could be at multiple tens of GHz, and the number of input and output signals may be larger than 3000. A simple analysis of what it means to switch this huge interconnect volume leads to some incredible power numbers. Even aggressive voltage scaling may not be sufficient to keep the dissipation within bounds. Hence, novel approaches on how to distribute signals on a chip are required.

Slide 6.4

Increasing Impact of Interconnect

- Interconnect is now exceeding transistors in
  - Latency
  - Power dissipation
  - Manufacturing complexity
- Direct consequence of scaling

In fact, the problem is already with us today. If we evaluate today's most advanced 65 nm devices with up to 8 interconnect layers, multiple hundreds of I/O pins, and clock frequencies (at least locally) of up to 5 GHz, we see that providing connectivity between the components poses an essential limitation on the latency levels that can be achieved. It also dominates the power dissipation — at least, if we also take the clock distribution network into account. Manufacturing the multiple layers of metal (mostly Cu and Al) and dielectric material in a reliable and predictable fashion is already a challenge in itself.

Slide 6.5

To drive the point home, this slide shows the power distribution over the different resources for a number of typical classes of integrated circuits. If I/O, interconnect, and clocks are lumped together, they constitute 50% or more of the budget for each class of devices. The worst case is the FPGA,

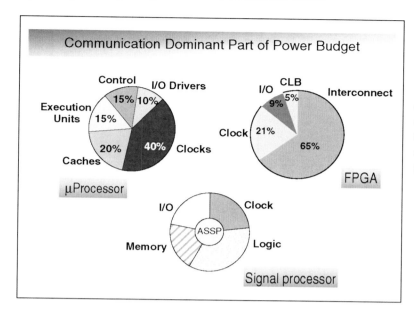

where interconnect power takes more than 80% of the power budget [Kusse'98]. Observe that these numbers represent designs of the late 1990s and that the pendulum has swung even more in the direction of interconnect in recent years.

Slide 6.6

To understand why this shift is happening, it is worthwhile examining the overall scaling behavior of wires. The ideal scaling model assumes that the two dimensions of the cross-section of the wire (W and H) are reduced with the same scaling factor S between process nodes (with S the same as the scaling factor of the critical dimensions of the process). How wire delay and energy dissipation per transition evolve depends upon how the wire length evolves with scaling. The length of local wires (e.g., those between gates) typically evolves the same way as the logic, whereas global wires (such as busses and clock networks) tend to track the chip dimensions (as is illustrated in the next slide). With the values of S and S_C (the chip scaling factor) typically being at 1.4 and 0.88, respectively, between subsequent processor nodes, we can derive the following scaling behavior:

- The delay of a local wire remains constant (in contrast to the gate delay which reduces by 1.4), whereas long wire gets 2.5 times slower!
- From an energy perspective, the picture does not seem too bad, and depends strongly on how the supply voltage scales (U). In the ideal model ($U = S$), things look quite good as the energy dissipation of a transition of a local and global wire reduces by 2.7 and 1.7, respectively. If the voltage is kept constant, the scaling factors are 1.4 and 0.88, respectively. Gate and wire energy exhibit approximately the same scaling behavior.

Unfortunately, the ideal model does not reflect reality. To address the wiring delay challenge, wire dimensions have not been scaled equally. For the layers at the bottom of the interconnect stack, where reducing the wire pitch is essential, wire heights have been kept almost constant between technology generations. This increases the cross-section, and hence decreases resistance – which is good for delay reduction. On the other hand, it increases capacitance (and hence energy) due to the increased contributions of the sidewalls of the wires.

Wires on the top of the stack are not scaled at all. These "fat" wires are used mostly for global interconnect. Their capacitance and energy now scales with the chip dimensions – which means that they are going up.

An important divergence between logic and interconnect is worth mentioning: though leakage has become an important component in the energy budget of logic, the same is not true (yet) in interconnect. The dielectrics used so far have been good enough to keep their leakage under control. This picture may change in the future though.

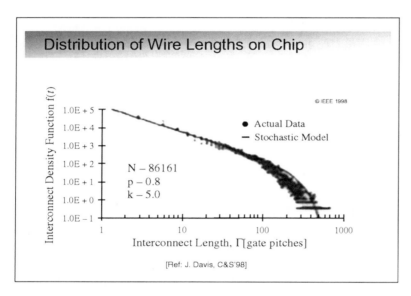

Slide 6.7
This slide plots a histogram showing the distribution of wire lengths in an actual microprocessor design, which contains approximately 90,000 gates. Though most of the wires are only a couple of gate pitches long, a substantial number of them are much longer, reaching lengths of up to 500 gate pitches, which is approximately the size of the die.

Slide 6.8
The obvious question is what technology innovations can do to address the problem. Research in novel interconnect strategies has been intense and is ongoing. In a nutshell, they can be summarized as follows:

- **Interconnect materials with *lower resistance*** – This only indirectly impacts energy. For the same delay, wires can be made thinner, thus reducing capacitance and in turn switching-energy. However, this avenue has led to a dead end. With copper in general use, there are no other materials in sight that can provide a next step.
- **Dielectrics with *lower permittivity* (so-called low-*k* materials)** – These directly reduce capacitance and hence energy. Advanced process technologies already use organic materials such as polyimides, which reduce the permittivity compared to the traditional SiO_2. The next step would be to move to aerogels ($\varepsilon_r = \sim 1.5$). This is probably as close as we will ever get to free space. A number of companies are currently researching ways to effectively deposit "air bubbles", by using self-assembly, for instance.

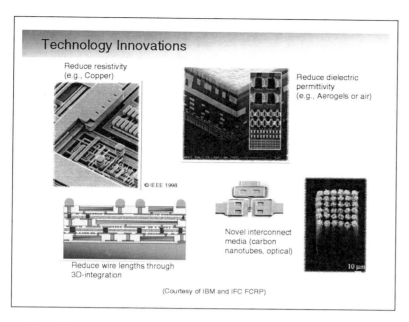

- **Shorter wire lengths** – One way to effectively reduce the wire lengths (at least those of the global wires) is to go the three-dimensional route. Stacking components vertically has been shown to have a substantial effect on energy and performance. The concept has been around for a long time, but recently has gathered a lot of renewed interest (especially in light of the perceived limits to horizontal scaling). The challenges still remain formidable – with yield and heat removal being the foremost ones.
- **Novel interconnect media** – optical interconnect strategies have long been touted as offering major performance and energy benefits. Although it is questionable if optical signaling ever will become competitive for on-chip interconnect due to the optical–electrical conversion overhead, recent advances have made off-chip optical interconnect a definitive possibility. Over the long term, carbon nanotubes and graphene offer other interesting opportunities. On an even longer time scale, we can only wish that we would be able to exploit the concept of quantum entanglement one day (tongue-in-cheek).

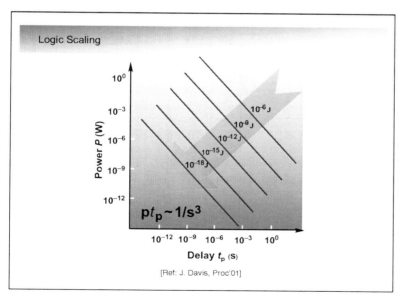

e 6.9
It is worthwhile to spend some time reflecting on the fundamental scaling differences between logic and wires. Under ideal scaling conditions, the Power–Delay product (i.e., the energy) of a digital gate scales as $1/S^3$. Hence gates get more effective with scaling.

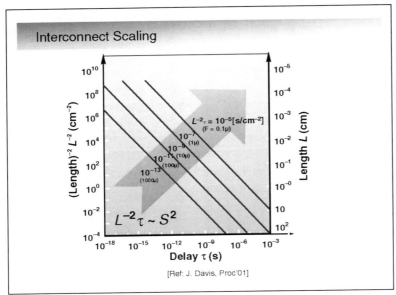

Slide 6.10
On the other hand, wires tend to become less effective. For a given technology, the product of wire delay and L^{-2} is a constant, assuming that the delay is dominated by the rc effect. It can hence be considered to be a figure of merit. Again assuming ideal scaling rules (i.e., all dimensions scale equally with the exception of the wire length), τL^{-2} scales as S^2:

$$\frac{\tau}{L^2} = rc = \frac{\rho\varepsilon}{HT} \propto S^2$$

In other words, the figure of merit of a wire gets worse with technology scaling, at least from a performance perspective. The only ways to change the scenario is to modify the material characteristics ($\rho\varepsilon$), or the propagation mechanism (for instance, by moving from rc-dominated diffusion to wave propagation).

Slide 6.11
Though we have established bounds on performance, it is valuable to know if there are bounds on the energy efficiency as well. And indeed there are – and they can be obtained by using nothing less than the *Shannon theorem*, famous in the communication and information theory communities. The theorem relates the available capacity of a link (in bits/sec) to the bandwidth and the average signal power. Using some manipulations and assuming that a link can take an infinite time to transmit a bit, we derive that the minimum energy for transmitting a bit over a wire equals $kT\ln(2)$ (where T is the absolute temperature, and k the Boltzmann constant) – a remarkable result as we will see in later chapters. At room temperature, this evaluates to 4 zeptoJoules (or 10^{-21} J – a unit worth remembering). As a reference, sending a 1 V signal over a

1 mm intermediate-layer copper wire implemented in a 90 nm technology takes approximately 200 fJ, or *eight orders of magnitude* more than the theoretical minimum.

Slide 6.12

Reducing Interconnect Power/Energy

- Same philosophy as with logic: reduce capacitance, voltage (or voltage swing), and/or activity
- A major difference: sending a bit(s) from one point to another is fundamentally a **communications/ networking problem**, and it helps to consider it as such
- Abstraction layers are different:
  - For computation: device, gate, logic, micro-architecture
  - For communication: wire, link, network, transport
- Helps to organize along abstraction layers, well-understood in the networking world: the OSI protocol stack

The techniques to make interconnect more energy efficient are in many ways similar to what we do for logic. In a way, they are somewhat simpler, as they relate directly to what people have learned for a long time in the world of (large-scale) communications and networking. Hence, it hence pays off to consider carefully what designers have come up with in those areas. We should keep the following caveat in mind however: what works on the macro scale, does not always scale well to the micro scale. Not all physical parameters scale equally. For example, at shorter wire lengths and lower energy levels, the cost of signal shaping and detection becomes more important (and often even dominant). Yet, over time we have seen more and more of what once was system- or board-level architecture migrate to the die.

Slide 6.13

OSI Protocol Stack

- Reference model for wired and wireless protocol design — Also useful guide for conception and optimization of on-chip communication
- Layered approach allows for orthogonalization of concerns and decomposition of constraints

Presentation/Application
Session
Transport
Network
Data Link
Physical

- No requirement to implement all layers of the stack
- Layered structure need not necessarily be maintained in final implementation

[Ref: M. Sgroi, DAC'01]

We had introduced the logical abstraction layers in Chapter 5. A similar approach can be taken for interconnect. Here, the layers are well understood, and have long been standardized as the OSI protocol stack (check http://en.wikipedia.org/wiki/OSI_model if you are not familiar with the concept). The top layers of the stack (such as the session and the presentation layers) are currently not really relevant for chip interconnects, and are more appropriate for the seamless communication between various applications over the internet. Yet, this picture may change over time when 100s to 1000s of processors get integrated on a single die. Today though, the relevant layers are the physical, link/MAC, and network layers. We organize the rest of the chapter along those lines. Before embarking on the discussion of the various techniques, it is worth pointing out that, just as for logic, optimizations at the higher layers of the abstraction chain often have more impact. At the same time, some problems are more easily and more cheaply addressed at the physical level.

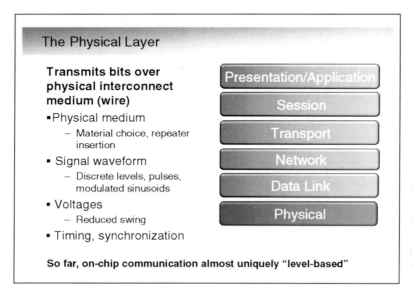

Slide 6.14
The physical layer of the interconnect stack addresses how the information to be transmitted is represented in the interconnect medium (in this case, the wire). Almost without exception, we are using voltage levels as the data representation today. Other options would be to use either currents, pulses (exploiting a wider bandwidth), or modulated sinusoids (as used in most wireless communication systems). These schemes increase the complexity of the transmitter and/or receiver, and hence have not been very attractive for integrated circuits. Yet, this may change in the future, as we discuss briefly at the end of the chapter.

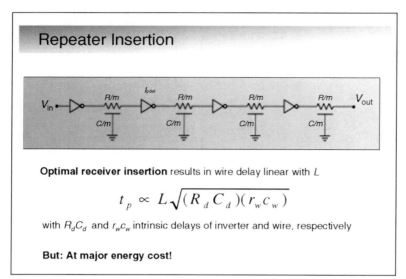

Slide 6.15
The majority of the wires on a chip can be considered either as being purely capacitive (for very short connections), or as distributed rc-lines. With the availability of thick copper lines at the top of the chip, on-chip transmission lines have become an option as well. They form an interesting option for the distribution of signals over longer distances.

Given their prominence, we focus most of our attention on the rc lines in this chapter. It is well-known that the delay of the wire increases quadratically with

Optimizing Power @ Design Time – Interconnect and Clocks

its length, whereas the energy dissipation rises linearly. The common technique to get around the delay concern is to insert repeaters at carefully selected intervals, which makes it possible to make the delay proportional to the length of the wire. The optimal insertion rate (from a performance perspective) depends upon the intrinsic delays of both the driver and the interconnect material.

The introduction of repeaters adds active components to an otherwise passive structure, and hence adds extra energy dissipation.

Repeater Insertion — Example

- 1 cm Cu wire in 90 nm technology (on intermediate layers)
  - r_w = 250 Ω/mm; c_w = 200 fF/mm
  - $t_p = 0.69 r_w c_w L^2$ = 3.45 ns
- Optimal driver insertion:
  - $t_{p\ opt}$ = 0.5 ns
  - Requires insertion of 13 repeaters
  - Energy per transition 8 times larger than just charging the wire (6 pJ versus 0.75 pJ)!
- It pays to back off!

Slide 6.16

The cost of optimal performance is very high (this should be of no surprise by now). Consider, for instance, a 1 cm copper line implemented in a 90 nm technology. The energy cost of the receiver is six times higher than what it takes to just charge the wire with a single driver. Again, just backing off a bit from the absolute minimum delay goes a long way in making the design more energy-efficient.

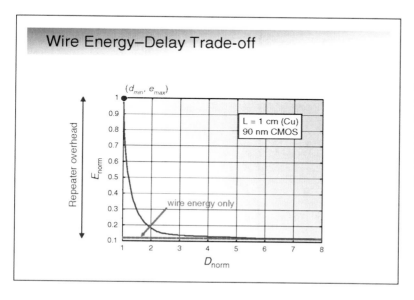

Slide 6.17

As always, the trade-off opportunities are best captured by the energy–delay curves. Doubling the allowable delay reduces the required energy by a factor of 5.5! Even just backing off 10% already buys a 30% energy reduction.

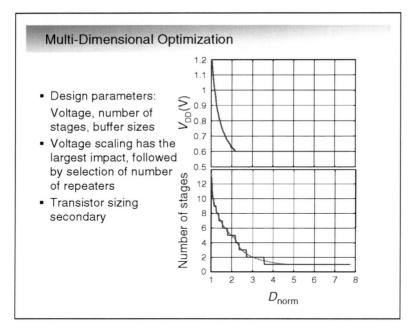

Slide 6.18
It is worth spending some time contemplating on how this pareto-optimal E–D curve was obtained. The design parameters involved include the supply (signal) voltage, the number of stages, and the transistor sizes (in the buffer/repeaters). From the results of the multi-dimensional optimization, it can be seen that the supply voltage has the biggest impact, followed by insertion rate of the repeaters. Observe that the width of the wire only has a secondary impact on the wire delay. Once the wire is wide enough to make the contribution of the fringing capacitance or sidewall capacitance ignorable, further increases in the wire width do nothing more than raising the energy dissipation, as is illustrated by the equations below.

$$c_w = w \cdot c_{pp} + c_f \quad r_w = r_{sq}/w$$
$$\tau_w = c_{pp}r_{sq} + c_f r_{sq}/w$$

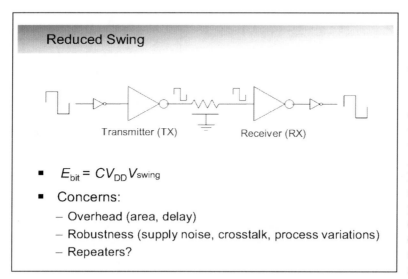

Slide 6.19
With reduction of the supply voltage (or more precisely, the signal swing) proven to be the most effective technique to save energy, some contemplation on how to accomplish this effectively is at hand. As we have observed earlier, sending a signal along a wire is a communication problem, and it is worth considering as such. A communication link consists of a transmitter (TX), a communication medium, and a receiver (RX). The generic configuration in CMOS is to have a driver (inverter) as TX, a stretch of aluminum or copper wire in between, and another inverter as a receiver.

Optimizing Power @ Design Time – Interconnect and Clocks

This changes once we reduce the signal swing. The TX acts as a driver as well as a level down-converter, whereas the RX performs the up-conversion. Though the energy savings are either linear or quadratic, depending upon the operational voltage of the TX, reducing the swing comes with an overhead in delay (maybe) and complexity (for sure). In addition, it reduces the noise margins and makes the design more susceptible to interference, noise, and variations. Yet, as we have learned from the communications community, the benefits of properly conditioning the signal can be quite substantial.

Slide 6.20

In Chapter 4, we had already touched on the topic of level conversion and multiple supply voltages (Slides 4.32, 4.33 and 4.34). It was concluded that down-conversion is relatively easy if multiple supply voltages are available. The challenge is in the up-conversion. In the logic domain, where the overhead penalty easily offsets the energy gains, we concluded that the level-conversion is best confined to the boundaries of the combinational logics (i.e., the flip-flops), where the presence of a clock helps to time when to perform the energy-hungry amplification. The availability of positive feedback in most of the latches/registers is another big plus.

Yet, synchronous or clocked conversion is not always an option in the interconnect space, and asynchronous techniques are worth examining. In this slide, a conventional reduced-swing interconnect scheme is presented. To reduce the signal swing at the transmit site, we simply use an inverter with a reduced supply voltage. The receiver resembles a differential cascade voltage switch logic (DCVSL) gate [Rabaey03], which consists of complementary pull-down networks and a cross-coupled PMOS load. The only difference is that the input signals are at a reduced voltage level, and that a low-swing inverter is needed to generate the complementary signal. The disadvantage of this approach is that it effectively needs two supply voltages.

Slide 6.21

One interesting way to create a voltage drop is to exploit implicit voltage references such as threshold voltages. Consider, for instance, the circuit presented in this slide. By swapping the NMOS and PMOS transistors in the driver, the logic levels on the wire are now set to $|V_{THp}|$ and $V_{DD} - V_{THn}$. For a supply voltage of 1 V and threshold voltages for NMOS and PMOS transistors around 0.35 V, this translates into a signal swing of only 0.3 V!

The receiver consists of dual cross-coupled pairs. The transistors N2 and P2 ensure full logic swing at the outputs, whereas N3 and P3 isolate the full-swing output nodes from the low-swing interconnect wires. A number of alternative level-conversion circuits, and a comparison of their effectiveness can be found in [Zhang'00]. From this, we learn that to be effective the

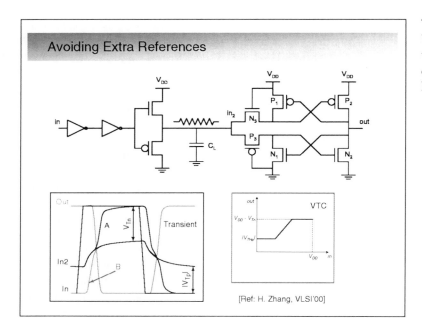

TX–RX overhead should be no more than 10% of the overall energy budget of the communication link.

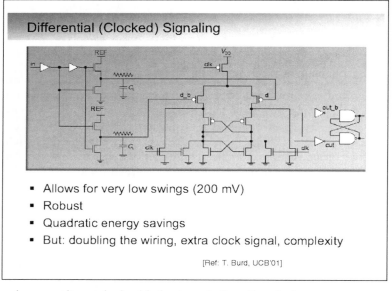

Slide 6.22
One of the concerns of the reduced-swing circuits is their increased sensitivity to interference and supply noise. Using a differential scheme not only offers a major increase in common-mode rejection, but also helps to reduce the influence of interference by 6 dB. Signaling schemes with reliable swing levels of as low as 200 mV have been reported and used. At the same time, differential interconnect networks come with a substantial overhead, as the overall wire capacitance is doubled – translating directly into extra energy dissipation. In addition, the differential detection scheme at the receiver consumes continuous power, and should be turned off when not in use. This most often (but not necessarily always) means that a clocked synchronous approach is required. Differential techniques are most effective when the wiring capacitance is huge, and the benefits of the extra small swing outweigh the overhead of the doubled capacitance and the extra clocking.

Optimizing Power @ Design Time – Interconnect and Clocks

Slide 6.23
At this point, it is worth wondering if there exists a lower bound on the signal swing that can be used in practice. A number of issues should be considered:

- Reducing the swing negatively impacts the delay, as it substantially increases the time it takes to reconstruct the signal level at the receiver end. In general, we may assume that the receiver delay is proportional to the swing at its input. This again leads to a trade-off. In general, the longer the wire the more the reduced swing makes sense.
- The smaller the signal, the larger the influence of parasitic effects such as noise, crosstalk, and receiver offset (if differential schemes are used). All of these may cause the receiver to make erroneous decisions. Besides the power supply noise of sender and receiver, the primary noise source in interconnect networks is the capacitive (and today even inductive) coupling between neighboring wires. This is especially a problem in busses, in which wires may run alongside each other for long distances, and crosstalk becomes substantial. This problem can be reduced by crosstalk repressing techniques such as proper shielding – which comes at the expense of area and wire folding.

So far, signaling swings that have been reported on busses hover around 200 mV. There is however no compelling reason to assume that this number can no further be reduced, and 100 mV swings have been considered in a number of designs.

Caveat: The usage of reduced signal swings on-chip definitely is incompatible with the standard design methodologies and flows, and hence falls into the realm of custom design. This means that the designer is fully responsible for the establishment of verification and test strategies – clearly not for the fainthearted ... Fortunately, a number of companies have brought modular intellectual-property (IP) solutions for energy-efficient on-chip communication on the market in recent years, thus hiding the complexity from the SoC designer.

Slide 6.24
A wide range of other energy-reducing on-chip data communication schemes have been published – few of which have made it onto industrial designs, though. A couple of those ideas are too compelling to omit from this text. The first one is based on the adiabatic charging approach we briefly touched on in Chapter 3. If delay is not of primary importance, we can extend the energy–delay space by using alternative charging techniques. Yet, the implementation of a truly adiabatic circuit requires the implementation of an energy-recovering clock generator, which typically requires a resonating network including inductors [L. Svensson, CRC'05]. The latter are expensive and low-quality when implemented on-chip. Off-chip inductors, on the other hand, increase the system cost.

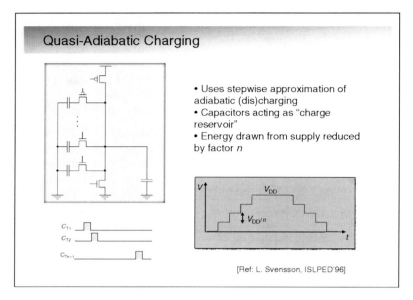

a factor of N over the single-step charging. The total energy is reduced by the same factor.

The $N-1$ intermediate voltage references are realized using a capacitor tank C_{Ti} (where $C_{Ti} \gg C_L$). During each charge-and-discharge cycle, each capacitor C_{Ti} provides and receives the same amount of charge, so the tank capacitor voltages are self-sustaining. Even more, it can be shown that during the start-up, the tank voltages automatically converge to an equal distribution.

In essence, this driver is not truly adiabatic. It rather belongs to the class of the "charge-redistribution" circuits: each cycle, a charge packet is injected from the supply, which then gradually makes its way from level to level during subsequent cycles, and is finally dumped into ground after N cycles. The following slide shows another circuit of the same class.

In this slide, a quasi-adiabatic driver for large capacitive busses is presented. A stepwise approximation of a ramp is produced by connecting the output in sequence to a number of evenly distributed voltages, starting from the bottom. From each reference, it receives a charge $C_L V/N$ before eventually being connected to the supply. To discharge the capacitance, the reverse sequence is followed. For each cycle, a charge equal to $C_L V/N$ is drawn from the supply, a reduction by

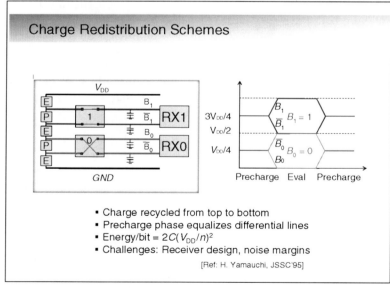

Slide 6.25

"Charge recycling" is another idea that is very intriguing, but has rarely been used. In a traditional CMOS scheme, charge is used only a single time: it is transferred from the supply to the load capacitor in a first phase, and dumped to the ground in a second. From an energy perspective, it would be great if we could use charge a couple of times before dumping it. This by necessity requires the use of multiple voltage levels. A simplified example of a chargerecycling bus with two levels is shown in this slide. Each bit i is present in differential form

(Bi and $\overline{Bi}$). During the precharge phase, the two differential lines for each bit are equalized by closing the switches P. During evaluation, one of the lines is connected to a line of a "higher-order" bit (representing a **1**), whereas the other is equalized with a "lower-order" bit (representing a **0**) using the switches E. This produces a differential voltage at each line pair, the polarity of which depending upon the logic value to be transmitted. Differential amplifiers at the end of the bus (one for each pair) reproduce the full swing signals.

Assuming that the capacitances of all lines are equal, we can see that the precharge voltage levels divide equally between V_{DD} and GND. The signal swing on each bus pair equals V_{DD}/N. The principle is quite similar to that of the quasi-adiabatic driver of the previous slide – a charge packet is injected from the supply, which sequentially drives every bit of the bus in descending order, until eventually being dumped on the ground.

The challenge resides in adequately detecting the various output levels in the presence of process variations and noise. Yet, the idea has enough potential that it is bound to be useful in a number of special cases.

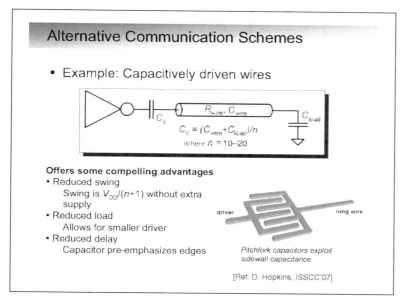

Slide 6.26
At the end of this discussion on physical-layer signaling, it is worth pointing out some other signaling strategies that may become attractive. In this slide, we show only one of them. Rather than connecting resistively into the interconnect network, drivers could also couple capacitively. The net effect is that the swing on the interconnect wire is reduced automatically without needing any extra supplies. In addition, driver sizes can be reduced, and signal transitions are sharper. The approach comes with a lot of challenges as well (one still needs a level-restoring receiver, for instance), but is definitely worth keeping an eye on.

Slide 6.27
So far, we have concentrated on the data representations of our signaling protocol, and have ignored timing. Yet, the interconnect network plays an important role in the overall timing strategy of a complex system-on-a-chip (SoC). To clarify this statement, let us consider the following simple observation: It takes an electronic signal moving at its fastest possible speed (assuming transmission-line conditions) approximately 66 ps to move from one side to the other of a 1 cm chip. When rc-effects dominate, the reality is a lot worse, as shown in Slide 6.16, where the minimum delay was determined to be 500 ps. This means that for clock speeds faster than 2 GHz, it takes more than one clock cycle for a signal to propagate across the chip! The situation is even worse, when the interconnect wire is loaded with a large distributed fan-out – as is always the case with busses, for instance.

There are a number of ways to deal with this. One commonly used option is to pipeline the wire by inserting a number of clocked buffer elements. This happens quite naturally in the

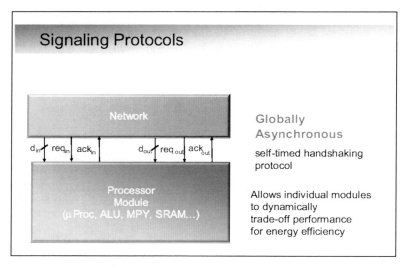

network-on-a-chip (NoC) paradigm, which we will discuss shortly. Yet, all this complicates the overall timing of the chip, and intrinsically links the timing of global interconnect and localized computation. This hampers the introduction of a number of power reduction techniques we have discussed earlier (such as multiple supply voltages and timing relaxation), or to be discussed in coming chapters (such as dynamic voltage and frequency scaling).

Hence it makes sense to decouple global interconnect and local compute timing through the use of asynchronous signaling. Along these lines of thinking, one methodology called GALS (Globally Asynchronous Locally Synchronous) has attracted a following in recent years [Chapiro'84]. The idea is to use a synchronous approach for the local modules (called synchronous islands), while communication between them is performed asynchronously. This approach dramatically relaxes the clock distribution and interconnect timing requirements, and enables various power-saving techniques for the processor modules.

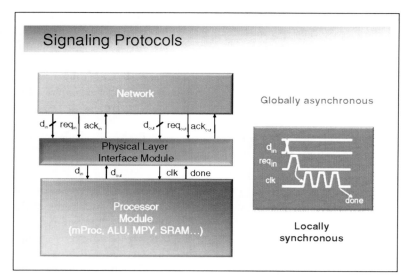

Slide 6.28
Asynchronous signaling opens the door for multiple optimizations. Yet it comes with the overhead of adding various extra control signals such as req(uest) and ack(nowledgement). Although that overhead can be shared by the N wires in the bus, it still is substantial. This is why the two-phase signaling protocol is preferred over the more robust four-phase protocol for large-interconnect networks [Rabaey'03].

The generation and termination of the control signals is most effectively performed by a standardized wrapper around the computational modules, which serves as the boundary between the synchronous and asynchronous domains. One of the very first designs that followed that concept is presented in [Zhang'00].

Optimizing Power @ Design Time – Interconnect and Clocks

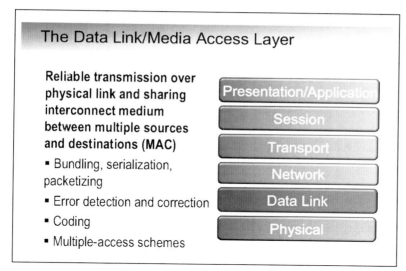

Slide 6.29
Whereas the physical layer deals with the various aspects of how to represent data on the interconnect medium, the function of the link layer is to ensure that data is reliably transmitted in the appropriate formats between the source and the destination. For example, in the wired and wireless networking world, packets of data are extended with some extra error-control bits, which help the destination to determine if the packet was not corrupted during its travel.

If the link connects to multiple sources and destinations, the media access control (MAC) protocol ensures that all sources can share the media in a fair and reliable fashion. Bus arbitration is a great example of a MAC protocol used extensively in SoCs.

Most designers still consider interconnect purely as a set of wires. Yet, thinking about them as a communication network opens the door for a broad range of opportunities, the scope of which will only increase with further scaling. As a starter, the link layer offers a great number of means of introducing energy-saving techniques for global interconnects. For example, adding error-correcting coding allows for a more aggressive scaling of the voltage levels used in a bus.

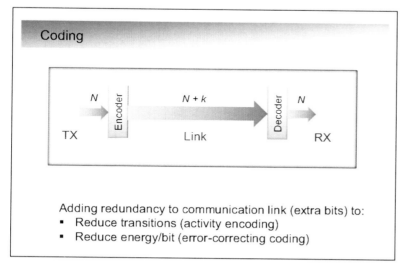

Slide 6.30
Coding is a powerful technique, which is extensively used in most wired and wireless communication systems. So far, its overhead has been too high to be useful in on-chip interconnects. With the growing complexity of integrated circuits, this may rapidly be changing. A number of coding strategies can be considered:

- Channel-coding techniques, which modify the data to be transmitted so that it better deals with imperfections of the channel.
- Error-correcting codes, which add redundancy to the data so that eventual transmission errors can be detected and/or corrected.

- Source codes, which reduce the communication overhead by compressing the data.

As the last one is application-dependent, we focus on the former two in this section. While coding may yield with a substantial energy benefit, it also comes with an overhead:

- Both channel and error-correcting codes require a redundancy in the representation, which most often translates into extra bits.
- Implementation of the coding requires an encoder (at the TX side) and a decoder (at the RX side).

As a result, coding is only beneficial today for interconnect wires with substantive capacitive loads.

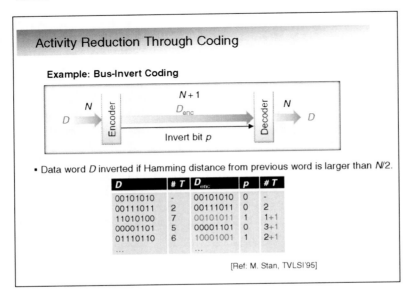

Slide 6.31

As stated in the beginning of the chapter, reducing activity on the interconnect network is an effective way of reducing the energy dissipation. Coding can be an effective means of doing exactly that. To demonstrate the concept, we introduce a simple coding scheme called "bus-invert coding" (BIC). If the number of bit transitions between two consecutive data words is high, it is advantageous to invert the second word, as is shown in the example. The BIC encoder computes the Hamming distance between the previous data transmission $D_{enc}(t\text{-}1)$ and the current $D(t)$. If the Hamming distance is smaller than $N/2$, we just transmit $D(t)$ and set the extra code bit p to 0. In the reverse case, $D_{enc}(t) = \overline{D(t)}$, and p is set to 1.

Optimizing Power @ Design Time – Interconnect and Clocks 169

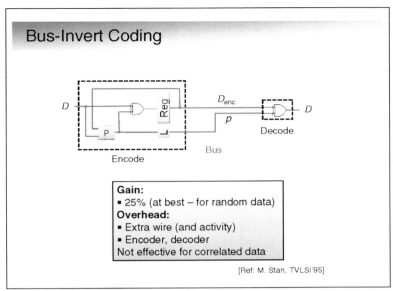

Slide 6.32
One possible realization of the BIC scheme is shown in this slide. The Encode module computes the cardinality of the number of bit transitions between $D_{enc}(t-1)$ and $D(t)$. If the result is larger than $N/2$, the input word is inverted by a bank of XORs, otherwise it is passed along unchanged. Decoding requires no more than another bank of XORs and a register. Observe that the BIC scheme comes with the overhead of one extra bit (p).

Under the best possible conditions, the bus-invert code may result in a 25% power reduction. This occurs when there is very little correlation between subsequent data words (in other words, data is pretty much random, and transitions are plentiful). When the data exhibits a lot of correlations, other schemes can be more effective. Also, the code is less effective for larger values of N (> 16).

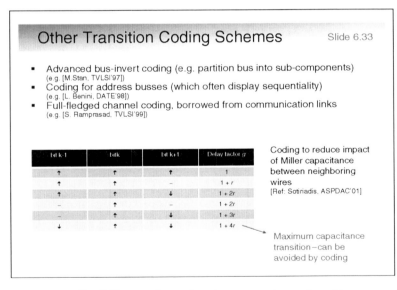

Slide 6.33
The idea of transition coding has gained quite some traction since the BIC scheme was introduced. For the sake of brevity, we just provide a number of references. One class of schemes further optimizes the BIC scheme by, for instance, partitioning the bus if the word length N gets too large. More generic channel-coding schemes have been considered as well.

In case the data exhibits a lot of temporal correlation, a totally different class of codes comes into play. For example, memory is often accessed sequentially. Using address representations that exploit the correlations, such as Gray coding, can help to reduce the number of transitions substantially.

Most transition-coding techniques focus on temporal effects. Yet spatial artifacts should not be ignored either. As we have observed before, bus wires tend to run alongside one another for long distances. The intra-wire capacitance can hence be as or even more important than the capacitance to ground. Under unfavorable conditions, as the table on the slide indicates, a wire can experience a capacitance that is many times larger than the most favorable case. Codes can be engineered to minimize the occurrence of "aggressions" between neighboring bits. Their overhead is manageable as well, as was established in [Sotiriades'01].

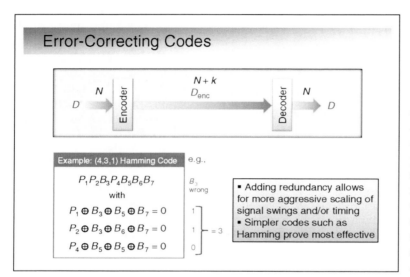

Slide 6.34
Error-correction codes present another interesting opportunity. Under the motto of "better-than-worst-case" design – a concept that receives a lot more attention in Chapter 10 – it often serves well to purposely violate a design constraint, such as the minimum supply voltage or clock period. If this only results in errors on an occasional basis, the energy savings can be substantial. Again, this requires that the overhead of encoding/decoding and transmitting a couple of extra bits is lower than the projected savings. This concept, which is instrumental to the operation of virtually all wireless communications, has been used extensively in various memory products such as DRAM and Flash over the last decades.

It is not too hard to envision that error-correcting coding (ECC) will play a major role in on-chip interconnect in the very near future as well. Our colleagues in the information theory community have done a wonderful job in coming up with a very broad range of codes, ranging from the simple and fast to the very complex and effective. A code is classified by the number of initial data bits, the number of parity bits, and the number of errors it can detect and/or correct. At the current time instance on the technology roadmap, only the simplest codes, such as Hamming, truly make sense. Most other schemes come with too much latency to be useful. An example of a (4,3,1) Hamming code is shown in the slide.

As we had established earlier, scaling causes the cost of communication and computation to increase and decrease, respectively. This will make coding techniques more and more attractive as time progresses.

Slide 6.35
Another aspect of the link layer is the management of media access control (MAC) – in case multiple senders and receivers share the media. A bus is a perfect example of such a shared medium. To avoid collisions between data streams from different sources, time-division multiplexing (TDM) is used. The overhead for sharing the media (in addition to the increased capacitance) is the scheduling of the traffic.

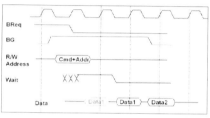

With an eye on the future, it is again worth pondering the possibility of borrowing ideas from the wireless- and optical-communication communities that utilize the diversity offered by the frequency domain. Today, all IC communications are situated in the baseband (i.e., from 0 Hz to 100s of MHz), similar to where optical communications used to be for a long time (before the introduction of wave-division multiplexing or WDM). Modulation of signal streams to a number of higher-frequency channels allow for the same wire to be used simultaneously by a number of streams. The overhead of modulation/demodulation is quite substantial – yet, it is plausible that frequency-division multiplexing (FDM) or code-division multiplexing (CDM) techniques may be used in the foreseeable future for high-capacity energy-efficient communication backplanes between chips, where the link capacitance is large. A great example of such a strategy is given in [Chang'08].

In light of this, TDM is the only realistic on-chip media-access protocol today. Even here, many different options exist, especially regarding the granting of access to the channel. The simplest option is for a source to just start transmitting when data is available. The chance for a collision with other sources is extremely high though, and the overhead of retransmission attempts dominates the energy and latency budgets. The other extreme is to assign each stream its own time slot, granted in advance. This works well for streams that need guaranteed throughput, but may leave some channels heavily under-utilized. Bus arbitration is the most common scheme: a source with data available requests the channel, and starts transmitting when access is granted. The overhead of this scheme is in the execution of the "arbitration" protocol.

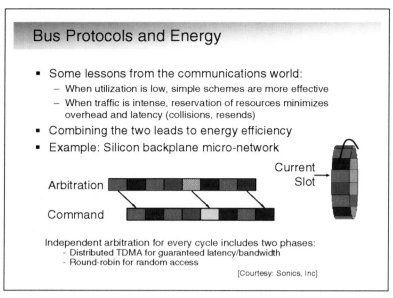

Slide 6.36

It is possible to combine energy efficiency, latency control, channel utilization, and fairness in a single scheme. Streams that are periodic and have strict latency requirements can be assigned their own time slots. Other time slots can be made available through arbitration. The OCP-IP protocol [Ref: Sonics, Inc] does exactly that. OCP (open-core protocol) creates a clean abstraction of the interconnect network. Modules access the network through a socket with a well-defined protocol (which is orthogonal to the actual interconnect implementation). Based on throughput and latency requirements, links are either granted a number of fixed slots, or compete for the others using a round-robin arbitration protocol.

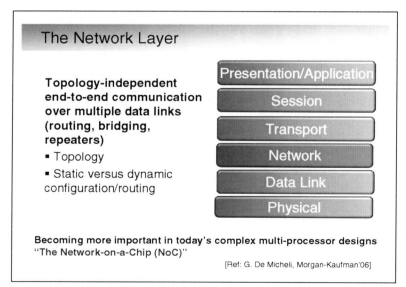

Slide 6.37

The network is the next layer in the OSI stack. With the number of independent processing modules on a chip growing at a fast rate, this layer – which was missing on chips until very recently – is rapidly gaining attention. The networking and parallel computing communities have provided us with an overwhelming array of options, a large number of which is not really relevant to the "network-on-a-chip" (NoC) concept. The choices can be classified into two major groups: (1) the network topology; and (2) the time of configuration.

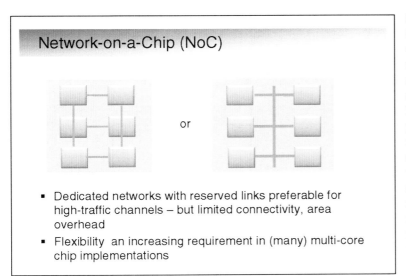

Slide 6.38

One may wonder if a network-on-a-chip approach truly makes sense. We are convinced that it is an absolute necessity. With a large number of modules, point-to-point connections rapidly become unwieldy, and occupy a disproportionate amount of area. A shared time-multiplexed resource, such as a bus, saturates, if the number of connected components becomes too high. Hence, breaking up the connection into multiple segments does more than make sense. In addition, the inserted switches/routers act as repeaters and help to control the interconnect delay.

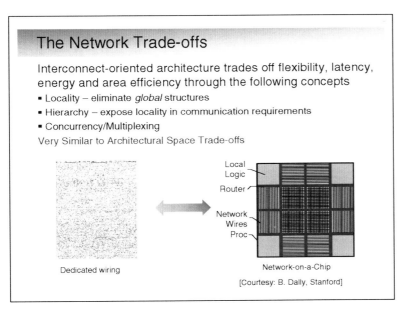

Slide 6.39

The architectural exploration of NoCs follows the lines of everything else in this text: it involves a trade-off between delay and energy, and in addition, flexibility and area. From an energy perspective, common themes re-emerge:

- Preserving locality – The advantage of a partitioned network is that communications between components that are near to each other – and which make up a lot of the overall traffic – are more energy-efficient.
- Building hierarchy – This creates a separation between local and global communications. Networks that work well for one do not work well for the other.
- Optimal re-use of resources – depending upon the energy and delay constraints there exists an optimum amount of concurrency and/or multiplexing that minimizes the area.

Sound familiar?

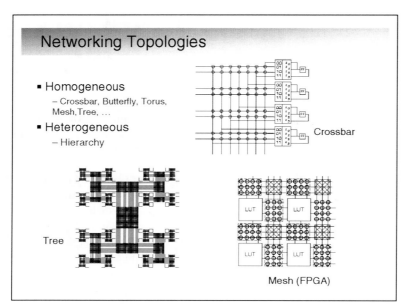

Slide 6.40

We could spend a lot of valuable book "real estate" on an overview of all the possible interconnect topologies that are known to mankind, but much of it would be wasted. For instance, a number of structures that excel in high-performance parallel computers do not map well on the two-dimensional surface of an integrated circuit; an example of such is the hyper cube. We therefore restrict our discussion to the topologies that are commonly used on-chip.

- The *crossbar* presents a latency-efficient way of connecting n sources to m destinations. However, it is expensive from both an area and energy perspective.
- The *mesh* is the most popular NoC architecture today. The FPGA was the first chip family to adopt this topology. The advantage of the mesh is that it uses only nearest-neighbor connections, thus preserving locality when necessary. For long-distance connections, the multi-hop nature of the mesh leads to large latencies. To combat this effect, FPGAs overlay meshes with different granularity.
- A *binary tree* network realizes a $\log_2(N)$ latency network (where N is the number of elements in the network) with relatively low wiring and capacitance costs. Other versions of this network vary the cardinality of the tree. In a fat tree, the cardinality is gradually increased for the higher levels. Trees offer an interesting counterpart to meshes as they are more effective in establishing long-distance connectivity.

Given that each network topology has its strengths and weaknesses, it comes as no surprise that many of the deployed NoCs pragmatically combine a number of schemes in a hierarchical fashion, presenting one solution for local wiring supplemented by another for the global connections. In addition, point-to-point connections are used whenever needed.

Slide 6.41

Navigating the myriad choices in an educated fashion once again requires an exploration environment that allows a study of the trade-off between the relevant metrics over the parameter set.

This is illustrated by the example in this slide, which compares the mesh and binary tree networks in the energy–delay space. As can be expected, the mesh network is the most effective solution for short connections, whereas the tree is the preferred choice for the longer ones. A solution that combines the two networks leads to a network with a merged pareto-optimal curve.

This combination is not entirely effective. If the goal is to make the latency between any two modules approximately uniform, straightforwardly combining the tree and the mesh topologies

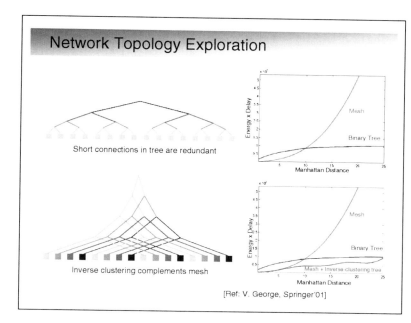

helps somewhat, but longer connections are still expensive. It would be better if the lower levels of the tree span nodes that are further away from each other, as shown in the lower diagram, picturing an "inverted-clustering" tree. The combination of the mesh with the inverse-clustering tree provides a superior solution, which only rises slightly with the Manhattan distance between the nodes.

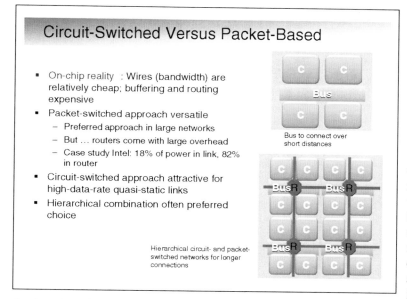

Slide 6.42

The other important dimension in the network exploration is the choice of the routing strategy, and the time at which it is established. The *static routing* is the simplest option. In this case, network routes are set up at design time. This is, for instance, the case in FPGAs, where the switches in the interconnect network are set at design time. A simple modification of this is to enable reconfiguration. This allows for a route to be set up for a time (for instance, for the duration of a computational task), and then be ripped up and rerouted. This approach resembles the *circuit-switched* approach of the traditional telephone networks. The advantage of both static- and circuit-switched routing is that the overhead is reasonable and solely attributable to the additional switches in the routing paths. (Note: for performance reasons, these switches are often made quite large and add a sizable amount of capacitance).

Packet-switched networks present a more flexible solution, where routes are chosen on a per-packet basis (as is done in some of the Internet routers). The overhead for this is large, as each router element has to support buffering as well as dynamic route selection. A major improvement is inspired by the realization that most data communications consist of a train of sequential packets. Under those conditions, the routing decision can be made once (for the first packet), with the other packets in the train just following. This approach is called "flit-routing" [Ref: Dally'01].

Experiments have shown that flit routing in NoCs is still quite expensive and that the energy cost of the dynamic routing is multiple times higher than the cost of the link. This realization is quite essential: *the energy cost of transmitting a bit over a wire on a chip is still quite reasonable compared to the implementation cost of routers and buffers.* As long as this is the case, purely dynamic network strategies may not be that attractive. Heterogeneous topologies, such as combining busses for short connections with mesh- or tree-based circuit- or packet-switched networks for the long-distance ones, most probably offer a better solution.

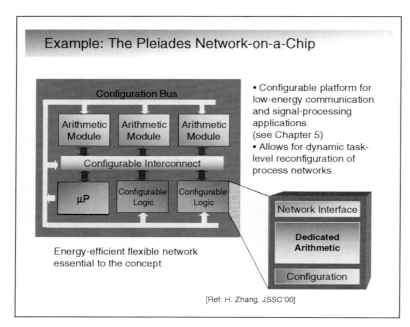

Slide 6.43

One of the earlier NoCs, with particular focus on energy efficiency, is found in the Pleiades reconfigurable architecture, already discussed in Chapter 5. In this platform, modules are wired together to form a dedicated computational engine on a per-task basis. Once the task is completed, the routes are ripped up and new ones established, reusing interconnect and compute modules for different functions. This approach hence falls into the "circuit-switched" class of networks.

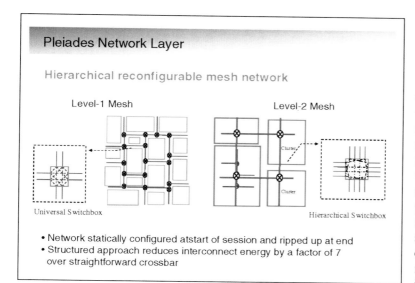

Slide 6.44

To maximize the energy efficiency, the Pleiades network consists of a two-layer heterogeneous mesh. Observe that the computational modules span a wide range of aspect ratios and sizes. The first-layer mesh follows the periphery of all the nodes with universal switchboxes at each crosspoint. Long-distance connectivity is supported by a second, coarser mesh (which couples into the lower-layer network). Nodes are divided into four clusters. The limited traffic requirements at this level allow for the use of a simpler and more restrictive switchbox. This topology, which was produced by an automated exploration tool, reduces the interconnect energy by a factor of seven over a straightforward crossbar, and is also substantially more area-efficient.

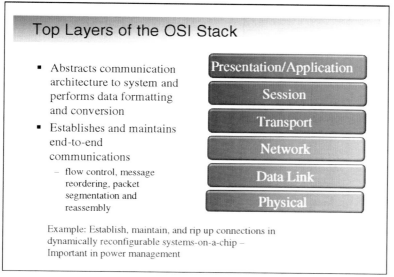

Slide 6.45

Traditional OSI network stacks support even more abstraction layers, such as *transport, session, presentation, and application*. Each of these layers serves to further abstract away the intricacies of setting up, maintaining, and removing a reliable link between two nodes. It will take some time before these abstractions truly make sense for a NoC. Yet, some elements are already present in today's on-chip networks. For example, the concept of a session is clearly present in the circuit-switched Pleiades network.

Although the impact on energy may not immediately be obvious, it is the existence of these higher abstraction levels that allow for a consistent, scalable, and manageable realization of energy-aware networks.

Slide 6.46

What about Clock Distribution?

- Clock easily the most energy-consuming signal of a chip
  - Largest length
  - Largest fan-out
  - Most activity ($\alpha = 1$)
- Skew control adding major overhead
  - Intermediate clock repeaters
  - De-skewing elements
- Opportunities
  - Reduced swing
  - Alternative clock distribution schemes
  - Avoiding a global clock altogether

At the end of this chapter on energy-efficient interconnect, it is worth spending some time on the interconnection that consumes the most: the *clock*. The clock network and its fan-out have been shown to consume as much as 50% of the total power budget in some high-performance processors.

When performance was the only thing that mattered, the designers of clock distribution networks spent the majority of their time on "skew management", and power dissipation was an afterthought. This explains, for instance, the usage of power-hungry clock meshes [Rabaey'03, Chapter 10]. A lot has changed since then. The clock distribution networks of today are complex hierarchical and heterogeneous networks, combining trees and meshes. In addition, *clock gating* is used to disable inactive parts of the network (more about this in Chapter 8). Mayhap, it is better to avoid using a clock altogether (Chapter 13). Detailed discussions on the design of clock networks and the philosophy of time synchronization are unfortunately out of the scope of this text.

Some interesting approaches at the physical level, however, are worth mentioning. Similar to what we had discussed with respect to data communications, it may be worthwhile to consider alternative clock signaling schemes, such as reduced signal swings.

Slide 6.47

Reduced-Swing Clock Distribution

- Similar to reduced-swing interconnect
- Relatively easy to implement
- But extra delay in flip-flops adds directly to clock period

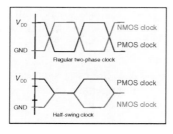
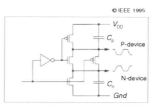

Example: half-swing clock distribution scheme
[Ref: H. Kojima, JSSC'95]

Reducing the clock swing is an attractive proposition. With the clock being the largest switching capacitance on the chip, reducing its swing translates directly into major energy savings. This is why a number of ideas on how to do this effectively popped up right away when power became an issue in the mid 1990s. An example of a "half-swing" clock generation circuit is shown in this slide. The clock generator uses charge redistribution over two equal capacitors to generate the mid voltage. The clock is distributed in two phases for driving the NMOS and PMOS transistors, respectively, in the connecting flip-flops.

The reduction in clock swing also limits the driving voltage at the fan-out flip-flops, which translates into an increase in clock-to-output delay. As this directly impacts the timing budget, reduced-swing clock distribution comes with a performance hit. Another challenge with the reduced clock swing is the implementation of the repeaters and buffers that are part of a typical clock distribution network. These need to operate from the reduced voltage as well. Because of these and other concerns, reduced-swing clock networks have been rarely used in complex ICs.

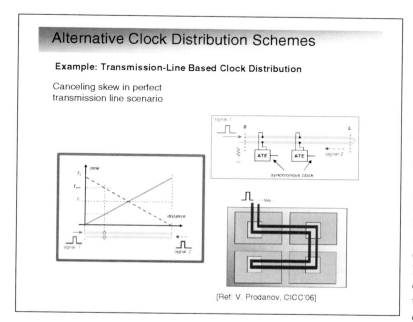

Slide 6.48

Another option is to consider alternative clock-distribution approaches. Over the years, researchers have explored a broad range of ideas on how to accurately synchronize a large number of distributed components on a chip. Ideas have ranged from coupled oscillator networks to generating standing waves in distributed resonant circuit elements (e.g., [Sathe'07]). Others have considered the idea of optical clock distribution. Given the importance of this topic, research is ongoing and a workable alternative to the paradigm of a centralized clock distributed with a "skewfree" network may emerge.

Given the limited space, we have chosen to present one single option in this slide (this does not presume a judgment on any of the other schemes). It is based on the assumption that virtually lossfree transmission lines can be implemented in the thick copper metal layers, which are available in all advanced CMOS processes. The transmission lines without a doubt present the fastest interconnect medium. Assume now that a pulse is transmitted over a folded transmission line (the contour of which is not important at all). At any point along the trajectory, the average between the early and late arrivals of the pulse is a constant – hence skewfree. By strategically positioning a number of "clock-extracting circuits" (which could be an analog multiplier) over the chip, a skewfree clock distribution network can be envisioned. The power dissipation of this network is very low as well. Though this scheme comes with some caveats, it is this form of disruptive technology that the energy-minded designer has to keep an eye on.

Summary

- Interconnect important component of overall power dissipation
- Structured approach with exploration at different abstraction layers most effective
- Lot to be learned from communications and networking community – yet, techniques must be applied judiciously
  - Cost relationship between active and passive components different
- Some exciting possibilities for the future: 3D integration, novel interconnect materials, optical or wireless I/O

Slide 6.49
The summary of this chapter is pretty much the same as that of the previous one: establishing clean abstractions and adhering to structured exploration methodologies is the key to low-energy interconnect networks. Borrowing ideas from the communication and networking communities is a good idea, but one must be watchful for some of the major differences between networking in the large and in the small.

References

Books and Book Chapters
- T. Burd, *Energy-Efficient Processor System Design*, http://bwrc.eecs.berkeley.edu/Publications/2001/THESES/energ_eff_process-sys_des/index.htm, UCB, 2001.
- G. De Micheli and L. Benini, *Networks on Chips: Technology and Tools*, Morgan-Kaufman, 2006.
- V. George and J. Rabaey, "Low-energy FPGAs: Architecture and Design", Springer 2001.
- J. Rabaey, A. Chandrakasan and B. Nikolic, *Digital Integrated Circuits: A Design Perspective*, 2nd ed, Prentice Hall 2003.
- C. Svensson, "Low-Power and Low-Voltage Communication for SoC's," in C. Piguet, *Low-Power Electronics Design*, Ch. 14, CRC Press, 2005.
- L. Svensson, "Adiabatic and Clock-Powered Circuits," in C. Piguet, *Low-Power Electronics Design*, Ch. 15, CRC Press, 2005.
- G. Yeap, "Special Techniques", in *Practical Low Power Digital VLSI Design*, Ch 6., Kluwer Academic Publishers, 1998.

Articles
- L. Benini et al., "Address Bus Encoding Techniques for System-Level Power Optimization," Proceedings DATE'98, pp. 861–867, Paris, Feb. 1998.
- T. Burd et al., "A Dynamic Voltage Scaled Microprocessor System," IEEE ISSCC Digest of Technical Papers, pp. 294–295, Feb. 2000.
- M. Chang et al., "CMP Network-on-Chip Overlaid with Multi-Band RF Interconnect", International Symposium on High-Performance Computer Architecture, Feb. 2008.
- D.M. Chapiro, "Globally Asynchronous Locally Synchronous Systems," PhD thesis, Stanford University, 1984.

Slides 6.50–6.52
Some references . . .

References (cont.)

- W. Dally, "Route packets, not wires: On-chip interconnect networks," *Proceedings DAC 2001*, pp. 684–689, Las Vegas, June 2001.
- J. Davis and J. Meindl, "Is Interconnect the Weak Link?," *IEEE Circuits and Systems Magazine*, pp. 30–36, Mar. 1998.
- J. Davis et al., "Interconnect limits on gigascale integration (GSI) in the 21st century," *Proceedings of the IEEE*, 89(3), pp. 305–324, Mar. 2001.
- D. Hopkins et al., "Circuit techniques to enable 430Gb/s/mm^2 proximity communication," *IEEE International Solid-State Circuits Conference*, vol. XL, pp. 368–369, Feb. 2007.
- H. Kojima et al., "Half-swing clocking scheme for 75% power saving in clocking circuitry," *Journal of Solid Stated Circuits*, 30(4), pp. 432–435, Apr. 1995.
- E. Kusse and J. Rabaey, "Low-energy embedded FPGA structures," *Proceedings ISLPED'98*, pp. 155–160, Monterey, Aug. 1998.
- V. Prodanov and M. Banu, "GHz serial passive clock distribution in VLSI using bidirectional signaling," *Proceedings CICC 06*.
- S. Ramprasad et al., "A coding framework for low-power address and data busses," *IEEE Transactions on VLSI Signal Processing*, 7(2), pp. 212–221, June 1999.
- M. Sgroi et al.,"Addressing the system-on-a-chip woes through communication-based design," *Proceedings DAC 2001*, pp. 678–683, Las Vegas, June 2001.
- P. Sotiriadis and A. Chandrakasan, "Reducing bus delay in submicron technology using coding," *Proceedings ASPDAC Conference*, Yokohama, Jan. 2001.

References (cont.)

- M. Stan and W. Burleson, "Bus-invert coding for low-power I/O," *IEEE Transactions on VLSI*, pp. 48–58, Mar. 1995.
- M. Stan and W. Burleson, "Low-power encodings for global communication in CMOS VLSI", *IEEE Transactions on VLSI Systems*, pp. 444–455, Dec. 1997.
- V. Sathe, J.-Y. Chueh and M. C. Papaefthymiou, "Energy-efficient GHz-class charg-recovery logic", *IEEE JSSC*, 42(1), pp. 38–47, Jan. 2007.
- L. Svensson et al., "A sub-CV^2 pad driver with 10 ns transition time," *Proc. ISLPED 96*, Monterey, Aug. 12–14, 1996.
- D. Wingard, "Micronetwork-based integration for SOCs," *Proceedings DAC 01*, pp. 673–677, Las Vegas, June 2001.
- H. Yamauchi et al., "An asymptotically zero power charge recycling bus," *IEEE Journal of Solid-Stated Circuits*, 30(4), pp. 423–431, Apr. 1995.
- H. Zhang, V. George and J. Rabaey, "Low-swing on-chip signaling techniques: Effectiveness and robustness," *IEEE Transactions on VLSI Systems*, 8(3), pp. 264–272, June 2000.
- H. Zhang et al., "A 1V heterogeneous reconfigurable processor IC for baseband wireless applications," *IEEE Journal of Solid-State Circuits*, 35(11), pp. 1697–1704, Nov. 2000.

Chapter 7
Optimizing Power @ Design Time – Memory

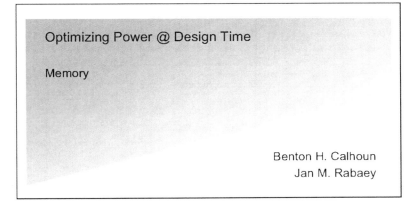

Slide 7.1

In this chapter, we discuss techniques for optimizing power in memory circuits. Specifically, we focus on embedded static random access memory (SRAM). Though other memory structures such as dynamic RAM (DRAM), Flash, and Magnetoresistive RAM (MRAM) also require power optimization, embedded SRAM is definitely the workhorse for on-chip data storage owing to its robust operation, high speed, and low power consumption relative to other options. Also, SRAM is fully compatible with standard CMOS processes, whereas the other memory options are technology-based solutions that usually require special tweaks to the manufacturing process (e.g., special capacitors for embedded DRAM).

This chapter focuses on design time approaches to reducing power consumption for an active SRAM. Although most of the cells in a large SRAM are not accessed at any given time, they must remain in a state of alert, so to speak, to provide timely access when required. This means that the total active power of the SRAM consists of both the switching power of the active cells and the leakage power of the non-active cells of the SRAM.

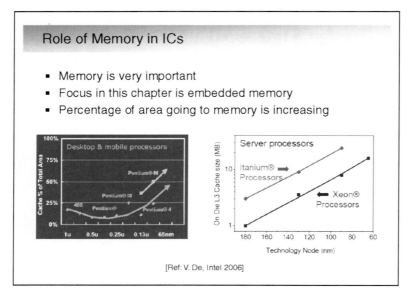

Slide 7.2
Almost all integrated chips of any substantial complexity require some form of embedded memory. This frequently means SRAM blocks. Some of these blocks can be quite large. The graphs on this slide establish a trend that the majority of processor area in scaled technologies is dedicated to SRAM cache. As higher levels of the cache hierarchy move on-chip, the fraction of die area consumed by SRAM will continue to balloon. Though large caches dominate the area, many recent processors and SoCs contain dozens or even hundreds of small SRAM arrays used for a variety of different purposes. From this, the importance of memory to the functionality and area (that is, cost) to future chip design is obvious.

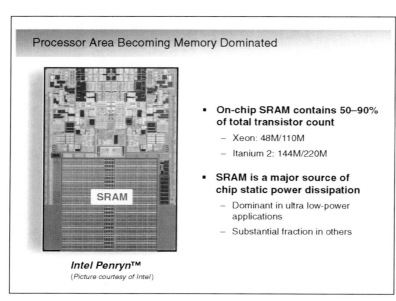

Slide 7.3
The die photo of Intel's Penryn™ processor makes the importance of SRAM even clearer. The large caches are immediately visible. We have circled in red just a few of the numerous other SRAM blocks on the chip. In addition to the impact on area, the power dissipation of the memory is growing relative to that of other components on the chip. This is particularly true for the leakage component of chip power. As the SRAM must remain powered on to hold its data, the large number of transistors in on-die SRAM will constantly draw leakage power. This leakage power can dominate the standby power and active leakage power budgets in low-power applications, and become an appreciable fraction of the total dissipation in others.

Slide 7.4

We begin our discussion of memory power optimization with an introduction to memory structures with a focus on embedded SRAM. Then we describe design time techniques for lowering power in the cell array itself, for reducing power during read accesses, and for decreasing power during write accesses. Finally, we present emerging devices that show promising results for reducing power in SRAM.

Given the limited space, an in-depth discussion on the operation of SRAM memories and the prevailing trends is unfortunately not an option. We refer the reader to specialized textbooks on the topic, such as [Itoh'07].

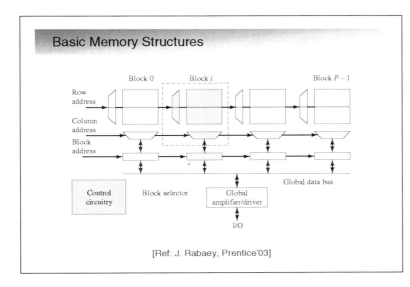

Slide 7.5

A two-dimensional array of SRAM bit-cells is the basic building block of large SRAM memories. The dimension of each cell array is limited by physical considerations such as the capacitance and resistance of the lines used to access cells in the array. As a result, memories larger than 64–256 Kb are divided into multiple blocks, as shown on this slide. The memory address contains three fields that select the block, the column, and the row of the desired word in the memory. These address bits are decoded so that the correct block is enabled and appropriate cells in that block are selected. Other circuits drive data into the cells for a write operation or drive data from the cells onto a data bus during a read. We may treat all of these peripheral circuits (e.g., decoders, drivers, control logic) as logic and apply to them the power-saving techniques from the preceding chapters of this book. The truly unique structure in an embedded SRAM is the array of bit-cells itself. In this chapter, we will focus on power-saving approaches specifically targeted at the bit-cell array.

Slide 7.6

In standard CMOS logic, the trade-off between power and delay tends to take precedence over other metrics. Although we certainly need combinational logic to function properly, static CMOS is sufficiently robust to make functionality relatively easy to achieve (at least for the time being). In

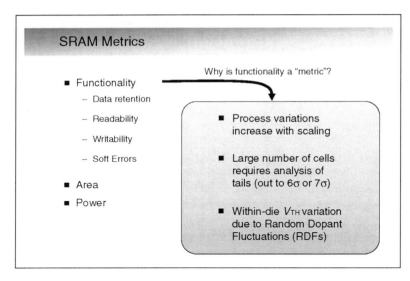

memory, this is not necessarily the case. The need for ever more storage density makes area the dominant metric – even though power is recently gaining ground in terms of its importance. SRAMs give up some of the important properties of static CMOS logic (e.g., large noise margins, non-ratioed circuits) to reduce the cell area. A typical cell is thus less robust (closer to failure) than typical logic. As we have discussed quite a number of times, the rapid increase in process variations that has accompanied CMOS process technology scaling causes circuit parameters to vary. Although process variations certainly impact logic, they have an even more profound impact on SRAM due to the tighter margins. One of the most insidious sources of variation is random doping fluctuation (RDF), which refers to the statistical variation of the number and position of doping ions in a MOSFET channel. RDF leads to significant variation in the threshold voltage of transistors with an identical layout. This means that physically adjacent memory cells exhibit different behaviors based on where their devices fall in the distribution of threshold voltages. As a result, important metrics related to the cell, such as delay and leakage, should be considered a distribution rather than a constant. When we further consider that embedded SRAMs may have many millions of transistors, we realize that some cells will necessarily exhibit behavior well out of the tail of the metric distribution (as far as 6σ or 7σ).

Although the power–delay trade-off certainly exists in memory, the more pressing issue in deeply scaled technologies is the trade-off between power and functional robustness (and area as a close second). Turning the circuit knobs to reduce SRAM power degrades the robustness of the array, so functionality is usually the limiting factor that prevents further power reduction. This means that, for SRAM, the primary goal when attempting to lower power is to achieve savings while maintaining correct operation across the entire array. The dominant aspects of functionality are readability, writability, data retention, and soft errors. In this chapter, we focus on the first two. We will look at data retention limits in detail in Chapter 9. Soft-error rates (SERs) for modern SRAM are increasing because each bit-cell uses less charge to store its data owing to smaller capacitance and lower voltage. As a result, the bits are more easily upset by cosmic rays and alpha particles. We will not discuss soft-error immunity in detail, although there is a variety of techniques that help to reduce the SER, such as error correction and bit interleaving.

The theme of this chapter centers around these threats to SRAM functionality: *To save power in SRAM, introduce new techniques to improve robustness, and trade off that gained robustness to lower power subsequently.*

Slide 7.7

Before we begin to reduce power in SRAM, we ought to ask ourselves, "Where does SRAM power go?" Unfortunately, this is a difficult question to answer. An expedition through the literature

Where Does SRAM Power Go?

- Numerous analytical SRAM power models
- Great variety in power breakdowns
- Different applications cause different components of power to dominate
- Hence: Depends on applications: e.g., high speed versus low power, portable

uncovers numerous analytical models for SRAM power consumption, and each one is complicated and different. Papers that report power breakdowns for SRAMs are equally inconsistent in their results. The reason for this variety goes back to our observation on Slide 7.2: SRAMs serve a huge variety of purposes. Even on the same chip, one large high-speed four-way cache may sit next to a 1 Kb rarely accessed look-up-table. From chip to chip, some applications require high-performance accesses to a cache nearly every cycle, whereas some portable applications need ultra low-energy storage with infrequent accesses. As a result, the optimal SRAM design for one application may differ substantially from that for another. As the constraints and specifications for each application determine the best SRAM for the job, we restrict ourselves to a survey of the variety of power-saving techniques that fall under the trade-off theme of this text book. Again, for SRAM, the trade-off is usually for functional robustness rather than for delay.

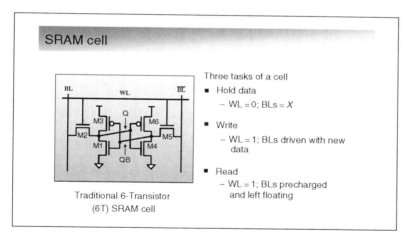

Slide 7.8

The circuit schematic on this slide shows the traditional topology for an SRAM bit-cell. It consists of six transistors, and is thus often called the 6T cell. The primary job of the bit-cell is to store a single bit of data, and it also must provide access to that data through read and write functions. The cell stores a single bit of data by using the positive feedback inherent in the back-to-back inverters formed by the transistors M1, M3 and M4, M6. As long as power is supplied to the cell and the wordline (WL) remains low (so that the transistors M2 and M5 are off), data at node Q will drive node QB to the opposite value, which will in turn hold the data at node Q. In this configuration, the voltages on the bitlines (BL and $\overline{BL}$ [or BLB]) do not impact the functionality of the bit-cell. To write the bit-cell (change the data in the bit-cell), we must overpower the positive feedback inside the cell to flip it to the opposite state. For example, if Q = 1 and QB = 0, we must drive Q to 0 and QB to 1 in order to write a new value into the cell. To accomplish this, we can drive the new data onto the BLs (e.g., BL = 0 and BLB = 1) and then assert the WL. This write operation is clearly ratioed, as it creates a fight between the devices inside the cell and the access transistors (M2 and M5). The NMOS access transistors are good at passing a 0, so we will rely on the side of the cell with a BL at ground to execute the write. To ensure that this works properly, we size M2 (M5) to win the fight with M3 (M6) so that we can pull the internal node that is high down to a 0 to flip the cell. We would also like to use the same access transistors (M2 and M5) to read the contents of the bit-cell to keep the size of the cell as small as possible. This means that we should be careful to avoid driving a BL to 0 during a read operation so that we do not inadvertently write the cell. To prevent this problem, we precharge both BLs to V_{DD}, and then allow them to float before asserting the WL. We can thus consider the BLs to be capacitors that are charged to V_{DD} at the onset of the read access. The side of the cell that stores a 0 will slowly discharge its BL – the read is slow because the cell transistors are small and the BL capacitance is relatively large – while the other BL remains near V_{DD}. By looking at the differential voltage that develops between the BLs, we can determine what value the cell is storing.

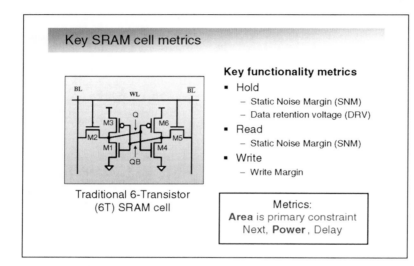

Slide 7.9

The traditional metrics of area, power, and delay apply to an SRAM. The driving metric has been area for a long time due to the large number of cells in SRAM arrays. However, power is becoming increasingly important to the point of rivaling area as the driving metric for the reasons that we described on Slide 7.3. Tuning the access delay of a memory is also of essence, but many embedded memories do not need to be super high-speed. Delay can thus be traded off to save area or power. As mentioned on Slide 7.6, robustness issues have floated to the top as a result of increasing process variations. This makes functionality a primary concern, and it limits the extent to which we can turn design knobs to lower power. A very useful metric to measure the robustness of a cell is the static noise margin (SNM), which is a measure of how well the cell can hold its data. An idle cell can generally hold its data quite well (i.e., the "hold SNM" is large), although the SNM decreases with a lowering of the supply voltage. V_{DD} *scaling* is a good knob for reducing leakage power, but the *hold SNM* places an upper limit on the achievable savings using this approach. We define *the data retention voltage* (DRV) as the lowest voltage at which a cell (or array of cells) can continue to hold its data. We will talk more about DRV in Chapter 9.

During a read access, the SNM is degraded due to the voltage-dividing effect that occurs between the access transistor and the drive transistor on the side of the bit-cell that holds a 0. This means that the bit-cell is most susceptible to losing its data during a read access. This type of read upset also becomes more likely as the power supply of the bit-cell is decreased.

As a successful write into an SRAM cell depends upon a sizing ratio, it also becomes more likely to fail in the presence of process variations. Specifically, variations that strengthen the PMOS transistors in the cell relative to the access transistors can be detrimental. An intended write may not occur if the access transistor cannot overpower the back-to-back inverters in the cell. The following slides discuss these metrics in more detail.

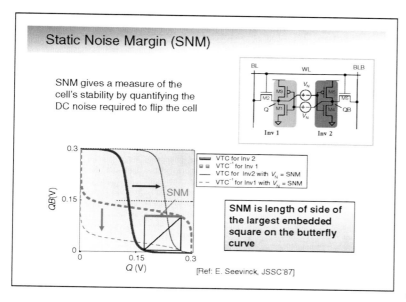

Slide 7.10

This slide provides a detailed illustration of the static noise margin (SNM) of a cell. The circuit schematic shows a cell with DC-voltage noise sources inserted into the cell. For now, let us assume that the value of these sources is $V_N = 0\,\text{V}$. The thick lines in the plot show the DC characteristics of the cell for the condition where there is no noise. The voltage transfer characteristic (VTC) curves cross at three points to make two lobes. The resulting graph is called the *butterfly plot* of the cell as the lobes resemble butterfly wings. The two crossing points at the tips of the lobes are the stable points, whereas the center crossing is a meta-stable point. Consider now the case where the value of the noise sources V_N start to increase. This causes the VTC of inverter 2 to move to the right, and the VTC of inverter 1 moves downward. The cell remains bistable (i.e., it holds its data) as long as the butterfly plot keeps its two lobes. Once the VTCs have moved so far that they only touch in two locations, one lobe disappears and any further increases in V_N result in a monostable bit-cell that has lost its data. This value of V_N is the static noise margin. The thin lines on the plot illustrate the VTCs in this condition. They touch at the corner of the largest square that can be inscribed inside the lobe of the original butterfly plot. The SNM is now defined as the length of the side of the largest square inside the butterfly plot lobe. If the cell is imbalanced (e.g., due to transistor sizing or process variations) – one lobe is smaller than the other in that case – then the SNM is the length of the side of the largest square that fits inside the smallest of the two lobes. This indicates that the bit-cell is more susceptible to losing one particular data value.

Slide 7.11

Process scaling causes the SNM of SRAM bit-cells to degrade. This slide shows simulations from predictive technology models (PTMs) of the SNM in 65 nm, 45 nm, and 32 nm. The upper plot shows that the typical SNM degrades with technology scaling and with voltage reduction. This means that it is harder to make a robust array in a scaled technology, and that lowering supply voltage to reduce power degrades the cell stability. Furthermore, this plot confirms that the read SNM is quite a bit smaller than the hold SNM.

If this story is not already bad enough, variations make it substantially worse. The bottom plots show distributions of the read SNM for the different technology nodes. Clearly, the tails of these distributions correspond to cells with vanishingly small noise margin, indicating that those cells will be quite unstable during a read access even in traditionally safe SRAM architectures. For the 32 nm technology, a substantial number of cells exhibit an SNM at (or below) 0, indicating a read upset even in the absence of other noise sources. This degradation of stability means that SRAM

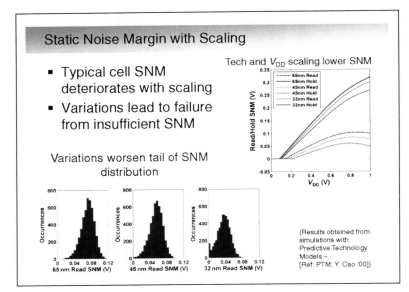

circuits/architectures must change if basic reading stability is to be maintained. For the power trade-off, this means that there is basically no room (negative room, actually!) to trade off robustness for power. Instead, we need to make fundamental changes to the SRAM to enable functionality and, hopefully, to lower power as well.

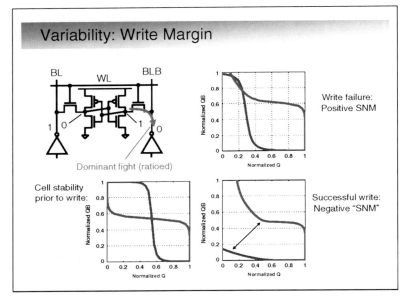

Slide 7.12

In CMOS sequential logic, the most-often-used latches simplify the write process by disabling the feedback loop between the cross-coupled inverters with the aid of a switch. The SRAM cell trades off this robust write method for area. The write operation now transgresses into a ratioed fight between the write driver and one of the inverters inside the cell. This "battle" is illustrated graphically on this slide. Write drivers assert the new data values onto the bitlines through a pass gate (not shown), then the WL goes high. This connects the internal nodes of the bit-cell with the driven bitlines, and a fight ensues between the cell inverters and the driver through the access transistor. As the NMOS access transistors pass a strong 0, the BL with a 0 is well-positioned to win its fight so long as the access transistor can overpower the PMOS pull-up transistor to pull down the internal node far enough to flip the cell. We can analyze the robustness of the cell's writability by looking at the equivalent of the butterfly plot during a write access. The bottom left-hand plot on this slide shows the butterfly plot of a bit-cell holding its data. For a successful write, the access transistors must drive the cell to a monostable condition. The lower right-hand plot shows a butterfly plot that no longer looks like a butterfly plot because it has successfully been made monostable (writing a 1 to Q). This corresponds to a negative SNM. In the upper right plot, the butterfly curve maintains bistability as indicated by the fact that both lobes of the butterfly plot persist during the write. This means that the write attempt has failed.

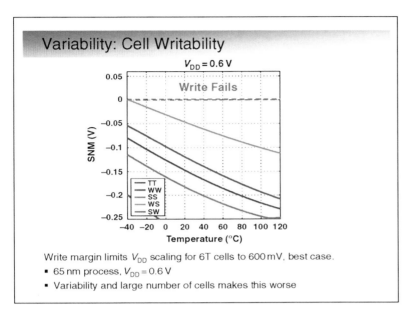

Slide 7.13
As with SNM, the write margin degrades in the presence of process variations. This is illustrated in this graph, which plots the onset of negative SNM for a write operation at the different global corners of a 65 nm process (e.g., typical NMOS, typical PMOS [TT]; weak NMOS, strong PMOS [WS]). Even before we account for local variations, 600 mV is the lowest voltage at which a 6T bit-cell allows for a successful write operation across global PVT (process, voltage, temperature) corners in this process. Local variations make the minimum operating voltage even higher. This indicates that successful write operation is compromised even for traditional 6T bit-cells and architectures in scaled technologies. As with read SNM, this limits the amount of flexibility that we have for trading off to save power. Hence approaches that improve functional robustness must be introduced. Only after this is accomplished can we start trading off robustness for power reduction.

Slide 7.14
Now that we have reviewed the traditional SRAM architecture, the bit-cell, and its important metrics, we take a look at the power consumption inside the bit-cell array when it is not being accessed. We assume for the moment that the array is still active in the sense that read or write accesses are imminent. In Chapter 9, we look at the standby case where no accesses are anticipated. Since the non-accessed array is merely holding its data, it does not consume switching power. Its power consumption is almost entirely leakage power.

Inactive cells leak current so long as the array is powered. This slide shows the primary paths for sub-threshold leakage inside the bit-cell. As we can assume that both the bitlines are precharged when the cell array is not being accessed, both BL and BLB are at V_{DD}. This means that the drain-to-source voltage across the access transistor on the "0"-side of the cell equals V_{DD}, causing that device to leak. Similar leakage occurs in the PMOS transistor on the same side of the cell, and the NMOS drive transistor on the opposite side. These three compose the dominant leakage paths inside the bit-cell.

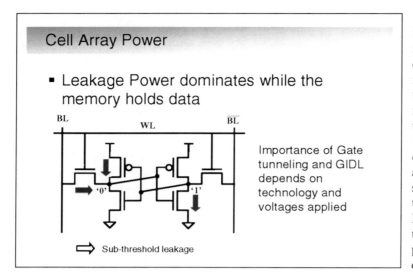

Other leakage mechanisms are also significant in modern technologies (see earlier chapters). Most notably, leakage through the gate terminal occurs for thin-gate-oxide transistors with a large V_{GD} or V_{GS}. Most scaled technologies have kept gate leakage to acceptable levels by slowing down the scaling of the gate oxide thickness. Emerging technologies at the 45 nm process node promise to include high-k dielectric material for the gate insulator, enabling further dielectric scaling. Assuming that this occurs, we postulate that sub-threshold leakage will continue to be the dominant leakage source in a CMOS cell. If not, the impact of gate leakage should be included in the design optimization process.

In the following slides, we examine two knobs for reducing the leakage power of an array: the threshold voltage and the peripheral voltages. These knobs can be set at design time such that SRAM leakage during active operation (e.g., when the array is ready to be accessed) decreases.

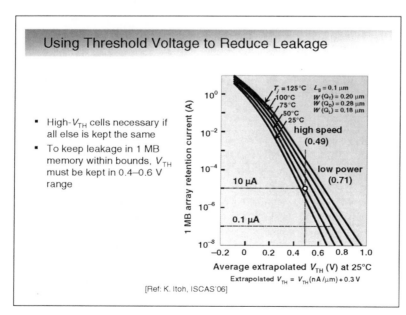

Slide 7.15

As established earlier, the sub-threshold leakage current equals $I_{SUB\,V_{TH}} = I_o \exp\left(\frac{V_{GS}-V_{TH}+\lambda_d V_{DS}}{nkT/q}\right)$. We can directly observe from this equation that the threshold voltage, V_{TH}, is a powerful knob, exponentially reducing the off-current of a MOSFET. Hence, one SRAM leakage reduction technique is to select a technology with a sufficiently high threshold voltage. The plot on this slide shows the leakage current for a 1 Mb array (log scale) versus the threshold voltage at different temperatures. Clearly, selecting a larger V_{TH} has an immediate and powerful effect on reducing the leakage current. If we assume that a high-speed memory application can tolerate 10 μA of leakage current at 50 °C, then this plot indicates that the V_{TH} of that array must be 490 mV. Likewise, the plot shows that a low-power array (0.1 µA of leakage current at 75 °C) needs a V_{TH} of over 710 mV if all other design parameters remain the same. This analysis not only indicates that threshold voltage can be used to control the leakage in an array, but also that it must remain fairly large if it is the lone knob used for controlling leakage power. On the negative side, the higher V_{TH} decreases the drive current of the bit-cells and limits the speed of the memory.

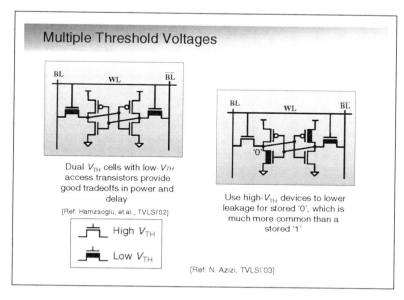

Slide 7.16

One alternative to using high-V_{TH} transistors for the entire cell is to selectively replace some high-V_{TH} devices with low-threshold transistors. Out of the large number of possible arrangements using transistors with two threshold voltages, only a few make sense. The best choice depends upon the desired behavior of the memory and the technology at hand.

A potential shortcoming of the multiple-threshold approach is that design rules may be imposed that force FETs with different V_{TH}s to be further apart. If area increases can be avoided or fall within an acceptable range, dual-V_{TH} cells can offer some nice advantages. One example is the cell shown on the left side of this slide (with low-V_{TH} devices shaded). The cross-coupled inverters in the cell are high-threshold, thereby effectively eliminating the leakage paths through the inverters. The access transistors are low-V_{TH}, along with the peripheral circuits. This translates into an improved drive current during read, minimizing the read-delay degradation due to the use of high-V_{TH} transistors.

The cell on the right side of the slide exploits the property that in many applications the majority of the cells in a memory store a "0". Selectively reducing the leakage of these cells hence makes sense. In fact, the leakage in a "0" cell is reduced by as much as 70% in this circuit. This obviously translates into a higher leakage for the "1" cells, but as these are a minority, the overall memory leakage is substantially reduced.

Slide 7.17

The use of multiple cell voltages provides another strong leakage reduction knob. Careful selection of the voltages inside and around the cell can decrease leakage in key devices by, for example, producing negative gate-to-source voltages. The sub-threshold current equation shows us that a negative V_{GS} has the same exponential impact on leakage current as raising V_{TH}. In the cell shown on this slide, the WL is at 0, but the source voltage of the cross-coupled inverters is increased to 0.5 V. This sets the V_{GS} for the access transistors to –0.5 V and –1.0 V for the sides of the cell holding a logical "0" and "1", respectively, producing a dramatic decrease in sub-threshold leakage. The supply voltage inside the cell must consequently be increased to maintain an adequate

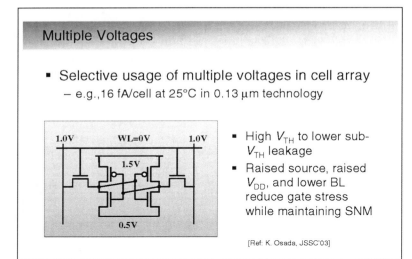

SNM. In this particular implementation, the author uses high-V_{TH} FETs in conjunction with voltage assignment to achieve a 16 fA/cell leakage current in a 130 nm technology.

In summary, increasing the threshold voltage is a strong knob for lowering sub-threshold leakage in the "non-active" cell, but higher-threshold devices translate into longer read or write latencies. Lowering the cell voltage or introducing multiple voltages also helps to reduce leakage power, but must be executed with care to avoid a degradation of the cell robustness. Be aware also of the hidden cost and area penalties associated with some of these techniques. For instance, extra thresholds mean extra masks – fortunately, most state-of-the-art processes already offer two thresholds. Providing extra supply voltages imposes a system cost due to the extra DC–DC converter(s) required, whereas routing multiple supply voltages incurs cell and periphery area overhead. All of these sources of overhead must be weighed against the power savings for a complete design.

In the next set of slides, we concentrate on how to impact power during a read access.

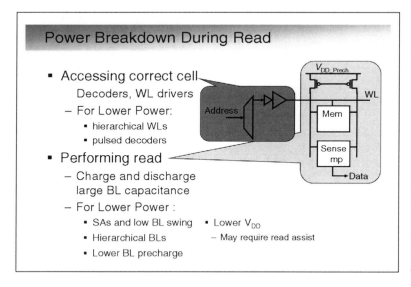

Slide 7.18
As the read access involves active transitions in the SRAM, the dominant source of power consumption during a read is switching power. This slide provides a general conceptual breakdown for where the switching power is dissipated during a read access. When the address is first applied to the memory, this address is decoded to assert the proper wordline. The decoded signals are buffered to drive the large capacitance of the wordline. The decoder and wordline drivers are nothing else but combinational logic, and techniques to manage power in this style of networks were treated in-depth in previous chapters.

We do not revisit these circuit-level techniques here, but restrict ourselves to techniques that specifically make sense in the memory context.

Once the decoded wordline signal reaches the cell array, the selected bit-cell selectively discharges one of the (precharged) bitlines. Owing to the large bitline capacitance, recharging it during the next precharge phase consumes a significant amount of power. Clearly, reducing the amount of discharge helps to minimize dissipation. It is traditional in SRAMs to use a differential sense amplifier (SA) to detect a small differential signal on the bitlines. This not only minimizes the read latency but also allows us to start precharging the bitline after it has only discharged a small fraction of V_{DD}, thus reducing power as well. On top of this, there are several other approaches for lowering power during this phase of the read access, some of which are described in the following slides.

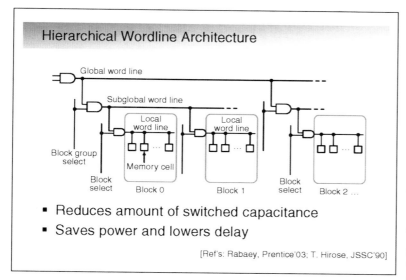

Slide 7.19

The capacitance of a wordline in an SRAM array can be quite large. It consists of the gate capacitance of two access transistors per bit-cell along the row in the array plus the interconnect capacitance of the wires. This capacitance gets even larger if a single wordline is deployed for accessing rows across multiple blocks in a large SRAM macro. To counter this, most large memories use a hierarchical wordline structure similar to the one shown on this slide. In this structure, the address is divided up into multiple fields to specify the block, block group, and column, for example. The column address is decoded into global wordlines, which are combined with select signals to produce sub-global wordlines. These in turn are gated with the block-select signals to produce the local wordlines. Each local wordline can thus be shorter and have less capacitance. This hierarchical scheme saves power and lowers delay by reducing the amount of capacitance that is switched on the wordline. The approach also allows for additional power savings by preventing wordlines in non-accessed blocks from being activated, which would cause dummy read operations in those blocks.

Slide 7.20

Dividing large lines into a hierarchy of smaller lines works for bitlines just the same way as it works for wordlines. The bitlines typically do not discharge all the way because of the sense amplifiers. Nevertheless, the large capacitance of these lines makes discharging them costly in terms of power and delay. Decreasing the number of cells on a local bitline pair reduces the delay and power consumption of the read access. The local bitlines can be recombined into global bitlines that provide the final data value from the read. The emergence of bitline leakage as a

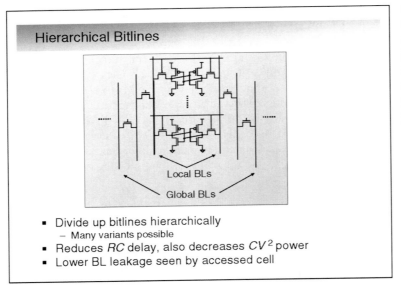

major issue in SRAM design for deep submicron technologies has made hierarchical bitlines more common, and the number of cells on a local bitline pair is decreasing to compensate for bitline leakage.

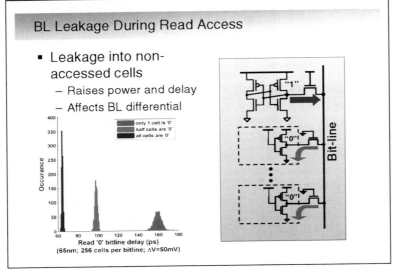

Slide 7.21
Bitline leakage refers to leakage current paths that flow from the bitlines into the bit-cells along an SRAM column. We have already identified this leakage path from a more local perspective on Slide 7.14. Bitline leakage is actually more problematic than described there, as it degrades the ability of the SRAM to read properly. This is illustrated in this slide, where a single cell tries to drive a "1" on the line while other cells on the column hold a "0". For this data vector, all of the non-accessed cells contribute leakage currents that oppose the (very small) drive current from the accessed bit-cell. As a result, the bitline, which should keep its precharged voltage at V_{DD}, may actually discharge to some lower voltage. As a consequence, the difference in its voltage from the voltage on the opposite bitline (not shown), which is supposed to discharge, is diminished. At the very least, this leakage increases the time for the sense amplifier to make its decision (hence raising the read latency).

The impact of the data distribution on the memory-access time is shown in the graph on the left. Clearly, if all of the non-accessed cells contain data opposite to that of the accessed cell, the delay increases dramatically. The variation of the access time also increases. In the worst case, bitline leakage into a large number of non-accessed cells can potentially become equal or larger than the drive current of the accessed cell, leading to a failure of the SRAM to read the cell. The number of bit-cells along a bitline pair must be carefully selected to prevent this type of bitline-leakage induced error.

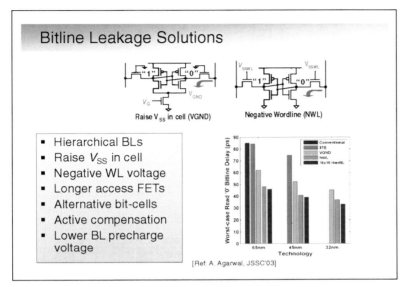

Slide 7.22

A number of possible solutions to combat bitline leakage are listed on this slide.

Hierarchical bitlines reduce the number of cells connected to a line, but increase complexity and area by requiring more peripheral circuits per bit-cell. Raising the virtual ground node inside the non-accessed bit-cells lowers leakage from the bitline at the cost of added area and reduced SNM. Reducing the wordline voltage below zero (negative WL) exponentially decreases the sub-threshold leakage through the access transistors, but this approach may be limited by gate current, which increases as a result of the large V_{DG}. Lengthening the access transistors lowers leakage at the cost of a decreasing drive current. Alternative bit-cells have been proposed, such as an 8-transistor (8T) cell that uses two extra access transistors (that are always off) to couple the same amount of leakage current to both bitlines. This cell successfully equalizes the leakage on both bitlines, but it does so by making the leakage worst-case. Hence, it is only successful in reducing the impact of bitline leakage on delay, not on power. Some active-compensation approaches have been proposed that measure the leakage on the bitline and then apply additional current to prevent the erroneous discharging of the high bitline. These sorts of schemes increase complexity and tend to focus on decreasing the delay at the expense of power. Reducing the precharge voltage is another approach (as it merits further discussion, we postpone its description to the next slide).

All of these techniques can help with the bitline leakage problem, but translate into some sort of trade-off. The best solution for a given application, as always, depends upon the specific circumstances and settings.

The bar graph on this slide provides a high-level comparison of the effectiveness of several of these techniques (obtained by simulations using the predictive modeling technology or PTM) with respect to read-access latency. The conventional approach and the 8T cell do not work all the way to the 32 nm node. Raising the virtual ground in non-accessed cells, using a negative wordline voltage, and subdividing the array with hierarchical bitlines all help to make the array less sensitive to bitline leakage. Again, the trade-offs that each approach makes to accomplish this must be carefully weighed, and included in the exploration process.

Lower Precharge Voltage

Lower BL precharge voltage decreases power and improves Read SNM

- Internal bit-cell node rises less
- Sharp limit due to accidental cell writing if access FET pulls internal '1' low

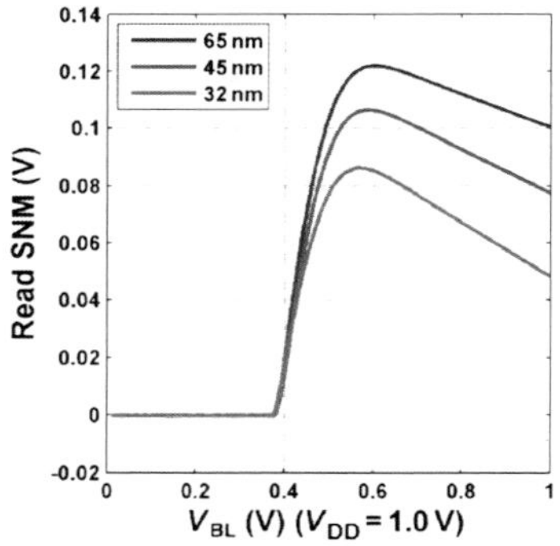

Slide 7.23
Reducing the precharge voltage on the bitlines below the traditional V_{DD} value helps to reduce bit-line leakage into the non-accessed bit-cells, because of the lower V_{DS} across the access transistors. Since the access transistor that drives a 1 onto the bit-line during a read does not turn on unless the bitline drops to $V_{DD}-V_{TH}$ any-way, the lower precharge voltage does not negatively affect the read itself. In fact, by weakening the access transistor on the 0 side of the cell, the lower precharge voltage actually makes the read more robust by improving the read SNM. The chart shows that a lower precharge value can improve the read SNM by over 10%, in conjunction with a lower leakage power. One simple way to implement this method is to precharge using NMOS devices instead of the traditional PMOS. The chart also indicates the major limitation to this approach: if the precharged bitline is at too low a voltage, the cell may be inadvertently written during a read access. This is indicated by a sharp roll-off in the read SNM.

V_{DD} Scaling

- Lower V_{DD} (and other voltages) via classic voltage scaling
  - Saves power
  - Increases delay
  - Limited by lost margin (read and write)
- Recover Read SNM with read assist
  - Lower BL precharge
  - Boosted cell V_{DD} [Ref: Bhavnagarwala'04, Zhang'06]
  - Pulsed WL and/or Write-after-Read [Ref: Khellah'06]
  - Lower WL [Ref: Ohbayashi'06]

Slide 7.24
A discussion of the power saving-approaches during read is incomplete without a closer look at "classic" V_{DD} scaling. Lowering the supply voltage of an SRAM array during active mode clearly decreases the switching power consumed by that array ($P = fCV^2$). It also decreases leakage power as $P = VI_{off}$, and I_{off} mostly decreases as a result of the DIBL effect. This double power-wins comes at the cost of increased access delays. We also know by now that the reduction in the operational V_{DD} is quite limited owing to functional barriers such as SNM and read/write margins.

There are two solutions to this problem. The first is to admit defeat so far as the array is concerned by using high-V_{TH} devices and maintaining a high V_{DD} to provide sufficient operating margins and speed. The peripheral circuits, on the other hand, can be scaled using traditional voltage scaling as they are essentially combinational logic. The complication to this approach is the need for level conversion at the interface between periphery and the array. The second solution is to recover the lost margin (read margin, as we are talking about the read access here) using *read-assist techniques*. These are circuit-level approaches that improve the read margin, which in turn can be used to reduce the V_{DD}. Examples of read-assist approaches include lowering the BL precharge voltage, boosting the bit-cell V_{DD}, pulsing the WL briefly, re-writing data to the cells after a read, and lowering the WL voltage. All of these approaches essentially work to sidestep the read-upset problem or to strengthen the drive transistor relative to the access transistor so as to reduce read SNM. The slide provides a number of references for the interested reader.

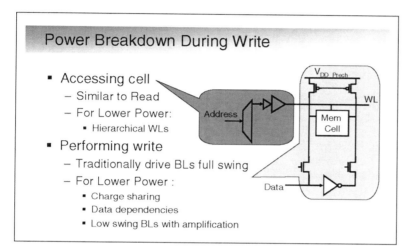

Slide 7.25

We now move on to look at the power consumed during the write access. We can partition the power consumed during a write access as belonging to two phases, similar to the way that we partitioned the read access. First, we must access the proper cells in the SRAM array, and second we must perform the write. The cell access is basically the same as for the read access. Once the correct local wordline is asserted, the new data must be driven into the accessed bit-cell to update the cell to the new value. The traditional mechanism for accomplishing this is to drive the differential value of the new data onto the bitlines in a full-swing fashion. As a subsequent write with a different data or a subsequent read (with precharge) will charge up the discharged bitline, this approach can be costly in terms of power. In fact, the power for a write access is typically larger than that for a read access owing to this full-swing driving of the bitlines. Fortunately write operations tend to occur less commonly than read operations. We examine techniques that use charge sharing, exploit data dependencies, and use low-swing bitlines to reduce the power consumption associated with the write access in the following three slides.

Slide 7.26

The full swing on the bitlines during a write operation seems particularly wasteful if successive writes are performed in the same block. In this case, the bitlines are charged and discharged according to the incoming data. The large capacitance of the bitlines causes significant CV^2 power consumption. If consecutive writes have different data values, then one bitline must discharge while the opposite bitline charges up for the next write. Instead of performing these operations separately, we can apply charge recycling to reduce the power consumption. This slide shows a simple example of how this works. The key concept is to introduce a phase of charge-sharing in between phases of driving data. Assume that the old values are 0 and V_{DD} on BL and BLB, respectively. During the charge-sharing phase, the bitlines are floating (e.g., not driven) and shorted together. If they have the same capacitance, then they will each settle at $V_{DD}/2$. Finally, the bitlines are driven to their new values. As BL only needs to be charged to V_{DD} from $V_{DD}/2$, the

Charge recycling to reduce write power

- Share charge between BLs or pairs of BLs
- Saves for consecutive write operations
- Need to assess overhead

Basic charge recycling – saves 50% power in theory

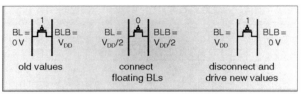

old values	connect floating BLs	disconnect and drive new values
BL = 0 V, BLB = V_{DD}	BL = $V_{DD}/2$, BLB = $V_{DD}/2$	BL = V_{DD}, BLB = 0 V

[Ref's: K. Mai, JSSC'98; G. Ming, ASICON'05]

power drawn from the supply equals $P = C_{BL} V_{DD} V_{DD}/2$. Hence, in theory this saves 50% of the power for this transition. In practice, the overhead of introducing the extra phase (both in terms of timing and power) needs to be weighed against the actual savings.

Memory Statistics

- 0's more common
  - SPEC2000: 90% 0s in data
  - SPEC2000: 85% 0s in instructions
- Assumed write value using inverted data as necessary [Ref: Y. Chang, ISLPED'99]
- New Bitcell:

1R, 1W port
W0: WZ = 0, WWL = 1, WS = 1
W1: WZ = 1, WWL = 1, WS = 0

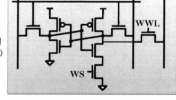

[Ref: Y. Chang, TVLSI'04]

Slide 7.27
A different type of approach to reducing write power is based on the earlier observation that one of the data values is more common. Specifically, for the SPEC2000 benchmarks, 90% of the bits in the data are 0, and 85% of the bits in the instruction memory are 0 [Chang'04]. We can take advantage of the predominance of 0s in a few ways.

First, we can use a write methodology that presets the BLs prior to each write based on the assumption that all of the bits will be 0. Then, as long as a word contains more 0s than 1s, the power consumed for driving the BLs to the proper values is reduced compared to the case in which both BLs are precharged to V_{DD}. In addition, words with more 1s than 0s can be inverted (keeping track of this requires one extra bit per word) to conform to the precharge expectation. This approach can reduce write power by up to 50% [Chang'99].

Second, an alternative bit-cell introduces asymmetry to make the power consumed when writing a 0 very low. As this is the common case, at least for some applications, the average write access power can be reduced by over 60% at a cost of 9% area increase. These approaches point out the intriguing concept that an application-level observation (i.e., the preponderance of 0s) can be exploited at the circuit level to save power. Of course, this is in accord with our discussion on Slide 7.7 regarding the close relationship between the intended application for a specific SRAM design and the memory design trade-offs.

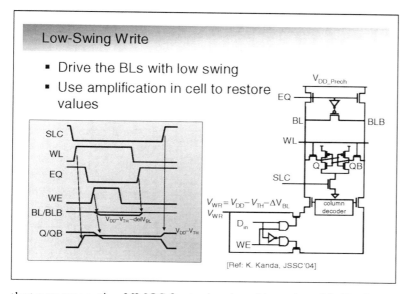

Slide 7.28
As the large bitline swing for a traditional write is the primary source of power dissipation, one seemingly obvious approach to is to reduce the swing on the bitlines. Doing so, of course, makes the access transistors less capable of driving the new data into the cell. This slide illustrates a solution that utilizes low-swing bitlines for writing along with an amplification mechanism in the cell to ensure successful write. The idea requires that a power gating NMOS footer be placed in series with V_{SS} for the bit-cell. This device (driven by SLC in the schematic) can be shared among multiple bits in the word. Prior to the onset of the write, this footer switch is disabled to turn off the NMOS driver FETs inside the bit-cell. The WL goes high, and the internal nodes of the bit-cell are set high. The (weakened) access transistors are able to do so, as the pull-down paths in the cell are cut off. Then, the bitlines are driven to $V_{DD}-V_{TH}$ and to $V_{DD}-V_{TH}-\Delta V_{BL}$, respectively, according to the input data. This bitline differential is driven into the bit-cell, and it is subsequently amplified to full swing inside the bit-cell after WL goes low and SLC goes high. This scheme can save up to 90% of the write power [Kanda'04].

Slide 7.29
As with read-power reduction techniques, the fundamental limit to most write-power saving approaches is the reduced functional robustness (i.e., the write margin becomes too small, and some cells become non-writable). Again the approach for pushing past this hurdle is to

Write Margin

- Fundamental limit to most power-reducing techniques
- Recover write margin with write assist, e.g.,
  – Boosted WL
  – Collapsed cell V_{DD} [Itoh'96, Bhavnagarwala'04]
  – Raised cell V_{SS} [Yamaoka'04, Kanda'04]
  – Cell with amplification [Kanda'04]

improve the write margin using some circuit innovations, and to trade off the improved robustness for power savings.

On this slide, we refer to a few of the many successful mechanisms for enabling this trade-off. Raising the voltage of the wordline during a write access relative to the V_{DD} does strengthen the access transistor relative to the cell pull-up transistors, creating a larger write margin and allowing for lower-voltage operation. Collapsing the V_{DD} or raising the V_{SS} inside of the bit-cell has the equivalent effect of reducing the strength of the cell relative to the access transistors. Finally, we have already described a method that provides amplification inside the cell. The references can help the interested reader to explore further.

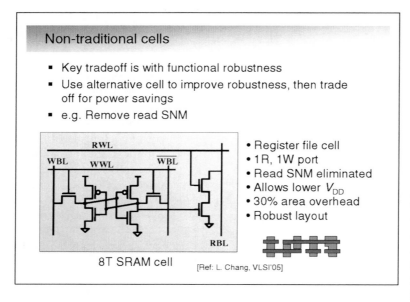

Slide 7.30
Most of the techniques described up to this point use the basic 6T as the basis. A more dramatic approach is to explore alternatives to the 6T bit-cell itself. These alternative bit-cells usually improve on the 6T cell in one or more ways at the expense of a larger area. A number of cells that may replace the 6T cell in some usage scenarios are proposed in the following slides. Even more dramatic (and much needed) changes to the SRAM cell could come from modifying the CMOS devices themselves (or even abandoning CMOS altogether). A number of new devices that offer enticing properties and may potentially change the way we design memories are therefore discussed as well. There is a huge amount of very creative activity going on in this field, and it will be no surprise to the authors if this leads one day to a very different approach of implementing embedded memory.

As we have repeatedly described, the key obstacle to power savings in SRAM is degraded functional robustness. Non-traditional bit-cells can provide improved robustness over the 6T bit-cell, which we can then trade off for power savings. In general, this leads to a larger area owing to the additional transistors.

One attractive alternative to the 6T with a number of interesting properties is the 8T, as shown on this slide. A 2T read buffer is added to the 6T cell. This extra read buffer isolates the storage node during a (single-ended) read so that the read SNM is no longer degraded. By decoupling the drive transistor from the storage node, this cell also allows for larger drive current and shorter read access times. In addition, the extra read buffer effectively enables separate read and write ports. This can improve the access rate to a memory by overlapping writes and reads. These improvements in read robustness allow the 8T to operate at lower supply voltages, and it does so without using extra voltage supplies.

Of course, these improvements come at a cost. The most obvious penalty is extra area, although a tight layout pattern keeps the array overhead down. Furthermore, the extra robustness of the cell may allow for the clustering of more cells along a single column, reducing the amount of peripheral circuitry required. The area overhead for the entire SRAM macro thus is less than the overhead in a single cell. The main challenge in using this cell is that it imposes architectural changes (i.e., two ports), which prevent it from acting as a direct replacement for 6T without needing a major macro redesign. However, the 8T cell is a wonderful example of how non-traditional bit-cells may offer improvements in robustness that can be exploited to improve power efficiency.

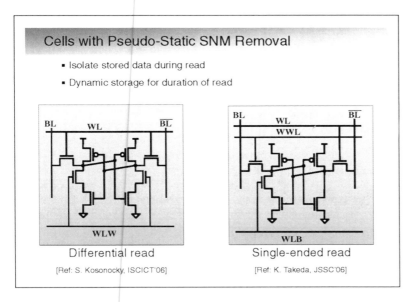

Slide 7.31

The 8T transistor statically isolates the storage node from the bitlines during a read operation. The two alternatives presented in this slide achieve the same effect using a pseudo-static approach. Both cells operate on the same principle, but the left-hand cell provides a differential read, whereas the right-hand cell uses a single-ended read. When the cells hold data, the extra wordlines (WLW, WLB) remain high so that the cell behaves like a 6T cell. During the read access, the extra wordline is dropped (WLW = 0, WLB = 0). This isolates the storage node, which holds its data dynamically while the upper part of the cell discharges the proper bitline. As long as the read access is sufficiently short to prevent the stored data from leaking away, the data is preserved. These cells each add complexity to the read operation by requiring new sensing strategies on the bitline(s).

Slide 7.32

A different tactic for reducing embedded SRAM power is to replace standard CMOS transistors with alternative devices. A slew of CMOS-replacement technologies are under investigation in labs around the world, and they range from minor modifications to CMOS, all the way to completely unrelated devices. Out of the many options, we take a brief look at one structure that is compatible with CMOS technologies, and which many people project as a likely direction for CMOS.

This slide shows a FINFET transistor (see also Chapter 2) that uses a vertical fin of silicon to replace the traditional planar MOSFET. Two types of devices can be constructed along this basic concept. The double-gated (DG) MOSFET is a vertically oriented MOSFET with a gate that wraps around three sides of the MOS channel. This allows the gate terminal to retain better control over the channel. In the back-gated (BG) MOSFET, the top part of the gate is etched away to leave

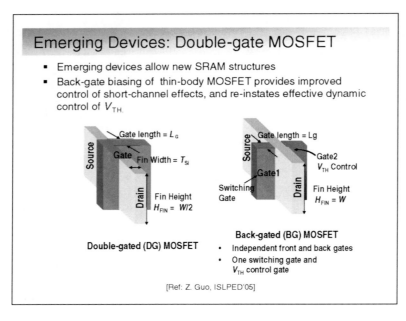

electrically disconnected gates along each side of the channel. This is analogous to having a planar MOSFET with an individual back-gate terminal (as, for instance, in an SOI process). If both of the gates are tied together, then the BG-MOS behaves like the DG-MOS. The BG-MOSFET retains the flexibility of allowing the back gate to serve as a control mechanism for modulating the threshold voltage of the transistor.

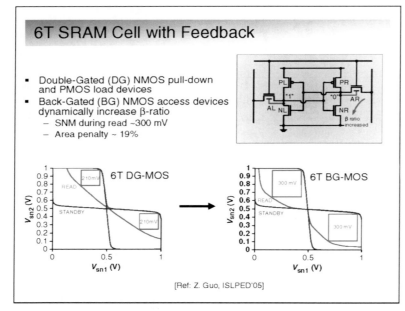

Slide 7.33
Using these two devices, we can re-engineer the 6T SRAM cell, so that butterfly diagrams as shown in this slide are obtained. The SNM for the DG-MOS bit-cell is quite similar to that of a traditional CMOS cell; the read SNM is degraded owing to the voltage-dividing effect between the access transistor and the drive transistor.

This can be remedied by connecting back-gate terminals of the BG-MOS access transistors as indicated by the red lines in the schematic, so that feedback is provided during a read access. When the storage node is high or low, the V_{TH} of the access transistor is raised or lowered, respectively. In the latter case, the access transistor becomes stronger, effectively increasing the -ratio of the cell. The bottom-right butterfly plot shows that this feedback results in a significantly improved read SNM for the cell that uses the BG-MOS devices.

This example demonstrates that device innovations can play a big role in the roadmap for embedded memory in the years to come. However, as always, the creation of a new device is only the first step in a long chain of events that ultimately may lead to a manufactureable technology.

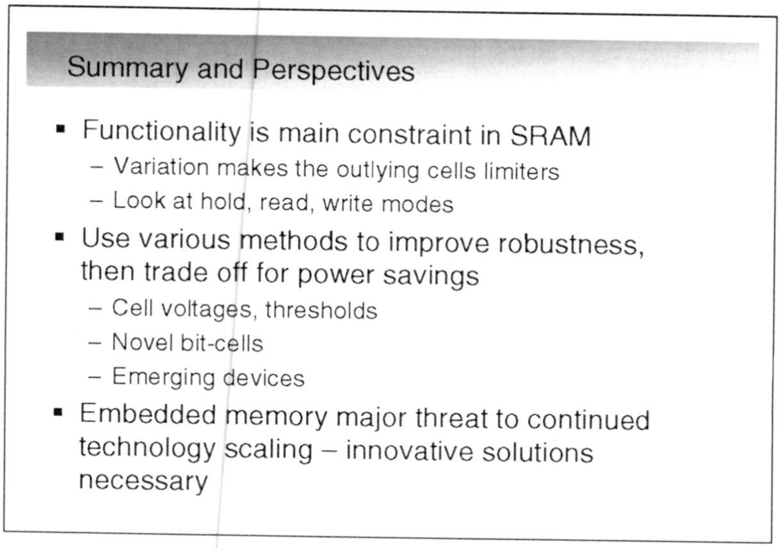

Summary and Perspectives

- Functionality is main constraint in SRAM
  - Variation makes the outlying cells limiters
  - Look at hold, read, write modes
- Use various methods to improve robustness, then trade off for power savings
  - Cell voltages, thresholds
  - Novel bit-cells
  - Emerging devices
- Embedded memory major threat to continued technology scaling – innovative solutions necessary

Slide 7.34

As we have seen repeatedly, process scaling and variations challenge the functionality of modern embedded SRAMs. The large sizes of embedded SRAM arrays, along with local variations, require us to examine the far tails ($>6\sigma$) of distributions to identify cells that will limit the array's functionality. Depending upon the application and the operating environment, the limiting conditions can occur during hold, read, or write operations. As robustness is so critical, the most effective method for saving power is to apply techniques to the memory that improve functional robustness. The resulting surplus of functional headroom can then be traded off for power savings. A number of techniques for doing so using device threshold voltages, cell and peripheral supply voltages, novel cells, and emerging devices have been presented.

In the long term, only novel storage devices can help to address the joined problem of power dissipation and reliability in memory. While waiting for these technologies to reach maturity (which may take some substantial amount of time and tax your patience), it is clear that in the shorter term the only solution is to take an area-penalty hit. Another option is to move large SRAM memories to a die different from that of the logic, and to operate it on larger supply and threshold voltages. Three-dimensional packaging techniques can then be used to reconnect logic and memory.

Slides 7.35–7.37
Some references . . .

References

Books and Book Chapters
- K. Itoh et al., *Ultra-Low Voltage Nano-scale Memories*, Springer 2007.
- A. Macii, "Memory Organization for Low-Energy Embedded Systems," in *Low-Power Electronics Design*, C. Piguet Ed., Chapter 26, CRC Press, 2005.
- V. Moshnyaga and K. Inoue, "Low Power Cache Design," in *Low-Power Electronics Design*, C., Piguet Ed., Chapter 25, CRC Press, 2005.
- J. Rabaey, A. Chandrakasan and B. Nikolic, *Digital Integrated Circuits*, Prentice Hall, 2003.
- T. Takahawara and K. Itoh, "Memory Leakage Reduction," in *Leakage in Nanometer CMOS Technologies*, S. Narendra, Ed., Chapter 7, Springer 2006.

Articles
- A. Agarwal, H. Li and K. Roy, "A Single-Vt low-leakage gated-ground cache for deep submicron," *IEEE Journal of Solid-State Circuits*, 38(2), pp. 319–328, Feb. 2003.
- N. Azizi, F. Najm and A. Moshovos, "Low-leakage asymmetric-cell SRAM," *IEEE Transactions on VLSI*, 11(4), pp. 701–715, Aug. 2003.
- A. Bhavnagarwala, S. Kosonocky, S. Kowalczyk, R. Joshi, Y. Chan, U. Srinivasan and J. Wadhwa, "A transregional CMOS SRAM with single, logic V_{DD} and dynamic power rails," in *Symposium on VLSI Circuits*, pp. 292–293, 2004.
- Y. Cao, T. Sato, D. Sylvester, M. Orshansky and C. Hu, "New paradigm of predictive MOSFET and interconnect modeling for early circuit design," in *Custom Integrated Circuits Conference (CICC)*, Oct. 2000, pp. 201–204.
- L. Chang, D. Fried, J. Hergenrother et al., "Stable SRAM cell design for the 32 nm node and beyond," *Symposium on VLSI Technology*, pp. 128–129, June 2005.
- Y. Chang, B. Park and C. Kyung, "Conforming inverted data store for low power memory," *IEEE International Symposium on Low Power Electronics and Design*, 1999.

References (cont.)

- Y. Chang, F. Lai and C. Yang, "Zero-aware asymmetric SRAM cell for reducing cache power in writing zero," *IEEE Transactions on VLSI Systems*, 12(8), pp. 827–836, Aug. 2004.
- Z. Guo, S. Balasubramanian, R. Zlatanovici, T.-J. King, and B. Nikolic, "FinFET-based SRAM design," *International Symposium on Low Power Electronics and Design*, pp. 2–7, Aug. 2005.
- F. Hamzaoglu, Y. Ye, A. Keshavarzi, K. Zhang, S. Narendra, S. Borkar, M. Stan, and V. De, "Analysis of Dual-V_T SRAM cells with full-swing single-ended bit line sensing for on-chip cache," *IEEE Transactions on Very Large Scale Integration (VLSI) Systems*, 10(2), pp. 91–95, Apr. 2002.
- T. Hirose, H. Kuriyama, S. Murakam, et al., "A 20-ns 4-Mb CMOS SRAM with hierarchical word decoding architecture," *IEEE Journal of SolidState Circuits-*, 25(5) pp. 1068–1074, Oct. 1990.
- K. Itoh, A. Fridi, A. Bellaouar and M. Elmasry, "A Deep sub-V, single power-supply SRAM cell with multi-V_T, boosted storage node and dynamic load," *Symposium on VLSI Circuits*, 133, June 1996.
- K. Itoh, M. Horiguchi and T. Kawahara, "Ultra-low voltage nano-scale embedded RAMs," *IEEE Symposium on Circuits and Systems*, May 2006.
- K. Kanda, H. Sadaaki and T. Sakurai, "90% write power-saving SRAM using sense-amplifying memory cell," *IEEE Journal of Solid-State Circuits*, 39(6), pp. 927–933, June 2004.
- S. Kosonocky, A. Bhavnagarwala and L. Chang, *International conference on solid-state and integrated circuit technology*, pp. 689–692, Oct. 2006.
- K. Mai, T. Mori, B. Amrutur et al., "Low-power SRAM design using half-swing pulse-mode techniques," *IEEE Journal of Solid-State Circuits*, 33(11) pp. 1659–1671, Nov. 1998.
- G. Ming, Y. Jun and X. Jun, "Low Power SRAM Design Using Charge Sharing Technique," pp. 102–105, *ASICON*, 2005.
- K. Osada, Y. Saitoh, E. Ibe and K. Ishibashi, "16.7-fA/cell tunnel-leakage- suppressed 16-Mb SRAM for handling cosmic-ray-induced multierrors," *IEEE Journal of Solid-State Circuits*, 38(11), pp. 1952–1957, Nov. 2003.
- PTM – Predictive Models. Available: http://www.eas.asu.edu/~ptm

References (cont.)

- E. Seevinck, F. List and J. Lohstroh, "Static noise margin analysis of MOS SRAM Cells," *IEEE Journal of Solid-State Circuits*, SC-22(5), pp. 748–754, Oct. 1987.
- K. Takeda, Y. Hagihara, Y. Aimoto, M. Nomura, Y. Nakazawa, T. Ishii and H. Kobatake, "A read-static-noise-margin-free SRAM cell for low-vdd and high-speed applications," *IEEE International Solid-State Circuits Conference*, pp. 478–479, Feb. 2005.
- M. Yamaoka, Y. Shinozaki, N. Maeda, Y. Shimazaki, K. Kato, S. Shimada, K. Yanagisawa and K. Osada, "A 300 MHz 25 µA/Mb leakage on-chip SRAM module featuring process-variation immunity and low-leakage-active mode for mobile-phone application processor," *IEEE International Solid-State Circuits Conference*, 2004, pp. 494–495.

Chapter 8
Optimizing Power @ Standby – Circuits and Systems

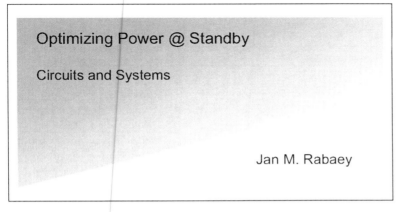

Slide 8.1

In Chapter 3, we observed that the optimal operation point in Energy–Delay space is a strong function of the activity – or, in other words, the operation mode of the circuit – and that there exists an optimal ratio between dynamic and static power dissipation. One special case is when there is no computational activity going on at all, that is, the standby mode. In an ideal world, this would mean that the *dynamic power consumption should be zero* or very small. Moreover (given the constant ratio), *static power dissipation should be eliminated as well*. Although the former can be achieved through careful management, the latter is becoming harder with advanced technology scaling. When all transistors are leaky, completely turning off a module is hard. In this chapter, we discuss a number of circuit and system techniques to keep both dynamic and static power in standby to an absolute minimum. As standby power is the main concern in memories (and as memories are somewhat special anyhow), we have relegated the discussion on them to Chapter 9.

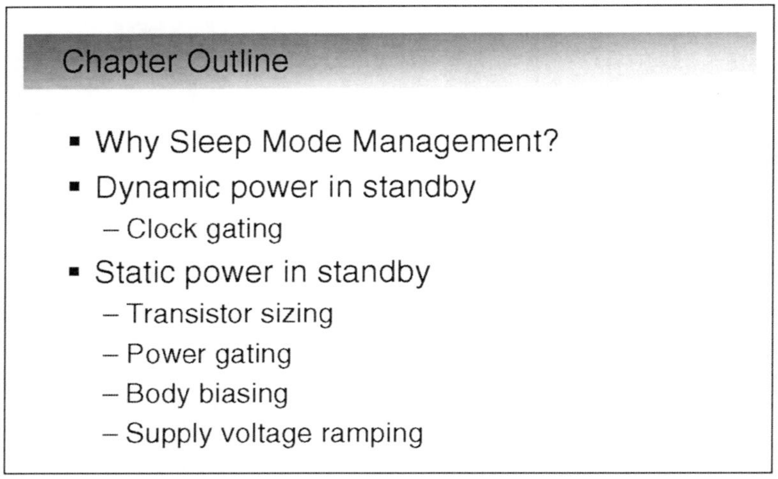

Slide 8.2

Chapter Outline

- Why Sleep Mode Management?
- Dynamic power in standby
  - Clock gating
- Static power in standby
  - Transistor sizing
  - Power gating
  - Body biasing
  - Supply voltage ramping

We start the chapter with a discussion on the growing importance of reducing standby power. Next, we analyze what it takes to reduce dynamic power in standby to an absolute minimum. The bulk of the chapter is devoted to the main challenge; that is, the elimination (or at least, minimization) of leakage during standby. Finally, some future perspectives are offered.

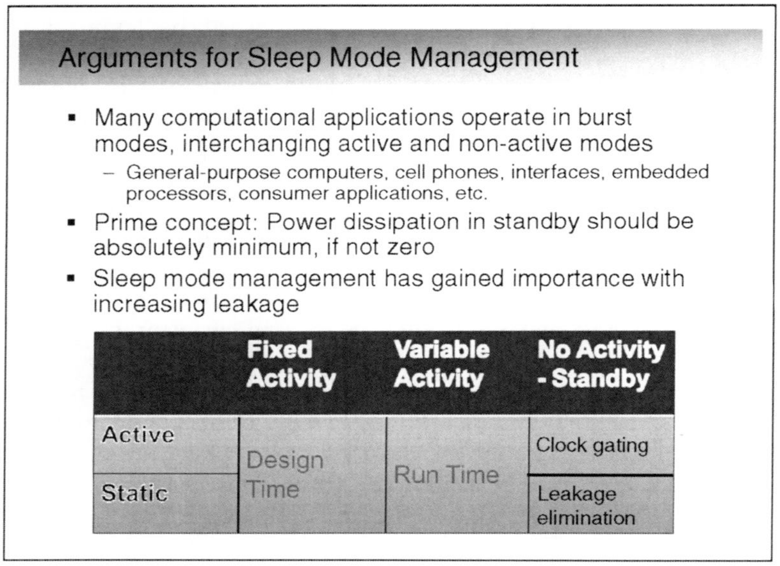

Slide 8.3

Arguments for Sleep Mode Management

- Many computational applications operate in burst modes, interchanging active and non-active modes
  - General-purpose computers, cell phones, interfaces, embedded processors, consumer applications, etc.
- Prime concept: Power dissipation in standby should be absolutely minimum, if not zero
- Sleep mode management has gained importance with increasing leakage

	Fixed Activity	Variable Activity	No Activity - Standby
Active	Design Time	Run Time	Clock gating
Static			Leakage elimination

With the advent of mobile applications, the importance of standby modes has become more pronounced, as it was realized that standby operation consumes a large part of the overall energy budget. In fact, a majority of applications tend to perform in a bursty fashion – that is, they exhibit short periods of intense activity interspersed with long intervals of no or not much activity. This is the case even in more traditional product lines such as microprocessors. Common sense dictates that modules or processors not performing any task should consume zero dynamic and also (preferably) zero static power.

Slide 8.4

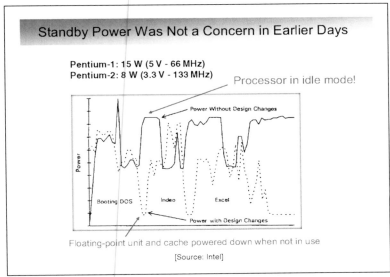

This was not a common understanding. In the long-gone days when power in CMOS designs did not rate very highly on the importance scale, designers paid scant attention to the power dissipation in unused modules. One of the (by now) classic examples of this neglect is the first Intel Pentium design, for which the power dissipation peaked when the processor was doing the least – that is, executing a sequence of NOPs. When power became an issue, this problem was quickly corrected as shown in the traces for the Pentium-2.

Slide 8.5

The main source of dynamic energy consumption in standby mode is the clock. Keeping the clock connected to the flip-flops of an idle module not only adds to the clock loading, but may cause spurious activity in the logic. In fact, as the data that is applied under those conditions is actually quite random, activity may be maximized as we have discussed earlier. This wasteful bit-flipping is avoided by two design interventions:

- Disconnect the clock from the flip-flops in the idle module through *clock gating*.
- Ensure that the inputs to the idle logic are kept stable. Even without a clock, changes at the inputs of a combinational block cause activity.

Clock gating a complete module (rather than a set of gates) makes the task a lot easier. However, deciding whether a module, or a collection of modules, is idle may not always be straightforward. Though sometimes it is quite obvious from the register-transfer level (RTL) code, normally it requires an understanding of the operational system modes. Also, clock gating can be more effective if modules that are idle simultaneously are grouped. What this basically says is that standby-power management plays at all levels of the design hierarchy.

Clock Gating

Turning off the clock to non-active components

[Diagram: Clk and Enable inputs to AND gate feeding Register File, which connects to Logic Module]

Disconnecting the inputs

[Diagram: Bus connected through Enable gate to Logic Module]

Slide 8.6
One possible way of implementing clock gating is shown in this slide. The clock to the register files at the inputs of an unused module is turned on or off using an extra AND gate controlled by an *Enable* signal. This signal is either introduced explicitly by the system- or RTL-designer, or generated automatically by the clock synthesis tools. Take for instance the case of a simple microprocessor. Given an instruction loaded in the instruction register (IR), the decoding logic determines which data path units are needed for its execution, and subsequently set their *Enable* signals to 1.

As the inputs of the logic module are connected to the register file, they remain stable as long as the clock is disabled. In the case that the inputs are directly connected to a shared bus, extra gates must be inserted to isolate the logic.

Observe that the gated clock signal suffers an additional gate delay, and hence increases the skew. Depending upon the time in the design process it is inserted, we must ensure that this extra delay does not upset any critical set-up and hold-time constraints.

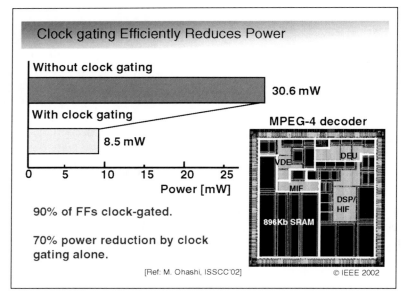

[Ref: M. Ohashi, ISSCC'02] © IEEE 2002

Slide 8.7
There is no doubt that clock gating is a truly effective means of reducing standby dynamic power. This is illustrated numerically with the example of an MPEG4 decoder [Ohashi'02]. Gating 90% of the flip-flops results in a straight 70% standby power reduction. This clearly indicates that there is NO excuse for not using clock gating in today's power-constrained designs.

Slide 8.8

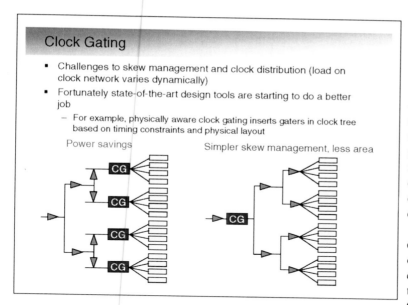

Yet, as mentioned, these gains do not come for free, and present an extra burden on the designers of the clock distribution network. In addition to the extra delay of the gating devices, clock gating causes the load on the clock network to vary dynamically, which introduces another source of clock noise into the system.

Let us, for instance, explore some different options on where to introduce the gating devices in the clock-tree hierarchy. One possible solution is to keep the *gaters* close to the registers. This allows for a fine-grain control on what to turn off and when. It comes at the expense of a more complex skew control and extra area. Another option is to move the gating devices higher up in the tree, which has the added advantage that the clock distribution network of the sub-tree is turned off as well – leading to some potentially large power savings. This comes at the expense of a coarser control granularity, which means that modules cannot be turned off as often.

Given the complexity of this task, it is fortunate that state-of-the-art clock synthesis tools have become more adept in managing the skew in the presence of clock gating. This will be discussed in more detail later, in the chapter on design methodology for power (Chapter 12).

Slide 8.9

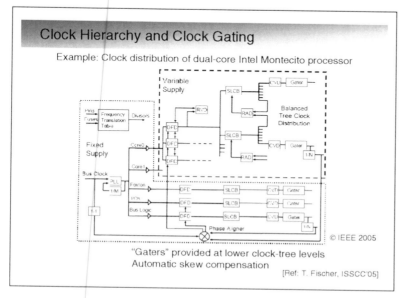

However effective these tools may be, it will be some time before they can handle the full complexity of the clock network of a modern microprocessor design. A bird's-eye view on the clock network of the dual-core Intel Montecito processor is shown here. Each core is allowed to run on variable frequencies (more about this in Chapter 10, when we discuss runtime optimization). The *digital frequency dividers* (DFDs) translate the central clock to the frequency expected for the different clock zones. The downstream clock network employs both *active deskew* (in the second-level clock buffers or SLCBs, and in the regional active deskew or RAD) and *fixed deskew*, tuned via scan (using the clock vernier devices or CVDs). The latter allow for final fine-tuning. *Gaters* provide the final stage of the network, enabling power saving and pulse shaping. A total of 7536 of those are distributed throughout the chip. Clock gating clearly has not simplified the job of the high-performance designer!

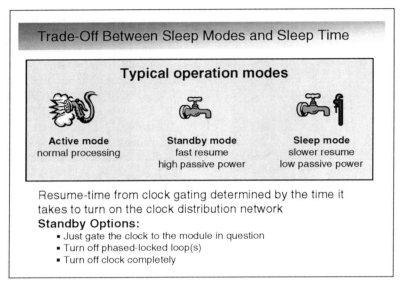

Slide 8.10
The introduction of clock gating succeeds in virtually eliminating the dynamic power dissipation of the computational modules during standby. However, although the ends of the clock tree have been disconnected, the root is still active and continues to consume power. Further power reductions would require that the complete clock distribution network and even the clock generator (which typically includes a crystal-driven oscillator and a phase-locked loop) are put to sleep. Although the latter can be turned off quite quickly, bringing them back into operation takes a considerable amount of time, and hence only makes sense if the standby mode is expected to last for considerable time.

Many processors and SoCs hence feature a variety of standby (or sleep) modes, with the state of the clock network as the main differentiator. Options are:

- Just clock gating
- Disabling the clock distribution network
- Turning off the clock driver (and the phase-locked loop)
- Turning off the clock completely.

In the latter case, only a wake-up circuit is kept alive, and the standby power drops to the microwatt range. Companies use different names for the various modes, with *sleep mode* typically reserved for the mode where the clock driver is turned off. It may take tens of clock-cycles to bring a processor back to operation from sleep mode.

Slide 8.11
The choice of the standby modes can be an important differentiator, as shown in this slide for a number of early-day low-power microprocessors. The Motorola PowerPC 603 supported four different operation modes, ranging from active, to doze (clocks still running to most units), nap (clock only to a timer unit), and sleep (clock completely shut off). The MIPS on the other hand did not support a full sleep mode, leading to substantially larger power dissipation in standby mode.

The MSP430^{TM} microcontroller from Texas Instruments shows the state-of-the-art of standby management. Using multiple on-chip clock generators, the processor (which is actively used in

Optimizing Power @ Standby – Circuits and Systems

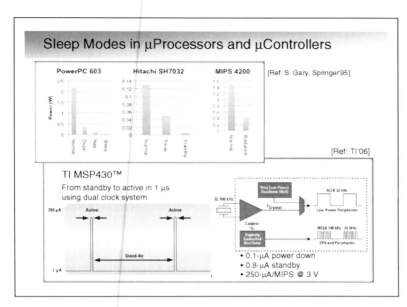

low-duty-cycle power-sensitive control applications) can go from standby (1 μA) to active mode (250 μA) in 1 μs. This rapid turnaround helps to keep the processor in standby longer, and makes it attractive to go into standby more often.

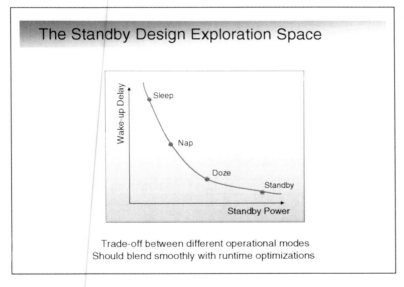

Slide 8.12

From the previous slides, a new version of our classic E–D trade-off curve emerges. The metrics to be traded off here are *standby power versus wake-up delay*.

Slide 8.13
Although standby modes are most often quoted for processors, they make just as much (if not more) sense for peripheral devices. Disks, wired and wireless interfaces, and input/output devices all operate in a bursty fashion. For instance, a mouse is in standby most of the time, and even when operational, data is only transmitted periodically. Clock gating and the support of different standby modes are hence essential. In this slide, the measured power levels and the transition times for two such peripheral devices are shown. Clearly the timing overhead associated with the wake-up from the standby mode cannot be ignored in each of these. Cutting down that time is crucial if standby is to be used more effectively.

Also the Case for Peripheral Devices

Hard disk

	P_{sleep} W	P_{active} W	T_{sleep} sec	T_{active} sec
IBM	0.75	3.48	0.51	6.97
Fujitsu	0.13	0.95	0.67	1.61

Wireless LAN Card

	TX	RX	Doze	Off
Power	1.65 W	1.4 W	0.045 W	0 W
Transitions			To Off: 62 ms	To Doze: 34 ms

[Ref: T. Simunic, Kluwer'02]

Slide 8.14
Given the effectiveness of clock gating, there is little excuse for dynamic power dissipation in standby. Eliminating or drastically reducing standby currents is a lot more problematic. The main challenge is that contemporary CMOS processes do not feature a transistor that can be turned off completely.

The Leakage Challenge – Power in Standby

- With clock gating employed in most designs, leakage power has become the dominant standby power source
- With no activity in module, leakage power should be minimized as well
  - Remember constant ratio between dynamic and static power …
- Challenge – how to disable unit most effectively given that no ideal switches are available

Slide 8.15

Standby Static Power Reduction Approaches

- Transistor stacking
- Power gating
- Body biasing
- Supply voltage ramping

A standby leakage control technique must be such that it has minimal impact on the normal operation of the circuit, both from a functional and performance perspective. Lacking a perfect switch, only two leakage-reduction techniques are left to the designer: increase the resistance in the leakage path, or reduce the voltage over that path. As the latter is harder to accomplish – you need either a variable or multiple supply voltages – most of the techniques presented in this chapter fall in the former category.

Slide 8.16

Transistor Stacking

- Off-current reduced in complex gates (see leakage power reduction @ design time)
- Some input patterns more effective than others in reducing leakage
- Effective standby power reduction strategy:
  - Select input pattern that minimizes leakage current of combinational logic module
  - Force inputs of module to correspond to that pattern during standby
- Pros: Little overhead, fast transition
- Con: Limited effectiveness

In Chapter 4 we established that the stacking of transistors has a super-linear leakage reduction effect. Hence, it pays to ensure that the stacking effect is maximized in standby. For each gate, an optimal input pattern can be determined. To get the maximum effect, one has to control the inputs of each gate individually, which is unfortunately not an option. Only the primary inputs of a combinational block are controllable. Hence, the challenge is to find the primary input pattern that minimizes the leakage of the complete block. Even though stacking has a limited impact on the leakage, the advantage is that it virtually comes for free, and that it has a negligible impact on performance.

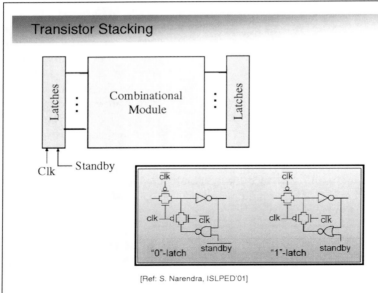

Slide 8.17

Standby leakage control using the stack effect requires only one real modification to the circuitry: all input latches or registers have to be presetable (either to the "0" or to the "1" state). This slide shows how this modification can be accomplished with only a minor impact on performance. Once the logic topology of a module is known, computer-aided design (CAD) tools can easily determine the optimal input pattern, and the corresponding latches can be inserted into the logic design.

Slide 8.18

Even when the technology-mapping phase of the logical-synthesis process is acutely aware of the stacking opportunity, it is unavoidable that some gates in the module end up with small fan-in. An inverter here or there is hard to avoid. And these simple gates contribute largely to the leakage. This can be remedied through the use of *forced stacking*, which replaces a transistor in a shallow stack by a pair (maintaining the same input loading). Although this transistor doubling, by necessity, impacts the performance of the gate – and hence should only be used in non-critical paths – the leakage reduction is substantial. This is perfectly illustrated by the leakage current (i.e., standby power) versus delay plots, shown on the slide for the cases of high- and low-threshold transistors. The advantage of forced stacking is that it can be fully automated.

Observe that this slide introduces another important trade-off metric: *standby power versus active delay*.

Slide 8.19

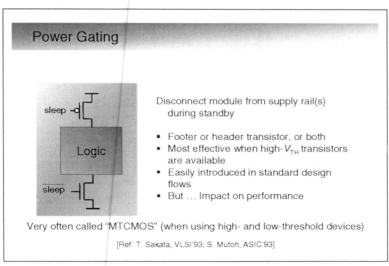

The ideal way to eliminate leakage current is to just disconnect the module from the supply rails – that is, if we could have perfect on–off switches available. The next best option is to use switches acting as "large resistors" between the "virtual" supply rails of the module and the global supply rails. Depending upon their position, those switches are called "headers" or "footers", connecting to V_{DD} or ground, respectively. This *power-gating* technique performs the best when the technology supports both high- and low-threshold transistors. The latter can be used for the logic, ensuring the best possible performance, whereas the others are very effective as power-gating devices. When multiple thresholds are used, the power-gating approach is often called MTCMOS.

Slide 8.20

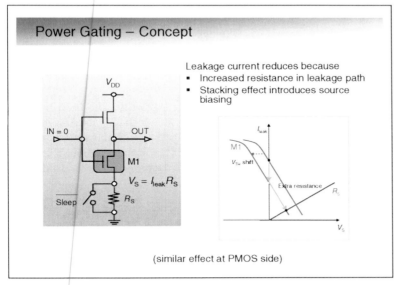

The headers/footers add resistance to the leakage path during standby. In addition, they also introduce a stacking effect, which increases the threshold of the transistors in the stack. The combination of resistance and threshold increase is what causes the large reduction in leakage current.

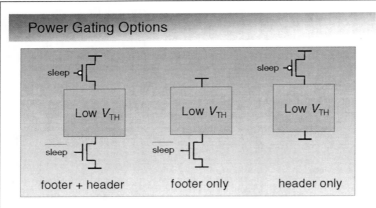

Slide 8.21

Obviously, introducing an extra transistor in the charge and discharge paths of a gate comes with a performance penalty, the effects of which we would like to mitigate as much as possible. In principle, it is sufficient to insert only a single transistor (either footer or header) for leakage reduction. The addition of the second switch, though far less dramatic in leakage reduction, ensures that the stacking effect is exploited independent of the input patterns. If one chooses a single power-gating device, the NMOS footer is the preferred option, because its on-resistance is smaller for the same transistor width. It can hence be sized smaller than its PMOS counterpart. This is the approach that is followed in a majority of today's power-conscious IC designs.

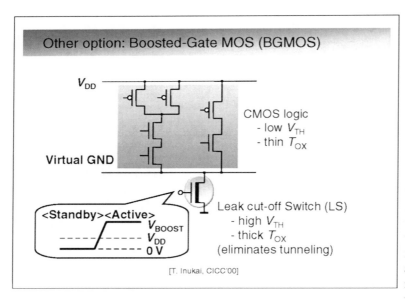

Slide 8.22

A number of modifications to the standard power-gating techniques can be envisioned, producing even larger leakage reductions, or reducing the performance penalty. The "boosted-gate" approach raises the gate voltage of the footer (header) transistors above the supply voltage, effectively decreasing their resistance. This technique is only applicable when the technology allows for high voltages to be applied to the gate. This may even require the use of thick-oxide transistors. Some CMOS processes make these available to allow for the design of voltage-converting input and output pads (Note: the core of a chip often operates at a supply voltage lower than the board-level signals to reduce power dissipation).

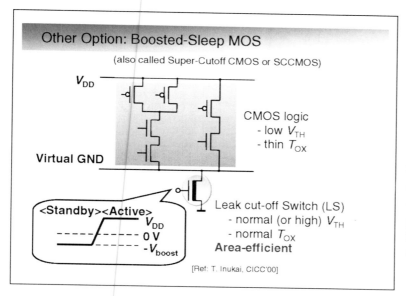

Slide 8.23

The reverse is also possible. Instead of using a high-V_{TH} device, the sleeper transistor can be implemented with a low-V_{TH} device, leading to better performance. To reduce the leakage in standby, the gate of the sleeper is reverse biased. Similar to the "boosted-gate" technique, this requires a separate supply rail. Be aware that this increases the latch-up danger.

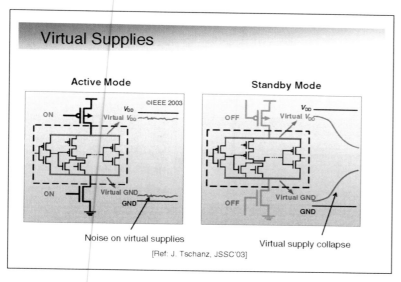

Slide 8.24

It is worth observing what happens with the virtual supplies in active and sleep modes. The extra resistance on the supply rail not only impacts performance, but also introduces extra IR-induced supply noise – impacting the signal integrity. During standby mode, the virtual supply rails start drifting, and ultimately converge to voltage levels determined by the resistive divider formed by the *on* and *off* transistors in the stack. The conversion process is not immediate though, and is determined by the leakage rates.

Slide 8.25

Reaching the standby operation mode is hence not immediate. This poses some interesting questions on where to put most of the decoupling capacitance (decap): on the chip supply rails, or on the virtual rails? The former has the advantage that relatively lower capacitance has to be (dis)charged when switching modes, leading to faster convergence and smaller overhead. The cost and overhead of going to standby mode is smaller. Also, the energy overhead for charging and discharging the decoupling capacitance is avoided. This approach

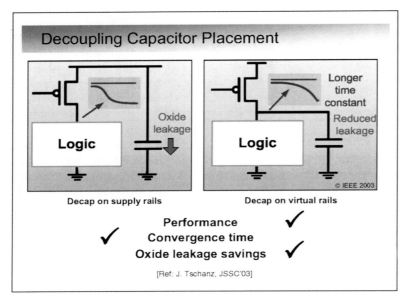

also has some important disadvantages: (1) the virtual supplies are more prone to noise, and (2) the gate-oxide capacitance that serves as decap stays under full voltage stress, and keeps contributing gate leakage current even in standby (Note that on-chip decoupling capacitance is often realized using huge transistors with their sources and drains short-circuited). Putting the decap on the chip supply rails hence is the preferred option if standby mode is most often invoked for a short time. The "decap on virtual supply" works best for long, infrequent standby invocations.

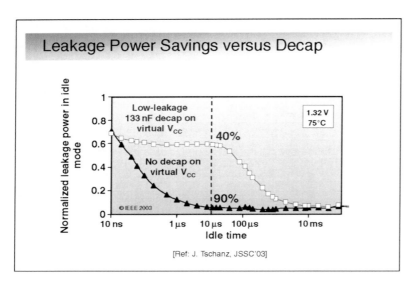

Slide 8.26
This trade-off is illustrated in the simulation of the virtual supply rails. After 10 ms, the leakage power of the "no decap on virtual rails" scenario has dropped by 90%. It takes 10 times as long for the "decap on virtual rails" to reach the same level of effectiveness.

Slide 8.27
As mentioned earlier, the sleep transistor does not come for free, as it impacts the performance of the module in active mode, introduces supply noise, and costs extra area. To minimize the area, a single switch is most often shared over a set of gates. An important question hence is how to size the transistor: making it wider minimizes performance impact and noise, but costs area. A typical target for sleep transistor sizing is to ensure that the extra ripple on the supply rail is smaller than 5% of the full swing.

How to Size the Sleep Transistor?

- Sleep transistor is not free – it will degrade the performance in active mode
- Circuits in active mode see the sleep transistor as extra power-line resistance
  - The wider the sleep transistor, the better
- Wide sleep transistors cost area
  - Minimize the size of the sleep transistor for given ripple (e.g., 5%)
  - Need to find the worst-case vector

If the designer has access to power distribution analysis and optimization tools, sizing of the sleep transistors can be done automatically – as we will discuss in Chapter 12. If not, it is up to her to determine the peak current of the module through simulations (or estimations), and size the transistor such that the voltage drop over the switch is no larger than the allowable 5%.

Slide 8.28

The table on this slide compares the effectiveness of the different power-gating approaches. In the MTCMOS approach, a high-V_{TH} device is used for the sleeper. To support the necessary current, the transistor must be quite large. When a *low-*V_{TH} transistor is used, the area overhead is a lot smaller at the expense of increased leakage. The boosted-sleep mode combines the best of both, that is small transistor width and low leakage, at the expense of an extra supply rail. The transistors were sized such that the supply bounce for each of them is approximately the same.

Sleep Transistor Sizing

- High-V_{TH} transistor must be very large for low resistance in linear region
- Low-V_{TH} transistor needs less area for same resistance.

	MTCMOS	Boosted Sleep	Non-Boosted Sleep
Sleep TR size	5.1%	2.3%	3.2%
Leakage power reduction	1450x	3130x	11.5x
Virtual supply bounce	60 mV	59 mV	58 mV

[Ref: R. Krishnamurthy, ESSCIRC'02]

Slide 8.29

The attentive reader must already have wondered about one important negative effect of power gating: when we disconnect the supply rails, all data stored in the latches, registers and memories of the module ultimately are lost. This sometimes is not a problem, especially when the processor always restarts from the same initial state – that is, all intermediate states can be forgotten. More often, the processor is expected to remember some part of its prior history, and rebooting from scratch after every sleep period is not an option. This can be dealt with in a number of ways:

- All essential-states are copied to memory with data retention before going to sleep, and is reloaded upon restart. Everything in the scratch-pad memory can be forgotten. The extra time for copying and reloading adds to the start-up delay.

Preserving State

- Virtual supply collapse in sleep mode causes the loss of state in registers
- Keeping the registers at nominal V_{DD} preserves the state
  - These registers leak ...
- Can lower the V_{DD} in sleep
  - Some impact on robustness, noise, and soft-error immunity

- The essential memory in the module is not powered down, but put in data retention mode. This approach increases the standby power, but minimizes the start-up and power-down timing overhead. We will talk more about memory retention in the next chapter.
- Only the logic is power-gated, and all registers are designed for data retention.

Latch-Retaining State During Sleep

Black-shaded devices use low-V_{TH} tranistors
All others are high-V_{TH}.
[Ref: S. Mutoh, JSSC'95]

Slide 8.30
The latter approach has the smallest granularity of control, but yields the smallest reduction in standby leakage, as all registers are still powered up. Also, the active performance of the latch/register may only be minimally impacted.

An example of a master slave register that combines high-speed active-mode performance with low-leakage data retention is shown in this slide. High-V_{TH} devices are used for all transistors with the exception of the transistors or gates that are shaded black. In essence, these are the transistors in the forward path, where performance is essential. The high-V_{TH} cross-coupled inverter pair in the second latch acts as the data retention loop and is the only leakage-producing circuitry that is *on* during standby.

Slide 8.31
The above represents only a single example of a data retention latch. Many others have been perceived since then.

A very different option for state retention is to ensure that the standby voltage over the registers does not drop below the retention voltage (i.e., the minimum voltage at which the register or latch still reliably stores data). This is, for instance, accomplished by setting the standby voltage to $V_{DD}-V_D$ (where V_D is the voltage over a reverse-biased diode), or by connecting it to a separate supply rail called V_{retain}. The former approach has the advantage that no extra supply is needed, whereas the latter allows for careful selection of the retention voltage so that leakage is minimized

Optimizing Power @ Standby – Circuits and Systems

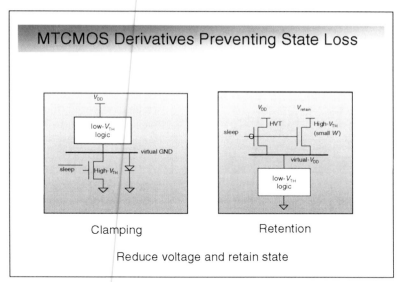

while retaining reliable storage. Both cases come with the penalty that the leakage through the logic may be higher than what would be obtained by simple power gating. The advantage is that the designer does not have to worry about state retention.

The topic of retention voltages and what determines their value is discussed in more detail in the next chapter on memory standby (Chapter 9).

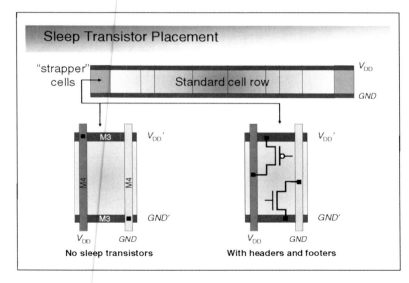

Slide 8.32

To conclude the discussion on power gating, it is worth asking ourselves how this impacts layout strategy and how much area overhead this brings with it. Fortunately power switches can be introduced in contemporary standard layout tools with only minor variations.

In a traditional standard-cell design strategy, it is standard practice to introduce "strapper" cells at regular intervals, which connect the V_{DD} and GND wires in the cells to the global power distribution network. These cells can easily be modified to include header and footer switches of the appropriate sizes. Actually, quite often one can determine the size of the switches based on the number of the cells they are feeding in a row.

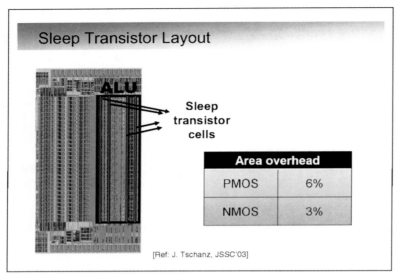

Slide 8.33

The area overhead of the power-gating approach was quantified in a study performed at Intel in 2003 [Ref: J. Tschanz, ISSCC'03], which compared the effectiveness of various leakage control strategies for the same design (a high-speed ALU). Both footers and headers were used, and all sleep transistors were implemented using low-threshold transistors to minimize the impact on performance. It was found that the area overhead of the power gating was 6% for the PMOS devices, and 3% for the NMOS footers. We will come back to the same study in a couple of slides.

Slide 8.34

An alternative to the power-gating approach is to decrease the leakage current by increasing the thresholds of the transistors. Indeed, every transistor has a fourth terminal, which can be used to increase the threshold voltage through reverse biasing. Recall that a linear change in threshold voltage translates into an exponential change in leakage current. Even better, this approach can also be used to decrease the transistor threshold in active mode through forward biasing! The alluring feature of dynamic biasing of the transistor is that it does not come with a performance penalty, and it does not change the circuit topology. The only drawback seems to be the need for a triple-well technology if we want to control both NMOS and PMOS transistors.

Although all this looks very attractive at a first glance, there are some other negatives that cannot be ignored. The range of the threshold control through dynamic biasing is limited, and, as we established in Chapter 2, it is rapidly decreasing with the scaling of the technology below 100 nm. Hence the effectiveness of the technique is quite small in nano-meter technologies, and will not get better in the future unless novel devices with much better threshold control emerge (for instance, the dual-gate transistors we briefly introduced in Chapter 2). Finally, changing the back-gate bias of the transistors requires the charging or discharging of the well capacitance, which adds a sizable amount of overhead energy and time.

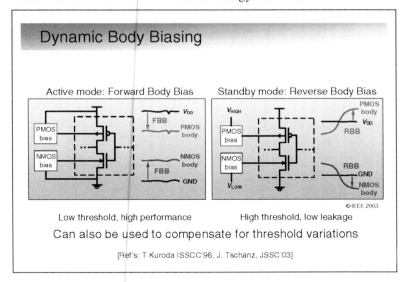

Slide 8.35

The concept of dynamic body biasing (DBB), as first introduced by Seta et al. in 1995, is illustrated pictorially in this slide. Obviously the approach needs some extra supply voltages that must be distributed over the chip. Fortunately, these extra supplies have to carry only little continuous current, and can be generated using simple on-chip voltage converters.

The technique of dynamic body biasing is by no means new, as it has been applied in memory designs for quite some time. Yet, it is only with leakage power becoming an important topic that it is being applied to computational modules. The attentive reader probably realizes that this technique has more to offer than just leakage management. It can, for instance, also be used for the compensation of threshold variations. To hear more about this, you have to wait for the later chapters.

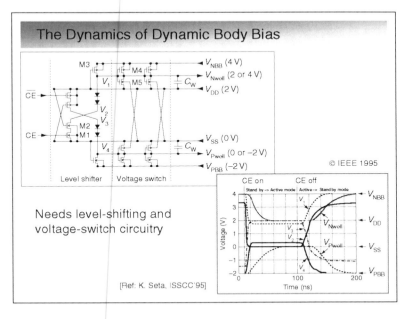

Slide 8.36

Though the adoption of DBB requires little changes in the computational modules, it takes some extra circuitry to facilitate the switching between the various biasing levels, which may extend above or below the standard voltage rails. Adapting the sleep control signals (CE) to the appropriate levels requires a set of level converters, whose outputs in turn are used to switch the well voltages. The resulting voltage waveforms, as recorded in [Seta95], are shown on the slide as well. Observe that in this early incarnation of the DBB approach it takes approximately the same time to charge and discharge the wells – for a total transient time of somewhat less than 100 ns.

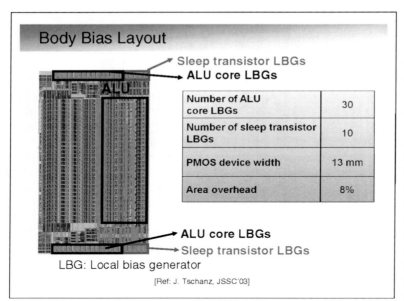

Slide 8.37

The area overhead of the dynamic-biasing approach mainly consists of the generation of the bias voltages, the voltage switches, and the distribution network for the bias voltages. To compare DBB with power gating, the example of Slide 8.33 is revisited. The body-bias circuitry consists of two main blocks: a central bias generator (CBG) and many distributed local bias generators (LBGs). The function of the CBG is to generate a process-, voltage-, and temperature-invariant reference voltage, which is then routed to the LBGs. The CBG uses a scaled-bandgap circuit to generate a reference voltage, which is 450 mV below the main supply — this represents the amount of forward bias to apply in active mode. This reference voltage is then routed to all the distributed LBGs. The function of the LBG is to refer the offset voltage to the supply voltages of the local block. This ensures that any variations in the local supplies will be tracked by the body voltage, maintaining a constant 450 mV of FBB.

To ensure that the impedance presented to the well is low enough, the forward biasing of the ALU required 30 LBGs. Observe that in this study only the PMOS transistors are dynamically biased, and that only forward biasing is used (in standby, zero bias is used). The total area overhead of all the bias units and the wiring turned out to be approximately 8%.

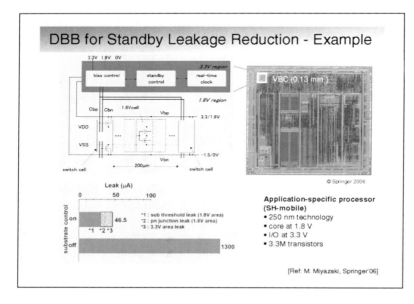

Slide 8.38

The effectiveness of the DBB approach is demonstrated with an example of an application-specific processor, the SH-mobile from Renesas (also called the SuperH Mobile Application Processor). The internal core of the processor operates at 1.8 V (for a 250 nm CMOS technology). In standby mode, reverse body-biasing is applied to the PMOS (3.3 V) and the NMOS (−1.5 V) transistors. The 3.3 V supply is already externally available for the

I/O pins, whereas the −1.5 V supply is generated on-chip. Similar to the power-gating approach, special "switch cells" are included in every row of standard cells, providing the circuitry to modulate the well voltages.

For this particular design, the DBB approach reduces the leakage by a factor of 28 for a fairly small overhead. Unfortunately, what works for 250 nm does not necessarily translate into similar savings in the future.

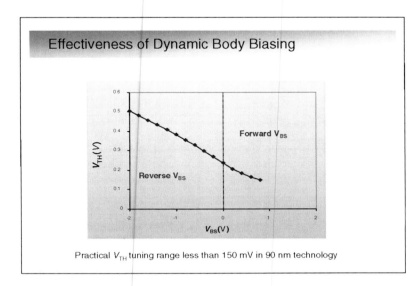

Slide 8.39

As we had already observed in Slide 2.12, the effectiveness of back biasing reduces with technology scaling. Although for a 90 nm technology, a combination of FBB and RBB may still yield a 150 mV threshold change, the effect is substantially smaller for 65 nm. This trend is not expected to change course substantially in the coming technology generations. The potential savior is the adoption of dual-gate devices, which may be adopted in the 45 nm (and beyond) technology generations. Hence, as for now, DBB is a useful technology up to 90 nm, but its future truly depends upon device and manufacturing innovations.

Slide 8.40

Ultimately, the best way to reduce leakage in standby mode is to ramp the supply voltage all the way down to 0 V. This is the only way to guarantee the total elimination of leakage. A controllable voltage regulator is the preferred way of accomplishing this *Supply Voltage Ramping* (SVR) scheme. With voltage islands and dynamic voltage scaling becoming common practice (see Chapter 10), voltage regulators and converters are being integrated into the SoC design process, and the overhead of SVR is negligible. In designs where this is not the case, switches can be used to swap the "virtual" supply rail between V_{DD} and GND. As the switches themselves leak, this approach is not as efficient as the ramping.

The overhead of the SVR scheme is that upon reactivation all the supply capacitance has to be charged up anew, leading to a longer start-up time. Obviously all state data are lost in this regime. If state retention is a concern, techniques discussed earlier in the chapter such as the transfer of essential state information to persistent memory or keeping the supply voltage of the state memory above the retention voltage (DRV), are equally valid.

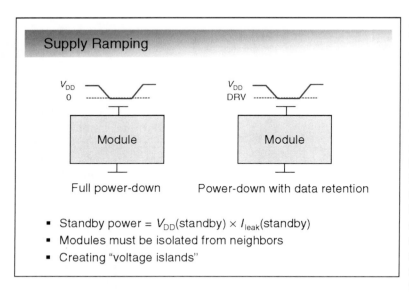

Slide 8.41
This slide shows a pictorial perspective of the supply ramping approach (both for ramping down to GND or to the data retention voltage DRV). SVR in concert with dynamic voltage scaling (DVS – see Chapter 10) is at the core of the "voltage island" concept, in which a chip is divided into a number of voltage domains that can change their values dynamically and independently. To have the maximum effect, it is important that signals crossing the boundaries of voltage islands are passed through adequate converters and isolators. This is in a sense similar to the boundary conditions that exist for signals crossing clock domains.

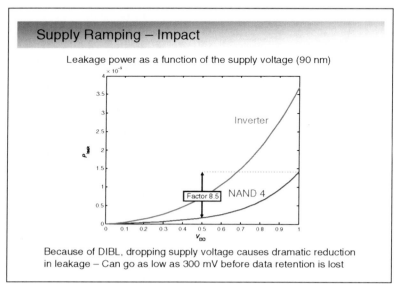

Slide 8.42
The impact of SVR is quite important. Because of the exponential nature of the DIBL effect, just reducing the supply voltage from 1 V to 0.5 V already reduces the static leakage power by a factor of 8.5 for a four-input NAND gate in a 90 nm CMOS technology. With the proper precautions (as we will discuss in Chapter 9), data retention may be ensured all the way to 300 mV. However, nothing beats scaling down all the way to ground.

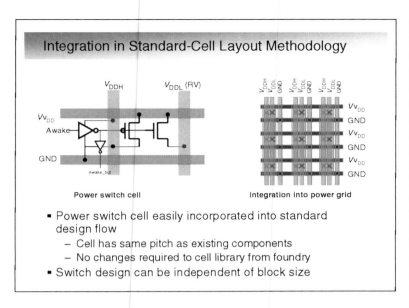

Slide 8.43

If voltage ramping is not an option, switching between different rails (from V_{DDH} to GND or V_{DDL}) is still a viable alternative, even though it comes with a larger leakage current in standby (through the V_{DDH} switch). The switch to the lower rail (V_{DDL} or GND) can be made small as it only carries a very small amount of current. The SVR approach can be incorporated in the standard design flows in a similar way as the power-gating and the dynamic body-biasing approaches discussed earlier. The only requirement is a number of extra cells in the library with appropriate sizing for the different current loads.

Slide 8.44

Though standby power reduction is a major challenge, it is also clear that a number of techniques have emerged addressing the problem quite effectively as shown in this overview slide. The main trade-offs are between standby power, invocation overhead, area cost, and runtime performance impact.

The main challenge the designer faces is to ensure that a module is placed in the appropriate standby mode when needed, taking into account the potential savings and the overhead involved. This requires a system-level perspective, and a chip architecture with power management integrated into its core.

Slide 8.45

In the end, what is really needed to deal with the standby power problem is an ideal switch, which conducts very small current when *off* and has a very low resistance when *on*. Given the importance of standby power, spending either area or manufacturing cost toward such a device seems to be worthwhile investment. In Chapter 2, we discussed some emerging devices that promise steeper sub-threshold slopes, such as, for instance, the dual-gate device. Some speculative transistors even promise slopes lower than 60 mV/dec.

Some Long-Term Musings

- Ideal power-off switch should have zero leakage current (S = 0 mV/decade)
- Hard to accomplish with traditional electronic devices
- Maybe possible using MEMS – mechanical switches have a long standing reputation for good isolation

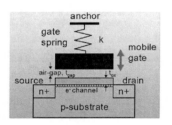 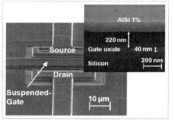

[Ref: N. Abele, IEDM'05]

Yet, a switch that can be fully turned off in standby mode would be the ultimate. This is why some of the current research into microelectromechanical systems (MEMS)-based switches is so appealing. A number of research groups are investigating the idea of a MOS transistor with a "movable" gate, where the thickness of the gate insulator is modified using electrostatic forces. This may lead to a switch with ignorable leakage in *off* mode, yet a good conductivity in the *on* mode. We believe that devices like this may ultimately play an important role in extending the life of CMOS design into the nanometer scale. Yet, as always, any modification in the manufacturing process comes at a considerable cost.

Summary and Perspectives

- Today's designs are not leaky enough to be truly power–performance optimal! Yet, when not switching, circuits should not leak!
- Clock gating effectively eliminates dynamic power in standby
- Effective standby power management techniques are essential in sub-100 nm design
  - Power gating the most popular and effective technique
  - Can be supplemented with body biasing and transistor stacking
  - Voltage ramping probably the most effective technique in the long range (if gate leakage becomes a bigger factor)
- Emergence of "voltage or power" domains

Slide 8.46
In summary, though leakage may not be such a bad thing when a circuit is switching rapidly, it should be avoided by all means when nothing is happening. It is fair to say that a number of effective techniques to manage standby power have emerged, especially for logic circuits. The most important challenge facing the designer today is how and when to invoke the different standby power saving techniques.

In the next chapter, we shall discuss how the inverse is true for memories, where controlling standby leakage while retaining storage has evolved into one of the most challenging problems for today and tomorrow.

References

Books and Book Chapters

- V. De et al., "Techniques for Leakage Power Reduction," in A. Chandrakasan et al., *Design of High-Performance Microprocessor Circuits*, Ch. 3, IEEE Press, 2001.
- S. Gary, "Low-Power Microprocessor Design," in *Low Power Design Methodologies*, Ed. J. Rabaey and M. Pedram, Chapter 9, pp. 255–288, Kluwer Academic, 1995.
- M. Miyazaki et al., "Case study: Leakage reduction in hitachi/renesas microprocessors", in A. Narendra, *Leakage in Nanometer CMOS Technologies*, Ch. 10., Springer, 2006.
- S. Narendra and A. Chandrakasan, *Leakage in Nanometer CMOS Technologies*, Springer, 2006.
- K. Roy et al., "Circuit Techniques for Leakage Reduction," in C. Piguet, *Low-Power Electronics Design*, Ch. 13, CRC Press, 2005.
- T. Simunic, "Dynamic Management of Power Consumption", in *Power Aware Computing*, edited by R. Graybill, R. Melhem, Kluwer Academic Publishers, 2002.

Articles

- N. Abele, R. Fritschi, K. Boucart, F. Casset, P. Ancey, and A.M. Ionescu, "Suspended-gate MOSFET: bringing new MEMS functionality into solid-state MOS transistor," Proc. Electron Devices Meeting, 2005. IEDM Technical Digest. *IEEE International*, pp. 479–481, Dec. 2005
- T. Fischer, et al., "A 90-nm variable frequency clock system for a power-managed Itanium® architecture processor," *IEEE J. Solid-State Circuits*, pp. 217–227, Feb. 2006.
- T. Inukai et al., "Boosted Gate MOS (BGMOS): Device/Circuit Cooperation Scheme to Achieve Leakage-Free Giga-Scale Integration," CICC, pp. 409–412, May 2000.
- H. Kam et al., "A new nano-electro-mechanical field effect transistor (NEMFET) design for low-power electronics, " IEDM Tech. Digest, pp. 463–466, Dec. 2005.
- R. Krishnamurthy et al., "High-performance and low-power challenges for sub-70 nm microprocessor circuits," *2002 IEEE ESSCIRC Conf.*, pp. 315–321, Sep. 2002.
- T. Kuroda et al., "A 0.9 V 150 MHz 10 mW 4 mm^2 2-D discrete cosine transform core processor with variable-threshold-voltage scheme," JSSC, 31(11), pp. 1770–1779, Nov. 1996.

References (cont.)

- S. Mutoh et al., 1V high-speed digital circuit technology with 0.5 mm multi-threshold CMOS, "*Proc. Sixth Annual IEEE ASIC Conference and Exhibit*, pp. 186–189, Sep. 1993.
- S. Mutoh et al., "1-V power supply high-speed digital circuit technology with multithreshold-voltage CMOS", *IEEE Journal of Solid-State Circuits*, 30, pp. 847–854, Aug. 1995.
- S. Narendra, et al., "Scaling of stack effect and its application for leakage reduction," *ISLPED*, pp. 195–200, Aug. 2001.
- M. Ohashi et al., "A 27MHz 11.1mW MPEG-4 video decoder LSI for mobile application," *ISSCC*, pp. 366–367, Feb. 2002.
- T. Sakata, M. Horiguchi and K. Itoh, Subthreshold-current reduction circuits for multi-gigabit DRAM's, *Symp. VLSI Circuits Dig.*, pp. 45–46, May 1993.
- K. Seta, H. Hara, T. Kuroda, M. Kakumu and T. Sakurai, "50% active-power saving without speed degradation using standby power reduction (SPR) circuit," *IEEE International Solid-State Circuits Conference*, XXXVIII, pp. 318–319, Feb. 1995.
- M. Sheets et al., J, "A Power-Managed Protocol Processor for Wireless Sensor Networks," Digest of Technical Papers 2006 Symposium on VLSI Circuits, pp. 212–213, June 15–17, 2006.
- TI MSP430 Microcontroller family, *http://focus.ti.com/lit/Slab034n/slab034n.pdf*
- J. W. Tschanz, S. G. Narendra, Y. Ye, B. A. Bloechel, S. Borkar and V. De, "Dynamic sleep transistor and body bias for active leakage power control of microprocessors," *IEEE Journal of Solid-State Circuits*, 38, pp. 1838–1845, Nov. 2003.

Slides 8.47 and 8.48

Some references.

Chapter 9
Optimizing Power @ Standby – Memory

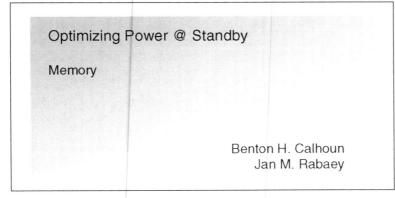

Slide 9.1
This chapter describes approaches for optimizing the power consumption of an embedded memory, when in standby mode. As mentioned in Chapter 7, the power dissipation of memories is, in general, only a fraction of the overall power budget of a design in active mode. The reverse is true when the circuit is in standby. Owing to the large (and growing) number of memory cells on a typical IC, their contribution to the leakage power is substantial, if not dominant. Reducing the standby power dissipation of memories is hence essential.

Slide 9.2
In this Chapter, we first discuss why the standby leakage current of large embedded SRAM memories is becoming a growing concern. When looking at the possible solution space, it becomes clear that static power in the memory core is best contained by manipulating the various voltages in and around the cell. One option is to reduce the supply voltage(s); another is to change the transistor bias voltages. Various combinations of these two can be considered as well. Bear in mind however that any realistic leakage power reduction technique must ensure that the data is reliably retained during the standby

period. Though the periphery presents somewhat of a lesser challenge, it has some special characteristics that are worth-examining. The Chapter is concluded with some global observations.

Slide 9.3

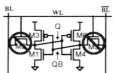

- SRAM is a major source of static power in ICs, especially for low-power applications
- Special memory requirement: need to retain state in standby
- Metrics for standby:
  - 1. Leakage power
  - 2. Energy overhead for entering/leaving standby
  - 3. Timing/area overhead

Memory Dominates Processor Area

During standby mode, an embedded memory is not accessed, so its inputs and outputs are not changing. The main function of the memory during standby is therefore to retain its data until the next transition to active operation. The retention requirement complicates the reduction of the leakage power. Although combinational logic modules can be disconnected from the supply rails using power gating, or their supply voltages reduced to zero, this is not an option for the embedded SRAM (unless it is a scratch-pad memory). Hence, minimizing the leakage, while reliably maintaining state, is the predominant requirement. Some of the techniques that are introduced in this chapter carry some overhead in terms of power and/or time to bring a memory in to and/or out of standby. A secondary metric is hence the energy overhead consumed during the transition, which is important because it determines the minimum time that should be spent in standby mode for the transition to be worthwhile. If the power savings from being in standby for some time do not offset the overhead of entering/leaving that mode, then standby should not be used. In addition, rapid transitions between standby and active modes are helpful in many applications. Finally, we also observe that reducing standby power often comes with an area overhead.

We begin this chapter by taking a brief top-level look at the operation of an embedded SRAM cell during standby. Next, we examine a number of standby power reduction techniques. The most effective techniques to date are based on voltage manipulation – either lowering the supply voltage, or increasing the bias voltages of the transistors inside the cell. The standby power of the peripheral circuits is briefly discussed before the chapter is summarized.

Slide 9.4

Reminder of "Design-Time" Leakage Reduction

- Design-time techniques (Chapter 7) also impact leakage
  - High-V_{TH} transistors
  - Different precharge voltages
  - Floating BLs
- This chapter: adaptive methods that uniquely address memory standby power

Some of the approaches described in Chapter 7 for lowering power at design time reduce leakage power in both active and standby modes. These approaches include using high-threshold-voltage transistors, lowering the precharge voltage, or allowing bitlines to float (they float to a voltage that minimizes leakage into the bitcells. Though these approaches do affect the leakage power during standby, this chapter focuses on approaches that uniquely address the standby leakage.

Slide 9.5

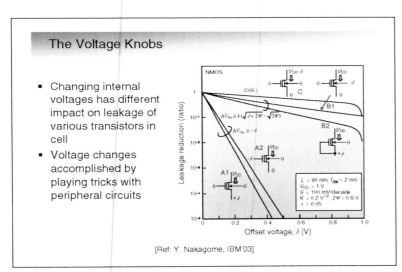

The Voltage Knobs

- Changing internal voltages has different impact on leakage of various transistors in cell
- Voltage changes accomplished by playing tricks with peripheral circuits

[Ref: Y. Nakagome, IBM'03]

Though there are many circuit-level knobs available for addressing leakage power, the various voltage levels in and around the bit-cell are the most effective. In Chapter 7, we discussed how these voltages can be assigned at design time to reduce power. Altering the voltages by manipulating the peripheral circuits during standby mode can decrease leakage power during standby mode. There is more flexibility to alter the voltages in standby mode because many of the functionality-limiting metrics are no longer relevant, such as read static noise margin and write margin. In standby mode, the primary functionality metric of concern is the hold static noise margin, as the bit-cells are only holding their data.

Slide 9.6

The most straightforward voltage scaling approach to lowering standby leakage power in a memory is reducing the supply voltage, V_{DD}. This approach lowers power in two ways: (1) voltage reduction ($P = VI$) and (2) leakage current reduction. The dominant mechanism behind the latter is the drain-induced barrier lowering (DIBL) effect. In addition, other contributors to leakage current drop off as well. Gate-induced drain leakage (GIDL) quickly decreases with V_{DD}, and

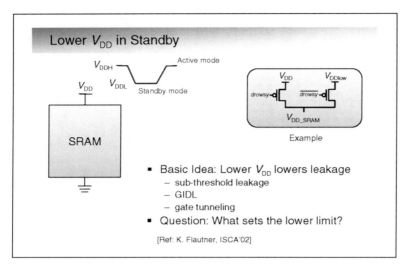

gate-tunneling current decreases roughly as V_{DD}. Junction leakage currents at the source and drain of the transistors also decrease rapidly with V_{DD}.

One approach to implementing standby voltage scaling is to switch to a lower supply voltage using PMOS header switches, as shown in the slide. The standby, or drowsy, supply provides a lower voltage to reduce leakage only for SRAM blocks that are in standby mode. During active mode, the power supply returns to the nominal operating voltage. As we have described before, the key limitation to the extent by which V_{DD} is lowered is that the data inside the cells must be protected. If the data are no longer required, then the power supply can simply be disconnected using power gating approaches like those that were described earlier for combinational logic, or by ramping the supply down to GND.

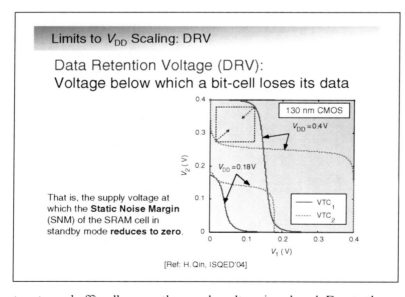

Slide 9.7
Given the effectiveness of voltage reduction in lowering the standby power of an SRAM memory, the ultimate question now is how much the supply voltage can safely be reduced. We define the minimum supply voltage for which an SRAM bit-cell (or an SRAM array) retains its data as the *Data Retention Voltage (DRV)*.

The butterfly plots shown on this slide illustrate how the noise margins of a 6T cell (with its access transistors turned off) collapse as the supply voltage is reduced. Due to the asymmetrical nature of a typical cell (caused by the dimensioning of the cell transistors as well as by variations), the SNM of the cell is determined by the upper lobe of the butterfly plot. Once the supply voltage reaches 180 mV, the SNM drops to zero and the stored value is lost. The cell becomes monostable at that point. In a purely symmetrical cell, the supply voltage could be lowered substantially more before the data is lost.

We can therefore also specify the DRV as the voltage at which the SNM of a non-addressed cell (or cell array) drops to zero.

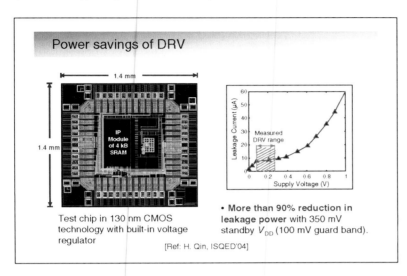

Slide 9.8
The advantages of scaling the V_{DD} during standby can be quite significant. A 0.13 μm test chip shows over 90% reduction in standby leakage by lowering the power supply to within 100 mV of the DRV. The reduction in the DIBL effect is one of the most important reasons behind this large drop in leakage current.

Hence, it seems that minimizing the DRV voltage of a memory is an effective means to further reductions in standby leakage power.

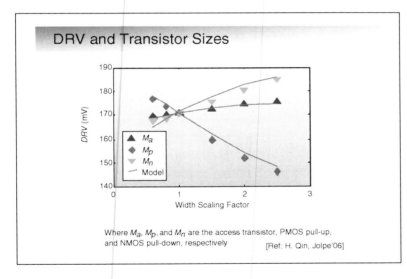

Slide 9.9
The DRV of a bit-cell depends upon a range of parameters. Intuitively we can see that the DRV would be minimized if its butterfly curve would be symmetrical – that is, that the upper and lower lobes should be of equal size. This is accomplished if the pull-up and pull-down networks (including the turned-off NMOS access transistors) are of equal strength.

From this, it becomes clear that the DRV must be a function of the sizes of transistors in the bit-cell. As the DRV voltage typically lies below the threshold voltage of the process, it means that all transistors operate in the sub-threshold mode. Under these operational conditions, the standard (strong-inversion) rationing rules between NMOS and PMOS transistors do not apply. In strong inversion, NMOS transistors are typically 2–3 times stronger than equal-sized PMOS devices owing to the higher electron mobility. In the sub-threshold region, the relative strength is determined by the leakage current parameter I_S, the threshold voltage V_{TH}, and the sub-threshold slope factor n of the respective devices. In fact, sub-threshold PMOS transistors may be substantially stronger than their NMOS counterparts.

238   Chapter #9

The influence of changing the respective transistor sizes on a generic 6T cell is shown in the slide. For this cell, increasing the size of the PMOS transistors has the largest impact on DRV. Given the strong pull-down/weak pull-up approach in most of the generic cells, this is not unexpected.

Note: Though a symmetrical butterfly curve minimizes the DRV voltage, it is most likely not the best choice from an active read/write perspective. SRAM memories provide fast read access through precharged bitlines and strong NMOS discharge transistors. This leads automatically to an asymmetrical cell.

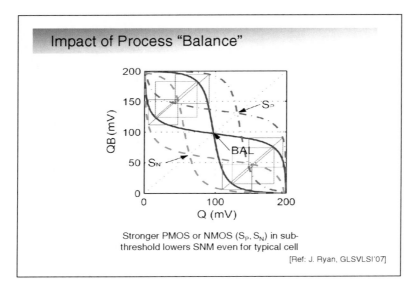

Slide 9.10
Any variation from the symmetrical bit-cell causes a deterioration of the DRV. This is illustrated in this slide where the impact of changing the relative strengths of the sub-threshold transistors is shown. Both strong NMOS (S_N) and strong PMOS (S_P) transistors warp the butterfly curves and reduce the SNM.

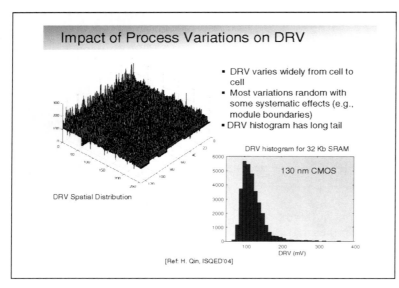

Slide 9.11
Given the high sensitivity of the DRV to the relative strengths of transistors, it should be no surprise that process variations have a major impact on the minimal operational voltage of an SRAM cell. Local variations in channel length and threshold voltages are the most important cause of DRV degradation. This is best demonstrated with some experimental results. This plot shows a 3-D rendition of the DRV of a 130 nm 32 Kb SRAM memory, with the x- and y-axis indicating the position of the cell in the array, and the z-axis denoting the value of the DRV. Local transistor variations seem to cause the largest DRV changes. Especially threshold variations play a major role.

The histogram of the DRVs shows a long tail, which means that only a few cells exhibit very high values of the DRV. This is bad news: the minimum operation voltage of a complete memory (that is, the DRV of the complete memory) is determined by the DRV of the worst-case cell, padded with some extra safety margin. This means that the DRV of this particular memory should be approximately 450 mV (with a 100 mV safety margin added), though most of the cells operate perfectly well even at 200 mV.

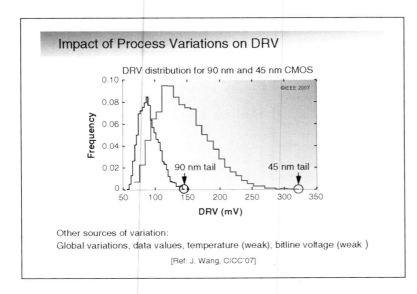

Slide 9.12

A similar picture emerges for memories implemented in the 90 nm and 45 nm (in this particular case, a 5 Kb memory). Clearly, local variations cause a DRV distribution with a long tail toward higher DRVs, and the influence of local variations increases with process technology scaling.

The DRV also depends on other factors (but less strongly so) such as global variations, the stored data values, temperature, and the bitline voltage.

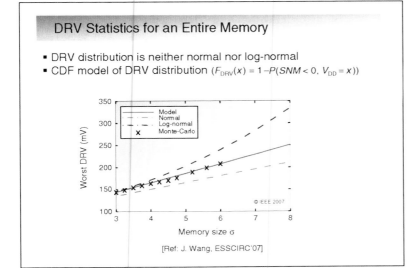

Slide 9.13

Understanding the statistical distribution of the DRV is a first and essential step toward identifying which circuit techniques would be most effective in lowering operational voltage and hence leakage. (This will be painstakingly made clear in Chapter 10, where we discuss runtime power reduction techniques).

Inspection of the DRV distribution shows that it follows neither a normal nor a log-normal model. A better match is presented by the equation shown in the slide. The resulting model matches true Monte-Carlo simulation along the DRV tail to 6σ – which means that outliers can be predicted quite effectively. The independent parameters of the model (μ_0 and σ_0 – the mean and variance of the SNM at a supply voltage V_0) can be obtained from a small Monte-Carlo simulation (at $V_{DD} = V_0$) of the SNM in the one lobe of the butterfly plot that is the most critical.

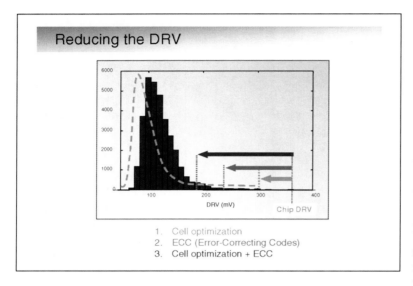

Slide 9.14
Building on the presented analysis of the DRV, its parameters and its statistics, we can devise a number of strategies to lower its value. The first approach is to use optimization. The available options are appropriate sizing of the transistors to either balance the cell or reduce the impact of variations, careful selection of the body-biasing voltages (for the same reasons), and/or playing with the peripheral voltages to compensate for unbalancing leakage currents. The net effect is to shift the DRV histogram to the left. Most importantly, the worst-case value is also lowered as is indicated in green on the chart. Nothing ever comes for free though – manipulating the DRV distribution means trading off some other metric such as area or access time. The designer must therefore weigh the importance of DRV for low-power standby mode with other design considerations.

A second approach is to lower the voltage below the worst-case value. This approach, which we will call "better-than-worst-case" design in the next chapter, may potentially lead to errors. As the tail of the distribution is long, the number of failing cells will be relatively small. The addition of some redundancy in the form of error detection can help to capture and correct these rare errors. Error-correcting (ECC) strategies have been exploited for a long time in DRAM as well as in non-volatile memories, but are not a common practice in embedded SRAMs. The potential benefits in leakage reduction and overall robustness are worth the extra overhead. From a DRV perspective, the impact of ECC is to lob off the tail of DRV distribution (as indicated in red).

Naturally, both cell optimization and ECC can be applied in concert resulting in a DRV with a lower mean and narrower distribution (indicated in blue).

Slide 9.15

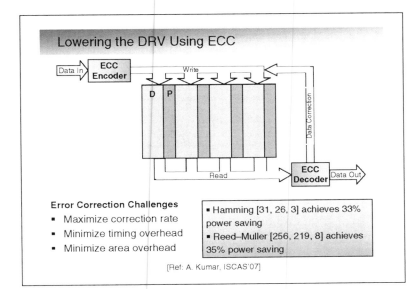

ECCs have been used in memories for a long time. Already in the 1970s, ECC had been proposed as a means to improve the yield of DRAMs. Similarly, error correction is extensively used in Flash memories to extend the number of write cycles. As indicated in the previous slides, another use of ECC is to enable "better-than-worst case", and lower the supply voltage during standby more aggressively.

The basic concept behind error detection and correction is to add some redundancy to the information stored. For instance, in a Hamming (31, 26) code, five extra parity bits are added to the original 26 data bits, which allows for the correction of one erroneous bit (or the detection of two simultaneous errors). The incurred overhead in terms of extra storage is approximately 20%. Encoder and decoder units are needed as well, further adding to the area overhead. The leakage current reduction resulting from the ECC should be carefully weighed against the active and static power of the extra cells and components.

Yet, when all is considered, ECC yields substantial savings in standby power. Up to 33% in leakage power reduction can be obtained with Hamming codes. Reed–Muller codes perform even a bit better, but this comes at the cost of a more complex encoder/decoder and increased latency.

Slide 9.16

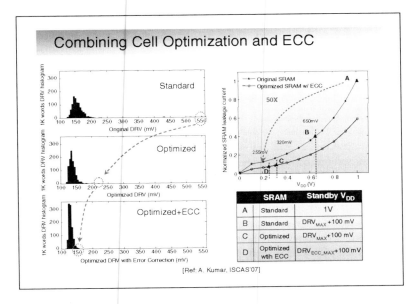

The impact of combining cell optimization and error correction is illustrated for a 26 Kb SRAM memory (implemented in a 90 nm CMOS technology). The use of a (31, 26, 3) Hamming code actually increases the total size of the memory to 31 Kb.

The optimized memory is compared with a generic implementation of the memory, integrated on the same die. In all scenarios, a guard band of 100 mV above the minimum allowed DRV is maintained. The DRV histograms illustrate how the combination of optimization and ECC both shifts the mean DRV to lower values and narrows the distribution substantially. The leakage current in standby is reduced by a factor of 50. This can be broken down as follows:

- Just lowering the supply voltage of the generic memory to its DRV + 100 mV reduces the leakage power by 75%.
- Optimizing the cell to lower the DRV yields another 90% reduction.
- Finally, the addition of ECC translates into an extra 35% savings.

For this small memory, the area penalty to accomplish this large leakage savings is quite substantial. The combination of larger cell area, additional parity bits, and encoders and decoders approximately doubles the size of the memory. Though this penalty may not be acceptable for top-of-the-line microprocessor chips with huge amounts of cache memory, the reduction in static power makes this very reasonable in ultra low-power devices with low duty cycles (such as those encountered in wireless sensor networks or implanted medical devices).

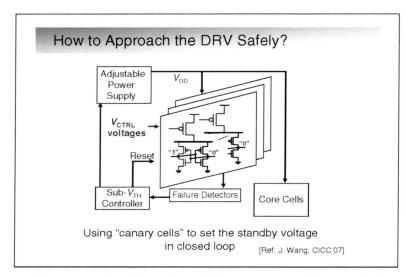

Slide 9.17

The standby voltage reduction techniques discussed so far lower the supply voltage to a value that is set at a guard band above the worst-case DRV. The latter is obtained by careful modeling, simulation, and experimental observation of the process variability. This open-loop approach means that all chips that do not suffer the worst-case DRV cannot take full advantage of the potential leakage savings. It has been widely reported that the difference in leakage current between the best- and worst-case instances of the same design can vary by as much as a factor of 30.

One way to get around this is to use a closed-loop feedback approach, which promises to increase the leakage savings for every chip. The idea is to measure the distributions on-line, and set the standby voltage accordingly. The measurements are provided by a set of "canary replica cells" added to the memory (as in "the canary in the coal mine" strategy used to detect the presence of noxious fumes in mines in older times).

The canary cells are intentionally designed to fail across a range of voltages above the DRV distribution of the core SRAM cells. Based on the knowledge of the shape underlying the SRAM-cell DRV distribution (using models such as the one presented in Slide 9.13), the feedback loop uses the measured data to dynamically set the supply voltage.

The diagram shows a prototype architecture that organizes the canary cells in banks. Each canary cell is structured and sized like the bit-cells in the core array, except that an additional PMOS header switch presents a lower effective V_{DD} to the cell. Controlling the gate voltage of the PMOS headers (V_{CTRL}) allows us to set the DRV of the canary cells across a wide range of voltages.

Note: The "canary" approach is a first example of a runtime power reduction technique, which is the topic of Chapter 10.

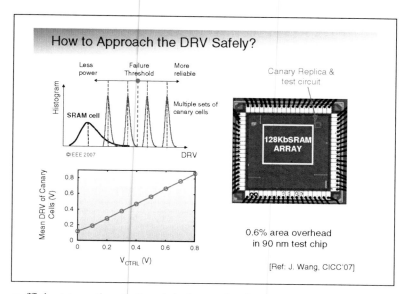

Slide 9.18

The concept of how canary cells can be used to estimate the "safe" operational voltage is illustrated in the top left drawing. The cells are divided into clusters, tuned to fail at regular intervals above the average DRV of the core cells. To reduce the spread of DRV distribution of the canary cells relative to the core, larger sizes are used for canary transistors. Of course, the small set of canary cells cannot track the local variations in the main array, but it is sufficient to estimate the global ones (such as systematic variations or the impact of temperature changes), and hence remove a large fraction of the guard band.

By varying V_{CTRL} (e.g., by providing different values of V_{CTRL} to different banks) and measuring the failure point, an estimate of the safe value of the minimum operational voltage is obtained. The measured relationship between the DRV of the canary cells and V_{CTRL} is shown in the lower-left plot, demonstrating clearly that V_{CTRL} is a good measure for the DRV value.

A 90 nm test chip implements the canary-based feedback mechanism at a cost of 0.6% area overhead. Measurements confirm that the canary cells reliably fail at voltages higher than the average core cell voltage and that this relationship holds across environmental changes. This approach helps to reduce leakage power by factors of up to 30 compared to a guard band approach.

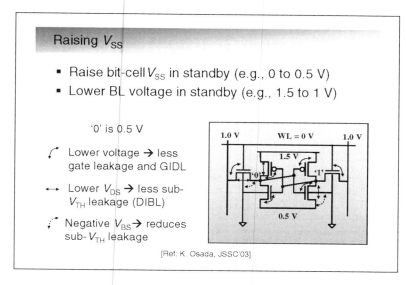

Slide 9.19

All standby power reduction techniques discussed so far are based on lowering the V_{DD}. An alternative approach is to raise the ground node of the bit-cells, V_{SS}. This approach decreases V_{DS} across a number of transistors, which lowers sub-threshold conduction (due to DIBL) as well as the GIDL effect. Furthermore, for bulk NMOS devices, the higher V_{SS} causes a negative V_{BS} that increases the threshold voltage of the transistors, and lowers the sub-threshold current exponentially. The cell presented in this slide exploits all of these effects. The choice between raising V_{SS} and lowering V_{DD} depends primarily on the dominant sources of leakage in a given technology and on the relative overhead of the different schemes.

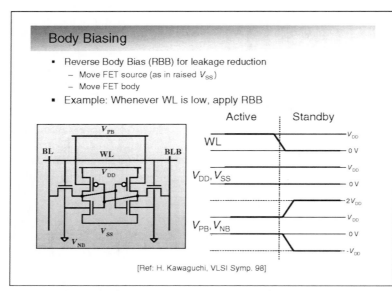

Slide 9.20

Another option is to intentionally apply reverse body biasing (RBB) to the transistors in the cell during standby mode. Again, an increase in threshold voltage translates into an exponential decrease in sub-threshold drain–source leakage current, which makes it a powerful tool for lowering standby currents.

To induce RBB, you can either raise the source voltage (as in raised-V_{SS} approach of Slide 9.19) or lower the body voltage for an NMOS. In traditional bulk CMOS, modulating the NMOS body node means driving the full capacitance of the P-type substrate. Transitioning in and out of standby mode hence comes with a substantial power overhead. Changing the body voltage of the PMOS is relatively easier because of the smaller-granularity control offered by the N-well. Many bulk technologies now offer a triple-well option that allows for the placement of NMOS transistors in a P-well nested inside an N-well. This option makes adjustable RBB for standby mode more attractive, but the energy involved in changing the voltage of the wells must still be considered.

This slide shows an RBB scheme that raises and lowers the PMOS and NMOS bulk voltages, respectively, whenever a row is not accessed. The advantage of this approach is that it operates at a low level of granularity (row-level), in contrast to all techniques discussed previously, which work on a per-block level. In general, at most a single row of a memory module is accessed at any given time. The penalty is an increase in read and write access times.

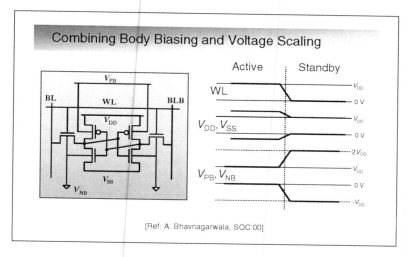

[Ref: A. Bhavnagarwala, SOC'00]

Slide 9.21

Body biasing is a technique that can easily be deployed in conjunction with other standby power reduction methods. This slide, for example, shows an SRAM that combines body biasing and supply voltage scaling. During active mode, the V_{DD} and V_{SS} rails for the accessed cells are set at slightly more positive and negative, respectively, than during standby. At the same time, the body terminals of the transistors are driven to 0 and V_{DD} such that the cell transistors have a slight forward body bias (FBB). The reduced V_{TH} improves the read/write access times. In standby mode, the power rails are pinched inward and RBB is applied. The combination of voltage scaling and body bias potentially provides for a dramatic reduction in standby power. However, one has to ensure that the double overhead of supply and body voltage scaling does not offset the gains. Also, one has to make sure that the source/drain diodes are not forward biased in FBB mode.

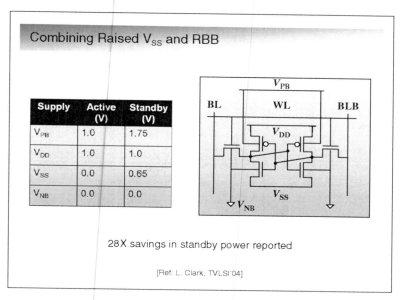

28X savings in standby power reported

[Ref: L. Clark, TVLSI'04]

Slide 9.22

Similarly we can combine the raised-V_{SS} approach with RBB. During standby, the raised-V_{SS} node reduces the effective supply voltage of the cell, while providing RBB for the NMOS transistors. A raised N-well voltage provides RBB to the PMOS devices. The advantage of this approach is that a triple-well technology is not required.

Slide 9.23

From Chapters 7 and 9 emerges a wide spectrum of choices in setting the voltages in SRAMs during active and standby modes. The design parameters include the choice of not only the supply and well voltages, but also the peripheral voltages such as wordline and bitline voltages. A literature survey illustrates the broad range of options available.

Voltage Scaling in and Around the Bitcell

Large number of reported techniques

Voltage	Approach	Source(s)
Bitcell V_{CC}	lower in active (e.g. DVS)	[1]
	lower in standby	[2][3][4][5][6][7]
	raise always	[8][9]
	raise for read access	[5][9]
	float or lower for write	[5][10]
	float for read access	[10]
	raise in standby	
Bitcell V_{SS}	raise in standby	[6][7][11][12][13][14][15]
	raise or float for write access	[16]
	lower for read access	[9]
Wordline (WL)	negative for standby	[4][10]
WL driver V_{DD}	lower in standby	[7]
Well-biasing	change with mode	[4][9]
Bitline V_{DD}	lower for standby	[12]

[1] K. Osada et al. JSSC 2001
[2] N. Kim et al. TVLSI 2004
[3] H. Qin et al. ISQED 2004
[4] K. Kanda et al. ASIC/SOC 2002
[5] A. Bhavnagarwala et al. SymVLSIC 2004
[6] T. Enomoto et al. JSSC 2003
[7] M. Yamaoka et al. SymVLSIC 2002
[8] M. Yamaoka et al. ISSC 2004
[9] A. Bhavnagarwala et al. ASIC/SOC 2000
[10] K. Itoh et al. SymVLSIC 1996
[11] H. Yamauchi et al. SymVLSIC 1996
[12] K. Osada et al. JSSC 2003
[13] K. Zhang et al. SymVLSIC 2004
[14] K. Nii et al. ISSCC 2004
[15] A. Agarwal et al. JSSC 2003
[16] K. Kanda et al. JSSC 2004

In essence, each of these approaches follows the same principles:

- For each operational mode, voltage values are selected to minimize power while ensuring functionality and reliability. The latter means that noise margin and DRV constraints must be met in active and standby mode, respectively. In addition, the impact on read and write access times as well as on memory area must be kept within bounds.
- Transition between modes often means that multiple voltages must be adopted. The overhead in time and power of these transitions should be carefully weighed against the gains.

Anyone who has ever designed SRAMs knows that the impact of a change in the cell or the periphery can be quite subtle. Although the different techniques presented here may seem to yield huge benefits, a careful analysis including intensive simulation and actual prototyping is absolutely essential in defining the ultimate benefits and pitfalls.

Periphery Breakdown

- Periphery leakage often not ignorable
  - Wide transistors to drive large load capacitors
  - Low-V_{TH} transistors to meet performance specs
- Chapter 8 techniques for logic leakage reduction equally applicable, but ...
- Task made easier than for generic logic because of well-defined structure and signal patterns of periphery
  - e.g., decoders output 0 in standby
- Lower peripheral V_{DD} can be used, but needs fast level-conversion to interface with array

Slide 9.24
As we had mentioned in Chapter 7, the peripheral circuits that go around the SRAM array primarily consist of combinational logic (examples are the write drivers, row and column decoders, I/O drivers). Most of these circuits can be disabled during standby mode, and their leakage can be reduced using the techniques from Chapter 8. However, there are some characteristics of the SRAM periphery circuits that differentiate them from generic logic and thus deserve mentioning.

- Although the majority of transistors in an SRAM are situated in the memory array, the SRAM periphery can still contribute a sizable amount of leakage. This can be attributed to the fact that most components of the periphery must be sized fairly large to drive the large capacitances inside the array (e.g., wordline and bitlines). These wide transistors come with large leakage currents.

- Whereas the SRAM bit-cells typically use high-threshold transistors, performance considerations dictate the use of low-threshold devices in the periphery.
- From our discussion, it had become clear that memory cells and logic are on somewhat different voltage-scaling trajectories. Logic supply voltages are expected to keep on scaling downward, whereas reliability concerns in the presence of increasing process variations force the voltages in memory to stay constant (if not increasing). Interfacing between periphery and memory core hence increasingly requires the presence of voltage-up and -down converters, which translates into a timing and power overhead. Moreover, this interface must be properly conditioned in standby mode. For example, floating wordlines caused by power gating of the periphery, could cause data loss in the bit-cells.

On the other hand, a sizable number of the generic standby power management techniques introduced in Chapter 8 perform even better when applied to memory periphery. This is largely due to the well-behaved repetitive structure of the peripheral circuits. In addition, many of the signal voltages during standby are known very well. For instance, we know that all of the wordlines must be 0 in standby. This knowledge makes it easier to apply power gating or forced stacking to maximally reduce the leakage power.

Summary and Perspectives

- SRAM standby power is leakage-dominated
- Voltage knobs are effective to lower power
- Adaptive schemes must account for variation to allow outlying cells to function
- Combined schemes are most promising
  - e.g., Voltage scaling and ECC
- Important to assess overhead!
  - Need for exploration and optimization framework, in the style we have defined for logic

Slide 9.25

In summary, SRAM leakage power is a dominant component of the overall standby power consumption in many SoCs and general-purpose processing devices. For components that operate at low duty cycles, it is often THE most important source of power consumption. In this chapter, we have established that the most effective knobs in lowering leakage power are the various voltages that drive the bit-cells. However, these voltages must be manipulated carefully so that data preservation is not endangered.

As with active operation, the large number of small transistors in an embedded SRAM means that the far tails of power and functionality distributions drive the design. This means that any worst-case or adaptive schemes must account for the outliers on the distributions to preserve proper SRAM functionality. The most promising schemes for leakage reduction combine several different voltage-scaling approaches (selected from the set of V_{TH}, V_{DD}, V_{SS}, and well and periphery voltages) along with architectural changes (e.g., ECC). In all of these approaches, the overhead requires careful attention to ensure that the overall leakage savings are worth the extra cost in area, performance, or overhead power.

All this having been said, one cannot escape the notion that some more dramatic steps may be needed to improve the long-term perspectives of on-chip memory. Non-volatile memory structures that are compatible with logic processes and that do not require high voltages present a promising venue. Their non-volatile nature effectively eliminates the standby power concern. However, their write and (sometimes) read access times are substantially longer than what can be obtained with SRAMs. It is worth keeping an eye on the multitude of cell structures that are currently trying to make their way out of the research labs.

Slides 9.26–9.28
Some references . . .

References

Books and Book Chapters:
- K. Itoh, M. Horiguchi and H. Tanaka, *Ultra-Low Voltage Nano-Scale Memories*, Springer 2007.
- T. Takahawara and K. Itoh, "Memory Leakage Reduction," in *Leakage in Nanometer CMOS Technologies*, S. Narendra, Ed, Chapter 7, Springer 2006.

Articles:
- A. Agarwal, L. Hai and K. Roy, "A single-V/sub t/low-leakage gated-ground cache for deep submicron," *IEEE Journal of Solid-State Circuits*, pp. 319–328, Feb. 2003.
- A. Bhavnagarwala, A. Kapoor, J. Meindl, "Dynamic-threshold CMOS SRAM cells for fast, portable applications," *Proceedings of IEEE ASIC/SOC Conference*, pp. 359–363, Sep. 2000.
- A. Bhavnagarwala et al., "A transregional CMOS SRAM with single, logic $V/sub\ DD$/and dynamic power rails," *Proceedings of IEEE VLSI Circuits Symposium*, pp. 292–293, June 2004.
- L. Clark, M. Morrow and W. Brown, "Reverse-body bias and supply collapse for low effective standby power," *IEEE Transactions on VLSI*, pp. 947–956, Sep. 2004.
- T. Enomoto, Y. Ota and H. Shikano, "A self-controllable voltage level (SVL) circuit and its low-power high-speed CMOS circuit applications," *IEEE Journal of Solid-State Circuits*, 38(7), pp. 1220–1226, July 2003.
- K. Flautner et al., "Drowsy caches: Simple techniques for reducing leakage power"., *Proceedings of ISCA 2002*, pp. 148–157, Anchorage, May 2002.
- K. Itoh et al., "A deep sub-V, single power-supply SRAM cell with multi-VT, boosted storage node and dynamic load," *Proceedings of VLSI Circuits Symposium*, pp. 132–133, June, 1996.
- K. Kanda, T. Miyazaki, S. Min, H. Kawaguchi and T. Sakurai, "Two orders of magnitude leakage power reduction of low voltage SRAMs by row-by-row dynamic Vdd control (RRDV) scheme," *Proceedings of IEEE ASIC/SOC Conference*, pp. 381–385, Sep. 2002.

References (cont.)

- K. Kanda, et al., "90% write power-saving SRAM using sense-amplifying memory cell," *IEEE Journal of Solid-State Circuits*, pp. 927–933, June 2004
- H. Kawaguchi, Y. Itaka and T. Sakurai, "Dynamic leakage cut-off scheme for low-voltage SRAMs," *Proceedings of VLSI Symposium*, pp. 140–141, June 1998.
- A. Kumar et al., "Fundamental bounds on power reduction during data-retention in standby SRAM," *Proceedings ISCAS 2007*, pp. 1867–1870, May 2007.
- N.Kim, K. Flautner, D. Blaauw and T. Mudge, "Circuit and microarchitectural techniques for reducing cache leakage power," *IEEE Transactions on VLSI*, pp. 167–184, Feb. 2004 167–184
- Y. Nakagome et al., "Review and prospects of low-voltage RAM circuits," *IBM J. R & D*, 47(516), pp. 525–552, Sep./Nov. 2003.
- K. Osada, "Universal-Vdd 0.65–2.0-V 32-kB cache using a voltage-adapted timing-generation scheme and a lithographically symmetrical cell," *IEEE Journal of Solid-State Circuits*, pp. 1738–1744, Nov. 2001.
- K. Osada et al., "16.7-fA/cell tunnel-leakage-suppressed 16-Mb SRAM for handling cosmic-ray-induced multierrors," *IEEE Journal of Solid-State Circuits*, pp. 1952–1957, Nov. 2003.
- H. Qin, et al., "SRAM leakage suppression by minimizing standby supply voltage," *Proceedings of ISQED*, pp. 55–60, 2004.
- H. Qin, R. Vattikonda, T. Trinh, Y. Cao and J. Rabaey, "SRAM cell optimization for ultra-low power standby," *Journal of Low Power Electronics*, 2(3), pp. 401–411, Dec. 2006.
- J. Ryan, J. Wang and B. Calhoun, "Analyzing and modeling process balance for sub-threshold circuit design" *Proceedings GLSVLSI*, pp. 275–280, Mar. 2007.
- J. Wang and B. Calhoun, "Canary replica feedback for Near-DRV standby VDD scaling in a 90 nm SRAM," *Proceedings of Custom Integrated Circuits Conference (CICC)*, pp. 29–32, Sep. 2007.

References (cont.)

- J. Wang, A. Singhee, R. Rutenbar and B. Calhoun, "Statistical modeling for the minimum standby supply voltage of a full SRAM array", *Proceedings of European Solid-State Circuits Conference (ESSCIRC)*, pp. 400–403, Sep. 2007.
- M. Yamaoka et al., "0.4-V logic library friendly SRAM array using rectangular-diffusion cell and delta-boosted-array-voltage scheme, *Proceedings of VLSI Circuits Symposium*, pp. 13–15, June 2002.
- M. Yamaoka, et al., "A 300 MHz 25 µA/Mb leakage on-chip SRAM module featuring process-variation immunity and low-leakage-active mode for mobile-phone application processor," *Proceedings of IEEE Solid-State Circuits Conference*, pp. 15–19, Feb. 2004.
- K. Zhang et al., "SRAM design on 65 nm CMOS technology with integrated leakage reduction scheme," *Proceedings of VLSI Circuits Symposium*, 2004, pp. 294–295, June 2004.

Chapter 10
Optimizing Power @ Runtime – Circuits and Systems

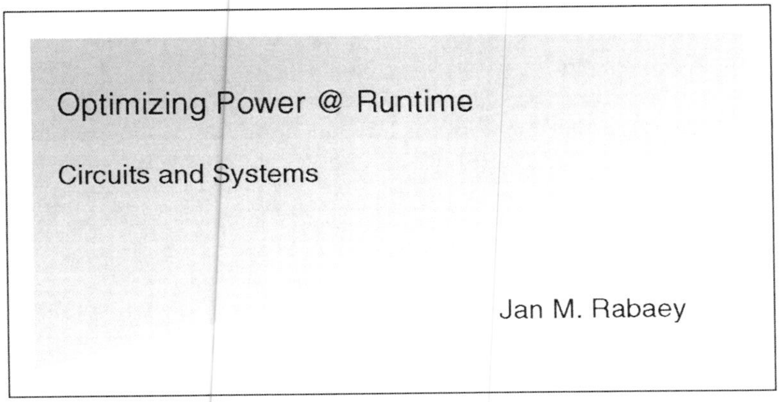

Slide 10.1

Optimizing Power @ Runtime

Circuits and Systems

Jan M. Rabaey

The computational load and hence the activity of a processor or an SoC may change substantially over time. This has some profound repercussions on the design strategy for low power, as this means that *the optimal design point changes dynamically*. The standby case, discussed in the previous chapters, is just a special case of these dynamic variations (with the activity dropping to zero). The concept of runtime optimization in the energy–delay space presents a fundamental departure from traditional design methodology, in which all design parameters such as transistor sizes and supply and threshold voltages were set by the designer or the technology, and remained fixed for the lifetime of the product. Though runtime optimization creates some unique opportunities, it also presents some novel challenges.

Slide 10.2

Chapter Outline

- Motivation behind runtime optimization
- Dynamic voltage and frequency scaling
- Adaptive body biasing
- General self-adaptation
- Aggressive deployment
- Power domains and power management

The Chapter starts by motivating the need for dynamic adaptation. A number of different strategies to exploit the variation in activity or operation mode of a design are then described in detail. Dynamic voltage- and body-bias scaling are the best-known examples. Increasingly, it becomes necessary to dynamically adjust a broad range of design parameters, leading to a self-adapting approach. In the extreme case, one can even adjust the design outside the safe operation zone to further save energy. This approach is called "aggressive deployment" or "better than worst-case" design. Finally, we discuss how the adoption of these runtime techniques leads to the need for a power management system, or, in other words, "a chip operating system".

Slide 10.3

Why Runtime Optimization for Power?

- Power dissipation strong function of activity
- In many applications, activity varies strongly over time:
  - Example 1: Operational load varies dramatically in general-purpose computing. Some computations also require faster response than others.
  - Example 2: The amount of computation to be performed in many signal processing and communication functions (such as compression or filtering) is a function of the input data stream and its properties.
- Optimum operation point in the performance–energy space hence varies over time
- Changes in manufacturing, environmental, or aging conditions also lead to variable operation points

Designs for a single, fixed operation point are sub-optimal

Activity variations and their impact on power are an important reason why runtime optimization had become an attractive idea in the late 1990s. Since then, other important sources of dynamic variations have emerged. Device parameters change over time owing to aging or stress effects, or owing to varying environmental conditions (e.g., temperature). Changes in current loads cause the supply rails to bounce up and down. These added effects have made runtime optimization over the available design parameters even more attractive. Sticking to a single operational point is just too ineffective.

Slide 10.4

To illustrate just how much workloads can vary over time, let us consider the case of a video compression module. A fundamental building block of virtually every compression algorithm is the motion compensation block, which computes how much a given video frame differs from the previous one and how it has changed. Motion compensation is one of the most computationally intensive functions in video compression algorithms such as MPEG-4 and H.264.

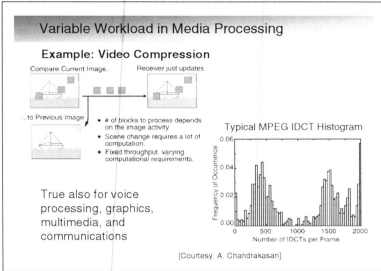

One can intuitively understand that the motion compensation module has to work a lot harder in a fast-moving car chase scene than in a slow pan of a nature landscape. This is clearly illustrated in the chart of the lower-right corner, which plots a histogram of the number of IDCTs (inverse discrete cosine transforms) that have to be performed per frame. The distribution is strongly bi-modal. It also shows that the computational effort per frame can vary over 2–3 orders of magnitude.

Slide 10.5

The same broad distribution holds for general-purpose computing as well. Just watch the "CPU Usage" chart of your laptop for a while. Most of the time, the processor runs at about 2–4% utilization, with occasional computational bursts extending all the way to 100% utilization. Identical scenarios can be observed for other computer classes, such as desktops, workstations, file servers, and data centers. When observing these utilization traces, it

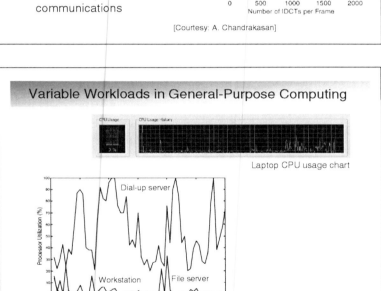

becomes quite obvious that there must be an opportunity to exploit the periods of low activity to reduce energy dissipation.

Slide 10.6

As stated in earlier chapters, the variation in activity moves or changes the optimal E–D curve. In addition, the delay expectation may change as well depending upon operating modes. The optimal operation point hence moves, which means that an energy-efficient design should adapt itself to the changing conditions. Unfortunately, the number of knobs that are available to the designer of the runtime system is restricted. Of the traditional design parameters, only supply and threshold voltages are truly available. Dynamically changing the transistor sizes is not very practical or

> **Adapting to Variable Workloads**
>
> - Goal: Position design in optimal operational point, given required throughput
> - Useful dynamic design parameters: V_{DD} and V_{TH}
>   - Dynamically changing transistor sizes non-trivial
> - Variable supply voltage most effective for dynamic power reduction

effective. One additional parameter we may consider is the dynamic adaptation of the clock frequency.

> **Adjusting Only the Clock Frequency**
>
> - Often used in portable processors
> - Only reduces power – leaves energy per operation unchanged
>   - Does not save battery life
>
> Compute ASAP — Excess throughput
> Always high throughput
>
> Clock Frequency Reduction
> f_{CLK} Reduced
>
> Energy per operation remains unchanged whereas throughput scales down with f_{clk}
>
> [Ref: T. Burd, UCB'01]

Slide 10.7
Consider, for instance, the case of the microprocessor embedded in a laptop computer. Assume that the computational tasks can be divided into high-performance tasks with short latency requirements, and background tasks, where the latency is not that important. A processor that runs at a fixed frequency and voltage executes both types of tasks in the same way – this means that the high-performance task is executed within specifications (as shown by the dotted lines), whereas the low-end task is performed way too fast. Executing the latter slower would still meet specifications, and offers the opportunity for power and energy savings.

One approach that was adopted by the mobile-computing industry early on is to turn down the clock frequency when the computer is operating on battery power. Lowering the clock frequency reduces the power dissipation ($P = CV^2f$) proportional to the frequency reduction (assuming that leakage is not a factor). However, it comes with two disadvantages:

- The reduced-frequency processor does fine with the high-latency tasks, but fails to meet the specifications for the high-performance functions;
- Though it scales power, this approach does not change the energy per operation. Hence, the amount of work that can be performed on a single battery charge remains the same.

Optimizing Power @ Runtime – Circuits and Systems

The first concern can be addressed by changing the frequency dynamically in response to the presented workload. *Dynamic Frequency Scaling* (or DFS) makes sure that performance requirements are always met ("just in time computing"), but misses out on the energy reduction opportunity.

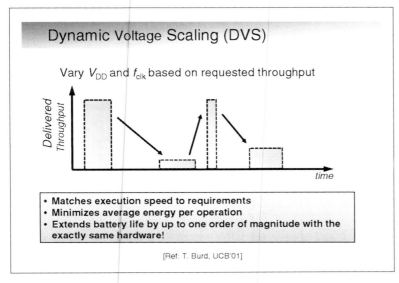

Slide 10.8

A more effective way of exploiting the reduced workload is to scale clock frequency and supply voltage simultaneously (called *dynamic voltage scaling*, or DVS). The latter is possible as frequency scaling allows for higher delays, and hence reduced supply voltages. Whereas pure frequency scaling does reduce power linearly, the additional voltage scaling adds a quadratic factor, and reduces not only the average power but also the energy per operation, while meeting all the performance needs.

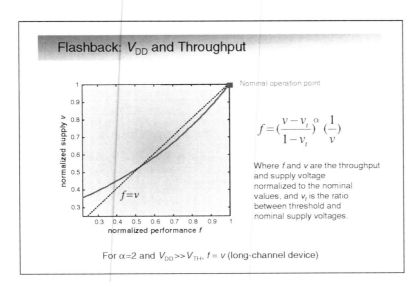

Slide 10.9

To analyze the effectiveness of DVS, it is worth revisiting the relationship between supply voltage and delay (or clock frequency). Using the α–delay expression introduced in Chapter 4, and normalizing supply voltage and frequency to their nominal values, an expression between the normalized frequency and required supply voltage can be derived. For long-channel devices, a linear dependency between frequency and voltage is observed. In short-channel transistors, the supply voltage initially scales super-linearly, but the effect saturates for larger reductions in clock frequency.

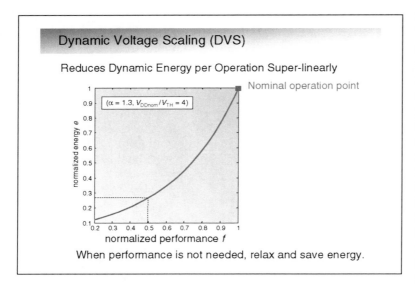

Slide 10.10
The results of the previous slide can now be used to compute the energy savings resulting from simultaneous supply and frequency scaling. The resulting chart clearly demonstrates that DVS reduces energy per operation super-linearly. Reductions of the clock frequency by factors of 2 and 4 translate into energy savings by factors of 3.8 and 7.4, respectively (in a 90 nm CMOS technology). Scaling only the frequency would have left the energy per operation unchanged – and taking leakage into account, it might even go up!

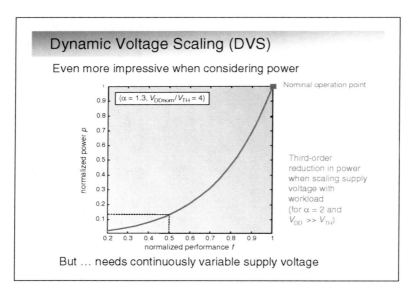

Slide 10.11
The impact of DVS is even more impressive when considering power dissipation, where an almost third-order reduction can be observed. More precisely, scaling of the clock frequency by factors of 2 and 4 translates into energy savings by factors of 7.8 and 30.6, respectively. Scaling only the clock frequency, leaving the voltage unchanged, would have led to a linear scaling – hence power would have been reduced by factors of 2 and 4, respectively.

The dynamic voltage scaling approach, though very attractive, comes with one major setback: it requires a supply voltage that can be adjusted continuously!

Slide 10.12
Having to adjust the supply voltage adaptively and continuously may or may not present a substantial overhead depending upon the system the processor is embedded in. Most microprocessor mother boards include sophisticated voltage regulators that allow for a range of

Optimizing Power @ Runtime – Circuits and Systems

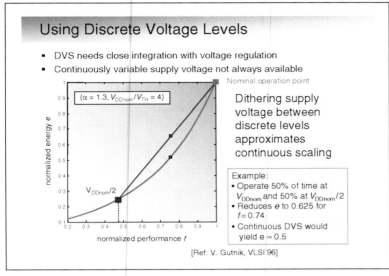

programmable output voltages. However, in other systems the cost of such a regulator might be just too high. Fortunately, the benefits of DVS can also be obtained when only a few discrete supply voltages are available. By *dithering* the module between the different voltages (i.e., flipping between them on a periodic basis), continuous voltage scaling can be emulated. The resulting energy dissipation per operation now lies on the linear interpolation between the different discrete-voltage operation points. The percentage of time spent at each voltage determines the exact operation point. Adding more discrete supply voltages allows for a closer approximation of the continuous DVS curve. This approach is often called *voltage hopping* or *voltage dithering*.

For example, if only one extra supply (at $V_{DD}/2$) is available in addition to the nominal supply, spending equal time at both supplies reduces the energy by a factor of 1.6 (instead of the factor 2 that would have resulted from continuous scaling).

Slide 10.13

A DVS system can be considered as a closed control system: based on the observed workload, the supply voltage and operational speed gets adjusted, which in its turn determines how fast the workload gets absorbed. The main challenge in the design of such a system is that the changes in supply voltages do not occur instantaneously – some delay is involved in ramping the large capacitance on the supply rail up or down – and that some energy overhead is incurred in changing the rails.

Hence, measuring and predicting the workload accurately is important. Misestimations can substantially reduce the efficiency of the DVS approach. How to perform workload estimation depends upon the application at hand.

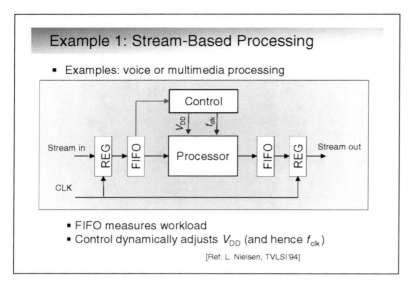

Slide 10.14

Stream processing is a particular class of applications where the workload estimation is relatively straightforward. The video-compression example of Slide 10.4 belongs to this class, and so do audio and voice compression, synthesis, and recognition. In stream processing, new samples are presented at a periodic rate. When buffering the incoming samples into a FIFO, the utilization of the FIFO is a direct measure of the presented workload. When it is close to full, the processor should speed up; when it empties, the processor can slow down. An output FIFO then translates the variable processing rate into the periodic signal that may be required by the playback device or the communication channel.

Buffer utilization is only one measure of the workload. Its disadvantage is that it comes with extra latency. More sophisticated estimators can be envisioned. For instance, many signal processing and communication algorithms allow for the construction of simple and quick estimators of the computational effort needed. The outcome of these can then be used to control the voltage–frequency loop.

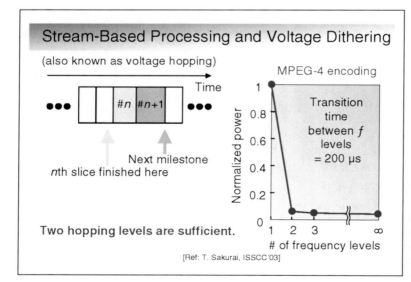

Slide 10.15

The effectiveness of this approach is shown for the case of an MPEG-4 encoder. Each time a new frame arrives, the "scheduler" estimates the amount of work to be performed before the next milestone, and the voltage is adjusted accordingly. In this particular example, the designers choose to use voltage dithering. Analysis showed that just two discrete voltages were sufficient to reap most of the benefits. A power reduction by a factor of 10 was obtained.

Relating V_{DD} and f_{clk}

- **Self-timed**
  - Avoids clock altogether
  - Supply is set by closed loop between V_{DD} setting, processor speed, and FIFO occupation
- **On-Line Speed Estimation**
  - Closed loop compares desired and actual frequency
  - Needs "dummy" critical path to estimate actual delay
- **Table Look-up**
  - Stores relationship between f_{clk} (processor speed) and V_{DD}
  - Obtained from simulations or calibration

Slide 10.16
Another challenge of the DVS approach is how to translate a given "request for performance" into voltage, and subsequently into frequency. One option is to use a self-timed approach (as was adopted in the first ever published DVS paper by Nielsen in 1994). This effectively eliminates the voltage-to-frequency translation step. The performance-to-voltage translation is performed by the closed control loop. Important design choices still need to be made: what voltage steps to apply in response to performance requests, and how fast to respond.

In the synchronous approach, the voltage-to-frequency translation can also be achieved dynamically using a closed-loop approach as well. A dummy delay line, mimicking the worst-case critical path, translates voltage into delay (and frequency).

A third approach is to model the voltage–frequency relationship as a set of equations, or as a set of empirical parameters stored in a look-up table. The latter can be obtained by simulation or from measurements when the chip is started up. To account for the impact of variations caused by temperature changes or device aging, the measurements can be repeated on a periodic base. Table look-up can also be used to translate computational requirements into frequency needs.

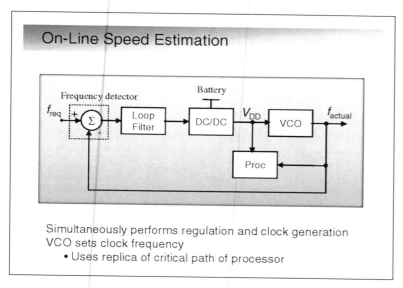

Simultaneously performs regulation and clock generation
VCO sets clock frequency
• Uses replica of critical path of processor

Slide 10.17
The closed-loop approach to set the voltage and the frequency is illustrated in this Figure. The difference between the desired and actual clock frequencies is translated into a control signal for the DC–DC converter (after being filtered to avoid rapid fluctuations). The voltage is translated into a clock frequency by the VCO, which includes a replica of the critical path to ensure that all the timing constraints in the processor or computational module are met.

Anyone familiar with phase-locked loops (PLLs) recognizes this scheme. It is indeed very similar to the PLLs commonly used in today's microprocessors and SoCs to set the clock phase and frequency. The only difference here is that the loop sets the supply voltage as well.

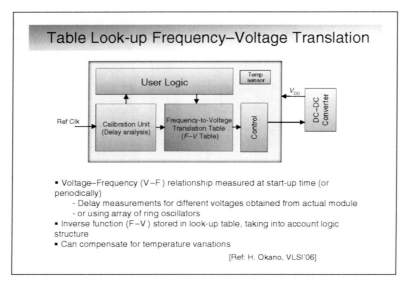

Slide 10.18
Although the replica critical path approach is an effective and simple way of relating supply voltage and clock frequency, it faces some important challenges in the deep-submicron age. With process variations causing different timing behavior at different locations on the chip, a single replica circuit may not be representative of what transpires on the die. One option is to combine the results of many distributed replica circuits, but this rapidly becomes complex.

An alternative approach is to calibrate the design at start-up time and record the voltage–frequency relationships (compared to an accurate reference clock) in a table. (It is also possible to create the table in advance using simulations.) One possible implementation of this calibration procedure is to apply a sequence of voltages to the logic module and measure the resulting delay. The inverse function, stored in a look-up table, can then be used to translate the requested frequency into the corresponding supply voltage. This approach can accommodate the impact of temperature changes, as the delay–temperature relationship is known or can be determined by simulation in advance. This information can be used to recalibrate the look-up table. A conceptual diagram of a table-based system is shown in this slide.

Another calibration option is to use an array of ring oscillators of different sizes, effectively measuring the actual process parameters. This process information can then be translated into a voltage using a $P-V$ (process–voltage) table, which is obtained in advanced using simulation. This approach, which was proposed in [Okano, VLSI06], has the advantage that it fully orthogonalizes design- and process-dependent factors.

Slide 10.19
All the above considerations are equally valid for general-purpose processors. The major difference is that the workload-to-voltage (or frequency) translation is typically performed in software. Actually, it becomes an operating-system function, the task of which is to translate a set of computational deadlines into a schedule that meets the timing requirements. The processor frequency is just one of the extra knobs that can be added to the set the OS has at hand.

A simple way of estimating the desired clock frequency is to divide the expected number of cycles needed for the completion of the task(s) by the allotted time. For the example of the MPEG

Optimizing Power @ Runtime – Circuits and Systems

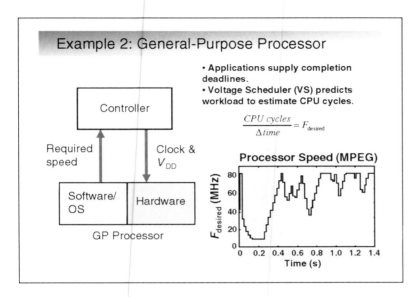

encoder, for instance, the desired clock frequency can be obtained empirically or through simulation.

Multi-tasking and the intricacies of modern microprocessors make this translation a lot more complex, however, and more complex workload and frequency estimates are needed.

Slide 10.20

The effectiveness of DVS in general-purpose processing is by no small means determined by the quality of the voltage-scheduling algorithm. Task scheduling has been studied intensively in a number of fields such as operations research, real-time operating systems, and high-level synthesis. Much can be learned from what has been developed in those fields.

The maximum savings would be obtained by a so-called Oracle scheduler, which has perfect foreknowledge of the future (as well as all the costs and overheads incurred by a change in voltage), and hence can determine the perfect voltage for every task at hand. The quality of any other scheduling algorithm can be measured by how close it gets to the Oracle schedule.

The worst-performing scheduler is the "ASAP" approach. This puts the processor in idle mode (and associated voltage) when there is no activity, and ramps to the maximum voltage whenever a computation has to be performed.

Most practical scheduling algorithms rely on a number of heuristics to determine the order of task execution as well as to select the design parameters. An example of such is the "zero" algorithm that was proposed in [Pering99].

From the results presented in the table, a number of interesting observations can be drawn.

- The savings that can be obtained by DVS and voltage scheduling are strongly dependent upon the applications at hand. The largest savings occur for applications that do not stress the processor, such as user-interface interactions or audio processing. On the other hand, only small gains are made for demanding applications such as video (de)compression. This is why it is worthwhile to farm these applications out to co-processors, as discussed in Chapter 5.
- Investing in a good scheduling algorithm is definitely worthwhile. Getting within a few percentiles from the Oracle scheduler is possible.

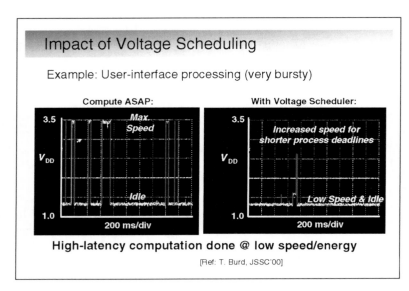

Slide 10.21
The impact of using different scheduling algorithms is illustrated in this slide for a "bursty" user-interface application. The supply voltage (and hence energy) levels as obtained by the "ASAP" and "zero" algorithms are compared. The latter raises the supply voltage only rarely above the minimum, does so only for latency-sensitive tasks, and never needs the maximum voltage.

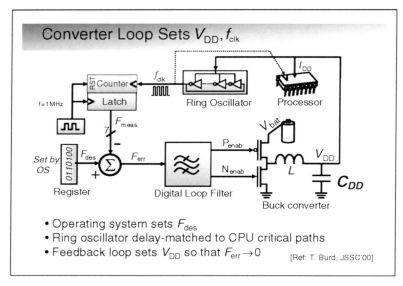

Slide 10.22
The voltage- and frequency-setting loop for general-purpose processors pretty much follows the scheme detailed in Slide 10.17. The actual clock frequency F_{CLK} is translated into a digital number by a counter–latch combination (sampled at a frequency of 1 MHz, in this example). The result is compared with the desired frequency, as set by the operating system. The goal of the feedback loop is to get the error frequency F_{ERR} as close as possible to zero. After filtering, the error signal is used to drive the DC–DC converter (which is an inductive buck converter in this particular case). The supply voltage is translated into the clock frequency F_{CLK} with the aid of a ring oscillator, which matches the critical path of the processor [Burd'00].

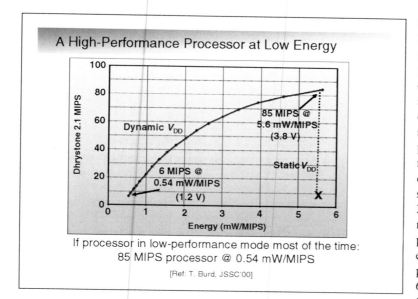

Slide 10.23

Instead of representing a single operational point, a DVS processor moves dynamically in the energy–delay space. This is illustrated quite nicely with the graph shown in this slide, which plots the operational performance versus the energy, for one of the first DVS processors, published at ISSCC in 2000. Implemented in a 600 nm technology(!), the same processor can implement either an 85 MIPS ARM processor operating at 6.5 µJ/instruction or a 6 MIPS processor at 0.54 µJ/instruction, or anything in between. If the duty cycle is low, which is often the case in embedded application processors, DVS creates a high-performance processor with a very low average energy per operation. The operational point just moves back and forth on the energy–delay (performance) curve.

Examples of DVS-Enabled Microprocessors

- Early Research Prototypes
  - Toshiba MIPS 3900: 1.3–1.9 V, 10–40 MHz [Kuroda98]
  - Berkeley ARM8: 1.2–3.8 V, 6–85 MIPS, 0.54–5.6 mW/MIPS [Burd00]
- Xscale: 180 nm 1.8 V bulk-CMOS
  - 0.7–1.75 V, 200–1000 MHz, 55–1500 mW (typ)
  - Max. Energy Efficiency: ~23 MIPS/mW
- PowerPC: 180 nm 1.8 V bulk-CMOS
  - 0.9–1.95 V, 11–380 MHz, 53–500 mW (typ)
  - Max. Energy Efficiency : ~11 MIPS/mW
- Crusoe: 130 nm 1.5 V bulk-CMOS
  - 0.8–1.3 V, 300–1000 MHz, 0.85–7.5 W (peak)
- Pentium M: 130 nm 1.5 V bulk-CMOS
  - 0.95–1.5 V, 600–1600 MHz, 4.2–31 W (peak)
- Extended to embedded processors (ARM, Freescale, TI, Fujitsu, NEC, etc.)

Slide 10.24

DVS has progressed immensely since its introduction by the research community. Today, a wide range of embedded, DSP, and notebook processors have embraced the concept. In typical applications, the energy per instruction can vary by as much as a factor of 10.

Slide 10.25

Although dynamic voltage scaling seems to be a no-brainer for a number of applications (if of course, continuous or discrete supplies can readily be made available), there existed a lot of resistance against its adoption in the early days. The main concerns were related to how it would be possible to guarantee that timing conditions and signal integrity are met under changing conditions. It is already a major challenge to verify that a processor functions correctly for a single

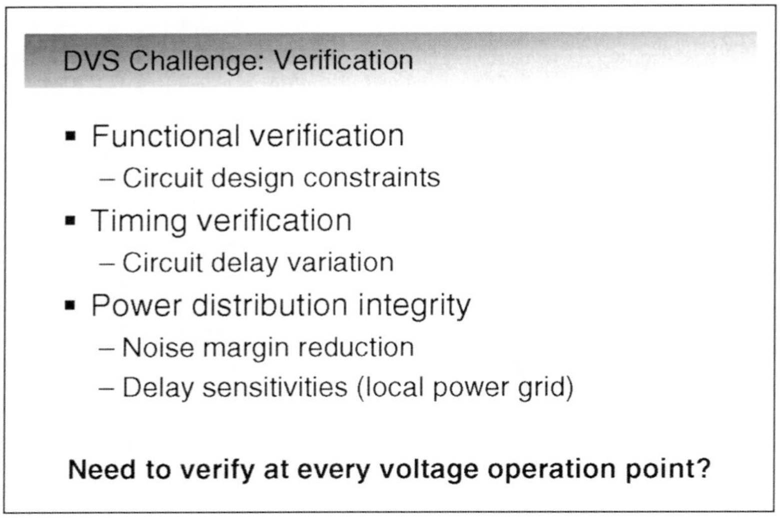

DVS Challenge: Verification

- Functional verification
  - Circuit design constraints
- Timing verification
  - Circuit delay variation
- Power distribution integrity
  - Noise margin reduction
  - Delay sensitivities (local power grid)

Need to verify at every voltage operation point?

supply voltage – a task that is getting more complicated with the increasing influence of process variations. Imagine now what it takes if the supply voltage is varied dynamically. Must one check correct functionality at every supply voltage within the operation range? What about the transient conditions while the voltage is being ramped? Must one halt the processor during that time, or can it keep on running?

All these questions are very pertinent. Fortunately, a number of key properties make the verification task a lot simpler than what would be expected.

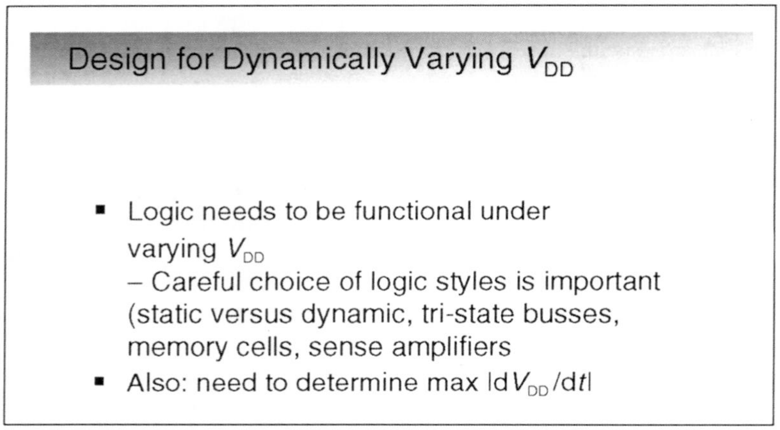

Design for Dynamically Varying V_{DD}

- Logic needs to be functional under varying V_{DD}
  - Careful choice of logic styles is important (static versus dynamic, tri-state busses, memory cells, sense amplifiers
- Also: need to determine max $|dV_{DD}/dt|$

Slide 10.26
Consider first the issue of ensuring functionality. It is important that the circuit does not fail as a result of the supply voltage changes. This depends upon a number of factors such as the logic style used, or the type of memory cell chosen. Most important is how fast the supply voltage is ramped during transitions.

Slide 10.27

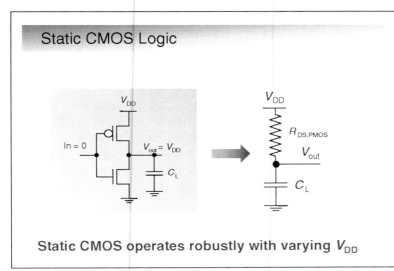

Let us consider the case of complementary CMOS logic. A positive property of this most popular logic style is that the output is always connected to either GND or V_{DD} through a resistive path (during a transition it may temporarily be connected to both). If the output is high and the supply voltage changes, the output of the gate just tracks that change with a short delay owing to the RC time constant. Hence, the functionality of the logic is by no means impacted by DVS. The same is true for a static SRAM cell. In fact, static circuits continue to operate reliably even while the supply voltage is changing.

Slide 10.28

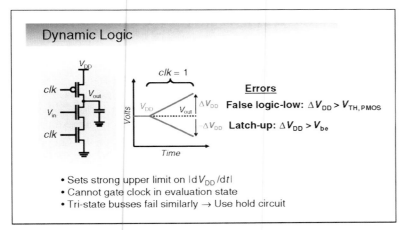

This is not the case for dynamic circuits. During evaluation, the "storage" node of the circuit may be at high impedance, and disconnected from the supply network. Ramping the supply voltage during that time period can lead to a couple of failure modes:

- When the supply voltage rises during evaluation, the "high" signal on the storage node drops below the new supply voltage. If the change is large enough ($> V_{TH,PMOS}$), it may be considered a logic-low by the connecting gate.
- On the other hand, when the supply voltage is ramped down, the stored node voltage rises above the supply voltage, and may cause the onset of latch-up, if the difference is larger than the V_{be} of the parasitic bipolar transistor.

These failure mechanisms can be avoided by either keeping the supply voltage constant during evaluation, or by ramping the rails slowly enough that the bounds, defined above, are not exceeded.

High-impedance tri-state busses should be avoided for the same reason.

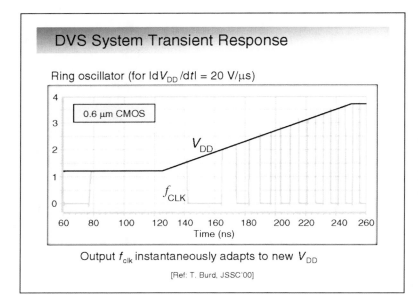

Slide 10.29

The simulated response of a CMOS ring oscillator, shown in this slide, amply serves to validate our argument that static CMOS keeps performing correctly while the voltage is ramped. The plot shows how the output clock signal f_{clk} keeps on rising while the supply voltage increases.

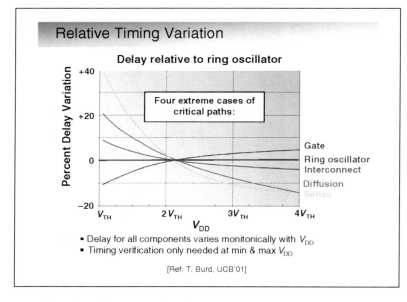

Slide 10.30

Even if a circuit works correctly at one voltage from a timing perspective, this by no means guarantees that it also does so at another one. The relative delays from modules in different logic styles may change owing to the voltage scaling. If so, it may be necessary to check the timing at every supply voltage in the operation range.

To evaluate what transpires during voltage scaling, the relative delay (normalized to the delay of a ring oscillator) versus supply voltage is plotted for four typical circuit elements. These include inverter chains, of which the loads are dominated by gate, interconnect, and diffusion capacitance (as each of these has a different voltage dependence). To model paths dominated by stacked devices, a fourth chain consisting of 4 PMOS and 4 NMOS transistors in series is analyzed as well. The relative delay of all four circuits is at a maximum at only the lowest or highest operating voltages, and is either monotonically falling or rising in between. This means that it is sufficient to ensure timing compliance at the extreme ends of the supply voltage range to guarantee compliance everywhere in between. This substantially reduces the timing verification effort.

Note: it may be possible to create a relative-delay curve with a minimum or a maximum occurring in-between the end points, by combining circuits of the different types. However, because the gate-capacitance-dominated delay curve is convex, whereas the others are concave, the combination typically results in a rather flat curve, and the observation above pretty much still holds.

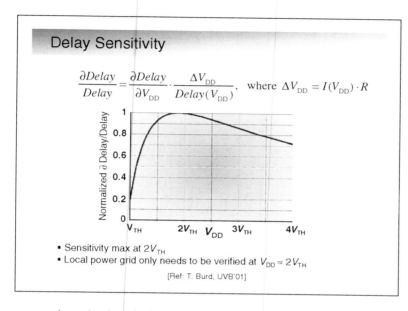

Slide 10.31

Another concern is the effect of supply bounce as it may induce timing variations and potential violations. We are not concerned about global supply voltage changes as they affect all timing paths equally and the clock period is adjusted as well – remember that the clock frequency is derived from the supply voltage in a DVS system.

Localized supply variations, however, may only affect the critical paths, and not the clock generator, and can lead to timing violations if the local supply drop is sufficiently large. As such, careful attention has to be paid to the local supply routing. As always, a certain percentage of the timing budget must be set aside to accommodate the impact of supply bounce. However, the question again arises as to the voltage at which the impact of supply noise is the largest and whether we should check it for the complete range.

The sensitivity of delay with respect to V_{DD} can be quantified analytically, and the normalized result is plotted as a function of V_{DD} in this slide. For a submicron CMOS process, the delay sensitivity peaks at approximately $2V_{TH}$. Thus, in the design of the local power grid, we only need to ensure that the resistive (inductive) voltage drop of the power distribution grid meets the design margins for one single supply voltage (i.e., $2V_{TH}$). This is sufficient to guarantee that they are also met at all other voltages.

All in all, though the DVS approach undoubtedly increases the verification task, the extra effort is bounded. In fact, one may even argue that the adaptive closed loop actually simplifies the task somewhat as some process variations are automatically adjusted for.

Slide 10.32

So far, we have only considered the dynamic adaptation of the supply voltage. In line with our discussions on design-time optimization, it seems only natural to consider adjusting the threshold voltages at runtime as well. This approach, called *Adaptive Body Biasing* or ABB, is especially appealing in light of the increasing impact of static power dissipation. Raising the thresholds when

Adapative Body Biasing (ABB)

- Similar to DVS, transistor thresholds can be varied dynamically during operation using body biasing
- Extension of DBB approach considered for standby leakage management
- Motivation:
  - Extends dynamic E–D optimization scope (as a function of activity)
  - Helps to manipulate and control leakage
  - Helps to manage process and environmental variability (especially V_{TH} variations)
  - Is becoming especially important for low V_{DD}/V_{TH} ratios

the activity is low (and the clock period high), and lowering them when the activity is high and the clock period short, seems to be a perfect alternative or complement to the DVS approach. It should be apparent that ABB is the runtime equivalent of the Dynamic Body Biasing (DBB) approach, introduced in Chapter 8 to address standby leakage.

In addition to dynamically adjusting the static power, ABB can help to compensate for some of the effects introduced by static or dynamic threshold variations – caused by manufacturing imperfections, temperature variations, aging effects, or all of the above. In fact, if well-executed, threshold variations can be all but eliminated.

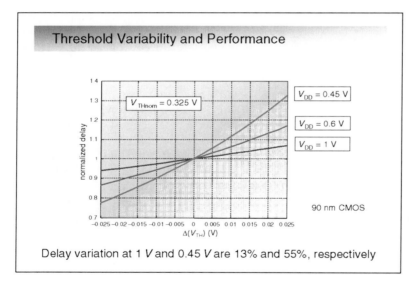

Delay variation at 1 V and 0.45 V are 13% and 55%, respectively

Slide 10.33
As can be observed, variations in thresholds may cause the performance of a module to vary substantially. This effect is more pronounced when the supply voltage is scaled down and the V_{DD}/V_{TH} ratio reduced. Though a 50 mV change in threshold causes a delay change of only 13% at a supply voltage of 1 V (for a 90 nm CMOS technology), it results in a 55% change when the supply is reduced to 0.45 V.

Slide 10.34
The idea of using ABB to address process variations was already introduced in 1994 [Kobayashi94] in a scheme called SATS (self-adjusting threshold voltage scheme). An *on-chip leakage sensor* amplifies the leakage current (the resistive divider biases the NMOS transistor for maximum gain). When the leakage current exceeds a preset threshold, the well bias generation circuit is turned on, and the reverse bias is increased by lowering the well voltage. The same bias is used for all NMOS

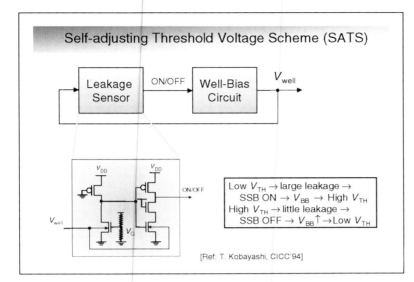

transistors on the chip. Though the circuit shown in the slide addresses the threshold adjustment of the NMOS transistors, the thresholds of the PMOS devices also can be controlled in a similar way.

Note that the overall goal of the SAT scheme is to set leakage to a specific value; that is, the transistor thresholds are set to the lowest possible value that still meets the power specifications.

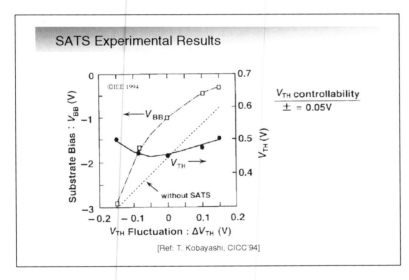

Slide 10.35
The effectiveness of the SATS is quite apparent from the measured results shown in this chart. Even with the raw threshold varying by as much as 300 mV, the control loop keeps the actual threshold within a 50 mV range.

Slide 10.36
This slide features a more recent study of the potential of adaptive body biasing. A test chip implemented by a group of researchers at Intel in a 150 nm CMOS technology [Tschanz02] features 21 autonomous body-biasing modules. The idea is to explore how ABB can be exploited to deal not only with inter-die, but also with intra-die variations. Both the reverse and forward body biasing options are available. Each sub-site contains a replica of the critical path of the circuit under test (CUT), a phase detector (PD) comparing the critical path delay with the desired clock period, and a phase-to-bias converter consisting of a counter, a D/A converter, and an op-amp driver. Only PMOS threshold control is implemented in this

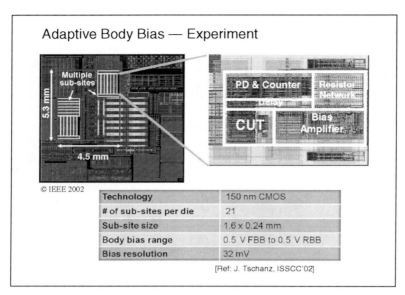

particular prototype circuit. The area overhead of the adaptive biasing, though quite substantial in this experimental device, can be limited to a couple of percents in more realistic settings.

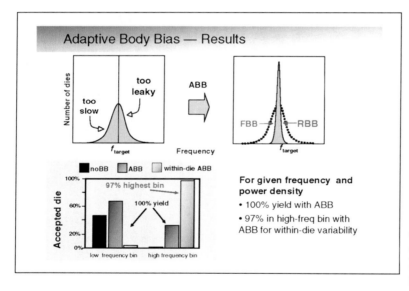

Slide 10.37

Measurement results collected over a large batch of dies are summarized in this slide. A design without adaptation shows a broad distribution with the fast, leaky modules with low thresholds on one side and the slow, low-leakage modules with high thresholds on the other. Applying RBB and FBB, respectively, tightens the frequency distribution, as expected. In addition, it helps to set the leakage current at a level that is both desirable and acceptable.

The economic impact of applying ABB should not be ignored either. In the microprocessor world, it is common to sort manufactured dies into frequency bins based on the measured performance (in addition, all of the accepted dies should meet both functionality and maximum power requirements). Without ABB, a majority of the dies ends up in the not-so-lucrative low-frequency bin, whereas a large fraction does not meet specifications at all. The application of per-die and (even more) within-die ABB manages to move a large majority to the high-frequency bin, while pushing parametric yield close to 100%.

From this, it becomes apparent that adaptively tuning a design is a powerful tool of the designer in the nanometer era.

Slide 10.38

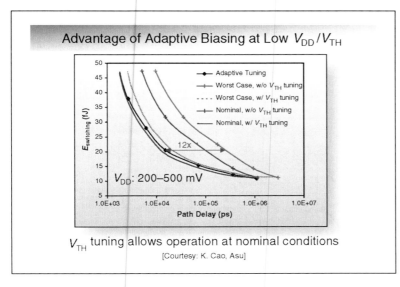

V_{TH} tuning allows operation at nominal conditions
[Courtesy: K. Cao, Asu]

ABB is even more effective in circuits operating at low supply voltages and low V_{DD}/V_{TH} ratios. Under those conditions, a small variation in the threshold voltage can cause either a large performance penalty or a major increase in energy consumption. This is illustrated by the distance between the "nominal" and "worst-case" E–D curves of a complex arithmetic logic function, implemented in a 130 nm CMOS technology. The supply voltage is variable and ranges between 200 and 500 mV, whereas the threshold voltage is kept constant. Observe that the delay is plotted on a logarithmic scale.

A substantial improvement is made when we allow the threshold voltages to be tuned. One option is to simultaneously modify all threshold voltages of a modules by adjusting the well voltage. Even then, the worst-case scenario still imposes a performance penalty of a factor of at least two over the nominal case. This difference is virtually eliminated if the granularity of threshold adjustment is reduced – e.g., allowing different body bias values for every logical path. Overall, introducing "adaptive tuning" allows a performance improvement by a factor of 12 over the worst-case un-tuned scenario, while keeping energy constant.

Slide 10.39

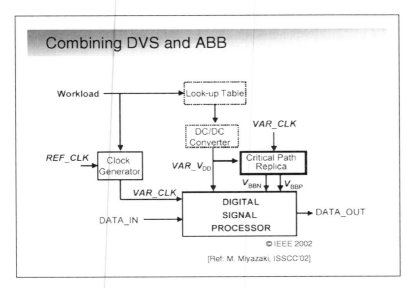

At this point, it is only a small step to consider the advantages of simultaneously applying DVS and ABB. Whereas DVS mainly addresses the dynamic power dissipation, ABB serves to set the passive power to the appropriate level. This combined action should lead to E–D curves that are superior to those obtained by applying the techniques separately.

An example of a circuit incarnation that adjusts both V_{DD} and V_{TH} is shown on this slide [Miyazaki02]. The requested workload is translated into a desired supply voltage using the table look-up approach. Given the selected V_{DD}, a replica of the critical path is used to set the well voltages for NMOS and PMOS transistors so that the requested clock frequency is met. This approach obviously assumes that the replica path shows the same variation behavior as the actual processor.

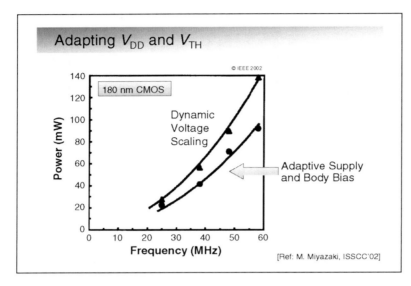

Slide 10.40
Actual measurements for the circuit of Slide 10.39 indeed show the expected improvements. Compared to DVS only, adding ABB improves the average performance of the circuit substantially for the same power level (and vice versa).

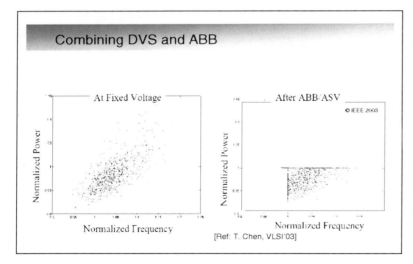

Slide 10.41
Although these performance improvements are quite impressive, one may question how effective the combined DVS/ABB approach is in suppressing the effects of (threshold) variations. The chart on the left side of the slide plots the measured clock frequency and power numbers for the same circuit as collected from a large number of dies (over different wafers). For these measurements, supply voltages were fixed to the nominal value, and no body biasing was applied. Though the measurements show a very broad and wide distribution (20% in both clock frequency and power), a general trend can be observed – that is, slower circuits consume less power (this obviously is not a surprise!).

With the introduction of DVS and ABB, circuits that do not meet the performance or power specification are adjusted and brought within the acceptable bounds (with the exception of some circuits that cannot be corrected within the acceptable range of supply and bias voltages, and hence should be considered faulty). The resulting distribution is plotted on the right, which indicates that dynamic adaptation and tuning is indeed a very effective means of addressing the impact of technology and device variations.

One *very important caveat* should be injected here: just when device variations are becoming a crucial design concern, one of the most effective means of combating them – that is body biasing – is losing its effectiveness. As we had already indicated in Chapter 2, the high doping levels used in sub-100 nm technologies reduce the body-effect factor: *at 65 nm and below, ABB maybe barely worth the effort (if at all)*. This is quite unfortunate, and is hopefully only temporary. The introduction of novel devices, such as dual-gate transistors, may restore this controllability at or around the 32 nm technology node.

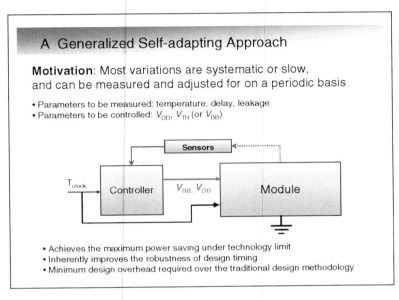

Slide 10.42

Another important and general observation is worth making. The DVS and ABB schemes, presented in the previous slides, are great examples of a new class of circuits (called *self-adaptive*) that deal with variations (be it caused by changes in activity, manufacturing, or the environment) by using a closed feedback loop. On-line sensors measure a set of indicative parameters such as leakage, delay, temperature, and activity. The resulting information is then used to set the value of design parameters such as the supply voltage and the body bias. In even more advanced schemes, functions might even be moved to other processing elements if performance requirements cannot be met.

The idea is definitely not new. In the 1990s, high-performance processors started to incorporate temperature sensors to detect over-heating conditions and to throttle the clock frequency when the chip got too hot. The difference is that today's self-adaptive circuits (as adopted in high-end products) are a lot more sophisticated, use a broad range of sensors, and control a wide range of parameters.

Slide 10.43

Although adaptive techniques go a long way in dealing with runtime variability, ultimately their effectiveness is limited by the "worst-case" conditions. These may be the voltage at which the timing constraints of a critical path cannot be met or when a memory cell fails. In a traditional-design approach, this is where the voltage scaling ends. However, on closer inspection, one realizes that these worst-case conditions occur only rarely. Hence, if we can cheaply detect the occurrence of such a condition and correct the resulting error when it occurs, we could over-scale the voltage, further reducing the energy dissipation.

Let us, for instance, consider the case of an SRAM memory. As we had discussed in Chapter 9, the minimum operational voltage of an SRAM cell (the DRV) varies cell-by-cell. Fortunately, the

> **Aggressive Deployment (AD)**
>
> - Also known as "Better-than-worst-case (BTWC) design"
> - Observation:
>   - Current designs target worst-case conditions, which are rarely encountered in actual operation
> - Remedy:
>   - Operate circuits at lower voltage levels than allowed by worst case, and deal with the occasional errors in other ways
>
>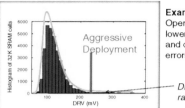
>
> Example:
> Operate memory at voltages lower than that allowed by worst case, and deal with the occasional errors through error correction
>
> Distribution ensures that error rate is low

measured distribution of the DRVs over a large memory block shows a long tail. This means that lowering the voltage below the worst case causes some errors, but only a few. The reduction in leakage currents far outweighs the cost of error detection and correction.

> **Aggressive Deployment – Concepts**
>
> - Probability of hitting tail of distribution at any time is small
>   - Function of critical-path distribution, input vectors, and process variations
> - Worst-case design expensive from energy perspective
>   - Supply voltage set to worst case (+ margins)
> - Aggressive deployments scales supply voltage below worst-case value
>   - "Better-than-worst-case" design strategy
>   - Uses error detection and correction techniques to handle rare failures

Slide 10.44

The basic concepts upon which this "better-than-worst-case" (BTWC) (first coined as such by Todd Austin) design is built are as follows:

- Over-scaling of the supply voltage leads only to rare errors, not to catastrophic breakdown. In the latter case, the overhead of dealing with the errors would dominate the savings. Hence, knowing the distribution of the critical parameters is important.
- The worst-case scenario leaves a large number of crumbs on the table. All circuitry in a module consumes way too much energy just because of a small number of outliers.
- Hence it pays to let some errors occur by over-scaling, and condone the small overhead of error detection and correction.

Optimizing Power @ Runtime – Circuits and Systems

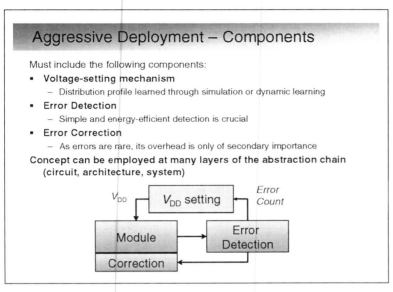

Slide 10.45

Like DVS and ABB, BTWC (very often also called *aggressive deployment*, or AD) relies on the presence of a feedback loop, positioning the system at its optimal operation point from a performance/energy perspective. A BTWC system consists of the following elements:

- A mechanism for setting the supply voltage based on the understanding of the trade-off between introducing errors and correcting them. In one way or another, this control mechanism should be aware of the error distribution (either by simulation in advance, or by adaptive learning).
- An error-detection mechanism – As this function is running continuously, its energy overhead should be small. Coming up with efficient error-detection approaches is the main challenge in the conception of BTWC systems.
- An error-correction strategy – As errors are expected to be rare, it is ok to spend some effort in correcting them. The correction mechanisms can vary substantially, and depend upon the application area as well as the layer in the abstraction chain where the correction is performed.

It is important to realize that the BTWC approach is very generic, and can be applied at many layers of the design abstraction chain (circuit, architecture, system) and for a broad range of application spaces, some of which are briefly discussed in the following slides.

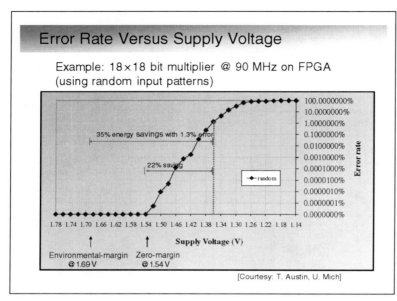

Slide 10.46

As a first example, let us consider what happens when we lower the supply voltage of a logical module such as a multiplier, which typically has a wide distribution of timing paths. In a traditional design, the minimum supply voltage is set by the worst-case timing path with an extra voltage margin added for safety. The example on the slide shows the results for an 18×18 multiplier, implemented on an FPGA. Including the safety margin, the minimal operational voltage equals 1.69 V.

Once we start lowering the supply voltages, some timing paths may not be met and errors start to appear. In the FPGA prototype, the first errors occur at 1.54 V. Observe that the y-scale of this plot is logarithmic. If we would allow for a 1.3% error rate (which means that one out of 75 samples is wrong), the supply voltage can be scaled all the way down to 1.36 V. This translates into a power reduction of 35%.

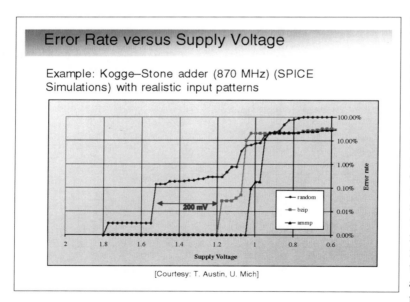

Slide 10.47

It is worth observing that error rate and the shape of the error histogram are functions of the data patterns that are applied. For the example of a multiplier, random data patterns tend to trigger the worst-case paths more often than the correlated data patterns that commonly occur in signal-processing applications. The same holds for many other computational functions, as is illustrated in this slide for a Kogge–Stone adder. When applying data patterns from applications such as *bzip* or *ammp*, the voltage can be scaled down by an extra 200 mV for the same error rate.

Hence, to be effective, the voltage-setting module must somehow be aware of the voltage-to-error function. As for DVS, this information can be obtained by simulation or during a training period and can be stored in a table.

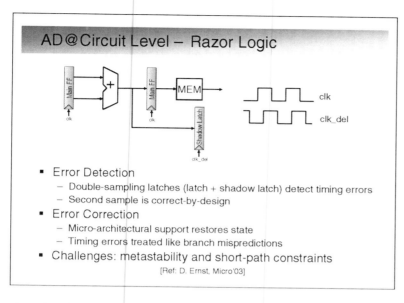

Slide 10.48

Based on these observations, it seems worthwhile to reduce the supply voltage below its worst-case value, assuming that a simple mechanism for detecting timing errors can be devised. One way of doing so (called RAZOR) is shown here. Every latch at the end of a critical timing path is replicated by a "shadow latch", clocked a certain time ΔT later. The value captured in the main latch is correct, if and only if the shadow latch shows an identical value. If on the other hand, the main clock arrives too early and a path has not stabilized yet, main latch and shadow latch will capture different values. A sole XOR is sufficient to detect the error.

The above description is somewhat over-simplifying, and some other issues need to be addressed for this approach to work. For instance, no shadow latches should be placed on "short paths" as this may cause the shadow latch to catch the next data wave. In other words, checking the set-up and hold-time constraints becomes more complicated. Also, the first latch may get stuck in a metastable state, leading to faulty or undecided error conditions. Extra circuitry can be added to get around this problem. For more detailed information, we refer the interested reader to [D. Ernst, MICRO'03].

Upon detection of an error, a number of strategies can be invoked for correction (depending upon the application space). For instance, since the original RAZOR targets microprocessors, it passes the correction task on to the micro-architectural level.

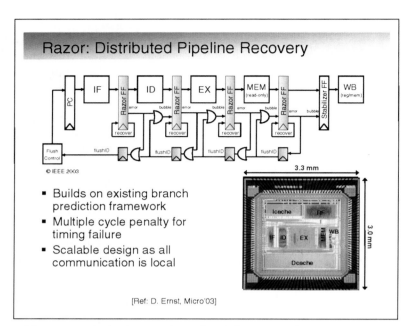

Slide 10.49

In a sense, an error in the data-path pipeline is similar to a branch misprediction. Upon detection of an error, the pipeline stalled, a bubble can be inserted, or the complete pipeline flushed. One interesting observation is that, upon error, the correct value is available in the shadow register. It can hence be re-injected in the pipeline at the next clock cycle, while stalling the next instruction. To state it simply, a number of techniques are available to the micro-architecture designer to effectively deal with the problem. It comes with some cycle (and energy) overhead, but remember: errors are expected to occur only rarely if the voltage-setting mechanism works correctly.

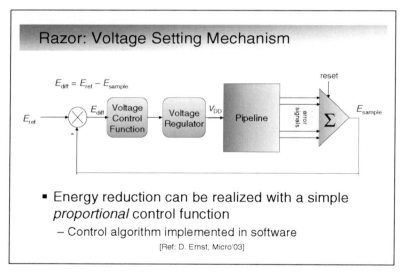

Slide 10.50

The voltage-setting mechanism is a crucial part of any AD scheme. In the RAZOR scheme, the errors per clock cycle occurring in the data path are tallied and integrated. The error rate is used to set the supply voltage adaptations. As mentioned earlier, knowledge of the voltage–error distribution helps to improve the effectiveness of the control loop.

Slide 10.51

Given our previous discussion of adaptive optimizations in the delay–energy space, it should come as no surprise that BTWC schemes converge to an optimal supply voltage that minimizes the energy per operation. Reducing the supply voltage lowers the energy, but at the same time increases the correction overhead. If the voltage–error relationship is gradual (such as the ones shown for the multiplier and the Kogge–Stone adder), the optimal operational point shows a substantial improvement in energy for a very small performance penalty.

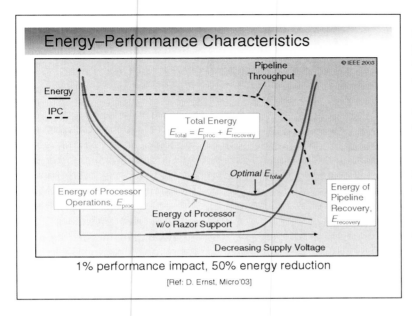

Trade-off curves such as the one shown are typical for any BTWC approach, as will be demonstrated in some of the subsequent slides.

Caveat: For aggressive deployment schemes to be effective, *it is essential that the voltage–error distribution has a "long tail"*. This means that the onset of errors should be gradual once the supply voltage is dropped below its worst-case value. The scheme obviously does not work if a small reduction leads to "catastrophic failures". Unfortunately, a broad range of the energy-reduction techniques, introduced earlier in this book, tend to create just this type of condition. Design-time techniques, such as the use of multiple supply and threshold voltages as well as transistor sizing, exploit the slack on the non-critical paths to minimize energy dissipation. The net result of this is that a larger percentage of timing paths become critical. Under such conditions, a small voltage reduction can lead to a catastrophic breakdown. This opens the door for an interesting discussion: wouldn't it be better to forgo the design-time optimizations and let the runtime optimizations do their job – or vice versa? The only way to get a relevant answer to this question is to exploit the systematic system-level design exploration framework, advocated in this book.

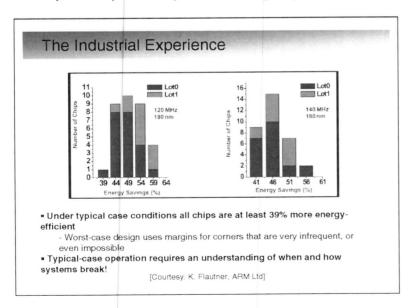

Slide 10.52

Although concepts such as runtime adaptation and BTWC design show great potential, it takes substantial effort to transfer them into producible artifacts. Similar to DVS, AD requires a re-evaluation of the standard design flows and a rethinking of traditional design concepts. As we had mentioned in the previous slide, concepts such as RAZOR require an understanding of how and when a chip breaks. To make the approach more effective, we may even want to rethink accepted design technologies. But the benefits of doing so can be very substantial.

In this slide, the impact of applying the RAZOR concept to an embedded processor of the ARM™-family is shown. An energy reduction of at least 39% over all processors is obtained, whereas average savings are at least 50%. Getting to this point required a major redesign not only of the data path but also of the memory modules. But the results show that that effort is ultimately very rewarding.

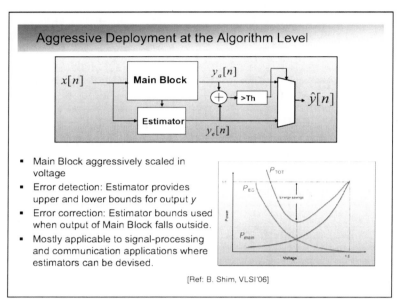

Slide 10.53

The RAZOR concept combines error detection at the circuit level with error correction at the micro-architecture level. Many other BTWC strategies can be envisioned. For example, in this slide we show an approach that employs both error detection and correction at the algorithm level.

One interesting property of many signal-processing and communication applications is that the theory community has provided us with simple ways to estimate the approximate outcome of a complex computation (based on the past input stream). The availability of such estimates provides us with a wonderful opportunity to reduce energy through BTWC.

The "Main Block" in the diagram represents some complex energy-intensive algorithm, such as for instant motion compensation for video compression. In normal operation, we assume this block to be error-free. Assume now that we aggressively scale the supply voltage of this block so that errors start to occur. In parallel with the "Main Block", a simple estimator is run which computes the expected outcome of the "Main Block". Whenever the latter ventures to values far from the prediction, an error condition is flagged (detection), upon which the faulty outcome is replaced by the estimation (correction). This obviously deteriorates the quality of the processor – in signal-processing speak, it reduces the signal-to-noise ratio (SNR). However, if the estimator is good enough, the increase in the noise level is masked by the noise of the input signal or by the added noise of the signal-processing algorithm, and hence barely matters. Also observe that "small errors" (errors that only effect the least-significant bits [LSBs]) may go undetected, which is ok as they only impact the SNR in a minor way.

As for RAZOR, algorithmic BTWC leads to an optimal supply voltage. If the error rate gets higher, the error correcting overhead starts to dominate (in addition, the deterioration in SNR may not be acceptable). For this scheme to work, clearly it is essential that the estimator does not make any errors itself. This requires that the "Estimate Module" be run at the nominal voltage. Since it is supposed to be a simple function, its energy overhead is small.

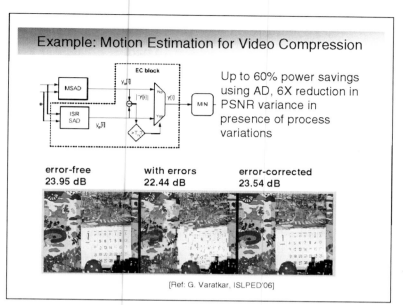

Slide 10.54

The effectiveness of algorithmic-level AD is demonstrated with a video compression example, more specifically for its motion estimation block, which is the most compute-intensive function. The main algorithm uses MSAD (main sum of absolute differences), whereas the estimator uses a simpler version called ISR-SAD (Input-sub-sampled replica of sum of absolute differences). The main simplifications used in the estimator are a reduced precision as well as a reduced sampling rate (through sub-sampling). Only the MSAD is voltage scaled. A pleasant surprise is that in the presence of process variations, the AD version performs better than the original one, from an SNR perspective. It turns out that this is not an exception – techniques that exploit the joined statistics of the application and the process often end up performing better than those that don't.

Slide 10.55

As mentioned, the concepts of runtime adaptation and AD are broad and far-reaching. We have only shown a couple of examples in this chapter. A number of other instantiations of the concept are enumerated on this slide. We also suggest that you consult the March 2004 issue of the IEEE Computer Magazine, which features a variety of BTWC technologies.

Slide 10.56

> **Power Domains (PDs)**
>
> **Introduction of multiple voltage domains on single die creates extra challenges:**
> - Need for multiple voltage regulators and/or voltage up–down converters
> - Reliable distribution of multiple supplies
> - Interface circuits between voltage domains
> - System-level management of domain modes
>   - Trade off gains of changing power modes with overhead of doing so
>   - Centralized "power management" very often more effective

This chapter has introduced a broad range of technologies that rely on the dynamic adaptation of supply and body bias voltages. A contemporary SoC features many partitions, each of which may need different voltage-scheduling regimes. In an integrated SoC for mobile communications, it is not unusual at any time for a number of modules to be in standby mode, requiring their supply voltage to be completely ramped down or at the data retention level, while other modules are active and require either the full supply or a dynamically varying one. Each chip partition that needs individual power control is called a *power domain* (PD).

The introduction of power domains in the standard design methodology comes with some major challenges. First of all, generating and distributing multiple variable supply voltages with a reasonable efficiency is not trivial. Many of the gains made by varying supply and well voltages could be lost if the voltage conversion, regulation, and distribution is not done efficiently. Another, often forgotten, requirement is that signals crossing power boundaries should be carefully manipulated. Level conversion, though necessary, is not sufficient. For instance, the output signals of an active module should not cause any activity in a connected module in standby; or, vice versa, the grounded output signals of a standby module should not result in any erroneous activity in a connected active block.

The most important challenge however is the *global power management* – that is, deciding what voltages to select for the different partitions, how fast and how often to change supply and well voltages, when to go in standby or sleep mode, etc. In the preceding slides and chapters, we had introduced voltage-setting strategies for individual modules. A distributed approach, in which each module individually chooses its preferred setting at any point in time, can be made to work. Yet, it is often sub-optimal as it lacks awareness of the global state of the system. A centralized *power manager* (PM) often can lead to far more efficient results.

Slide 10.57

There are a couple of reasons for this. First of all, the PM can examine the global state of the system, and may have knowledge of the past state. It is hence in a better position to predict when a block will become active or inactive, or what the level of activity may be. Furthermore, transferring the state to a centralized module allows a sub-module to go entirely dormant, reducing leakage power. For instance, many sleep strategies often employ timers to set the next wake-up time (unless an input event happens earlier). Keeping the timers running eliminates the possibility of complete power-down of the unit. Hence, transferring the time-keeping to a centralized "scheduler" makes clear sense.

Although many SoCs employ some form of a power management strategy, most often it is constructed ad hoc and after the fact. Hence, a methodological approach such as the one advocated

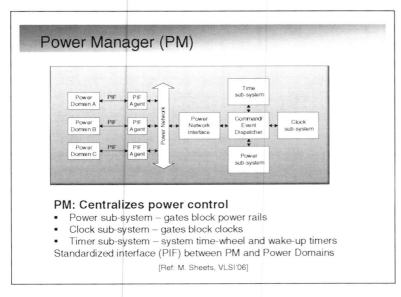

PM: Centralizes power control
- Power sub-system – gates block power rails
- Clock sub-system – gates block clocks
- Timer sub-system – system time-wheel and wake-up timers
Standardized interface (PIF) between PM and Power Domains

[Ref: M. Sheets, VLSI'06]

in this slide is advisable. A coordinated PM contains the following components: a central control module (called event/command dispatcher), and time, power, and clock sub-systems. The latter contain the necessary knowledge about past and future timing events, power- and voltage-setting strategies for the individual modules, and the voltage–clock relationships, respectively. The dispatcher uses the information of the three sub-systems to set a voltage-scheduling strategy for the different power domains on the chip. Inputs from PDs (such as a request to shut down, or a request to set up a channel to another PD), as well as scheduling decisions and power-setting commands are interchanged between the PDs and the PM over a "power network" with standardized interfaces.

In a sense, the PM takes on some of the tasks that typically would be assigned to a scheduler or an operating system (OS) (which is why another often-used name for the PM is the "chip OS"). However, the latter normally runs on an embedded processor, and consequentially that processor could never go into standby mode. Dedicating the PM functionality to a specialized processor avoids that problem, with the added benefit that its energy-efficiency is higher as well.

Managing the Timing

- Basic scheduling schemes
  - Reactive
    - Sleep when not actively processing
    - Wake up in response to a pending event
  - Stochastic
    - Sleep if idle and probably not needed in near future [Simunic'02]
    - Wake up on account of expected event in the near future
- Metrics
  - Correctness – PD awake when required to be active
  - Latency – time required to change modes
  - Efficiency – minimum total energy consumption [Liao'02]
    - Minimum idle time – time required for savings in lower-power mode to offset energy spent for switching modes

$$\text{Min. Idle Time} = \frac{E_{lost}}{P_{savings}} = \frac{E_{overhead} - P_{idle} t_{switch\_modes}}{P_{sleep} - P_{idle}}$$

Slide 10.58
In the literature, a number of PM-scheduling strategies have been proposed. The main metrics to judge the quality of a scheduling strategy are the "correctness" – that is, for instance, having a PD in inactive mode, when another PD attempts to communicate with it might be catastrophic – latency and energy efficiency. The scheduling strategies can be roughly divided into two classes: reactive (based on events at the signaling ports) and proactive. The PhD theses of Tajana Simunic (Stanford) and Mike Sheets (UCB) probably present the most in-depth treatments on the topic so far.

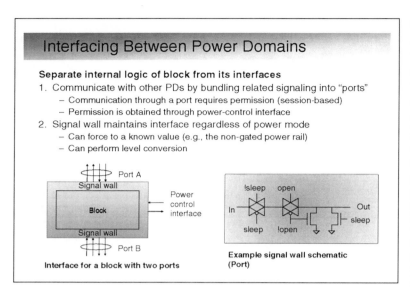

Slide 10.59
A structured approach to the construction of PDs can also help to address the challenge of the proper conditioning of signals crossing power domains. Putting a wrapper around every PD supporting only standardized interfaces makes the task of composing a complex SoC containing many PDs a lot simpler. For instance, the interface of each PD should support a port to communicate with the PM through the "Power Network" and a number of signaling ports to connect to other PDs. The signaling ports can contain a number of features such as level conversion or signal conditioning, as shown in the slide.

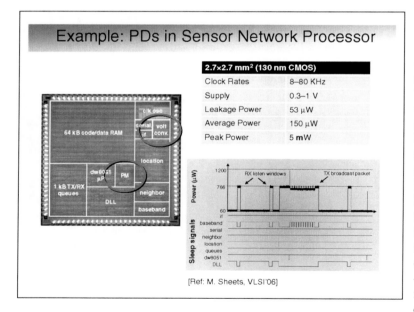

Slide 10.60
An example of a structured power management approach is shown on this slide. This integrated protocol and application processor for wireless sensor networks combines a broad range of functions such as the baseband, link, media-access and network-level processing of the wireless network, node locationing, as well as application-level processing. These tasks exhibit vastly different execution requirements – some of them are implemented in software on the embedded 8051 micro-controller, whereas the others are implemented as dedicated hardware modules – as well as dissimilar schedules. It is rare for all functions to execute simultaneously. To minimize standby power (which is absolutely essential for this low duty cycle application), an integrated power manager assumes that any module is in power-down mode by default. Modules transition to active mode as a result of either timer events (all timers are incorporated in the PM), or events at their input ports. For a module in standby, the supply voltage is ramped down either to GND if there is no state, or to a data retention

voltage of 300 mV. The latter is the case for the embedded micro-controller, whose state is retained in between active modes. To minimize overhead, the retention voltage is generated by an on-chip voltage converter. When the chip is in its deepest sleep mode, only the PM running at an 80 kHz clock frequency is still active.

The logic analyzer traces show how all modules are in standby mode by default. Power is mostly consumed during a periodic RX cycle, when the node is listening for incoming traffic, or during a longer TX cycle. Modules servicing the different layers of the protocol stack are only fired up when needed. For instance, it is possible to forward a packet without waking up the micro-controller.

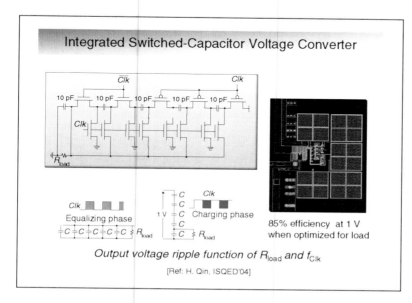

Slide 10.61

For low-power applications, such as wireless sensor networks, using off-the-shelf components to generate the various voltages that are needed on the chip turns out to be very inefficient. Most commercial voltage regulators are optimized for high-power applications drawing Amperes of current. When operated at mW levels, their efficiency drops to the single-digit percentage level (or even lower). Hence, integrating the regulators and converters on-chip is an attractive solution. The fact that the current demands for these converters are very low helps substantially in that respect. An additional benefit of the integrated approach is that the operational parameters of the converter can be adapted to the current demand, maintaining a high level of efficiency over the complete operation range.

The "switched-capacitor" (SC) converter, shown in this slide, works very well at low current levels, and can be easily integrated on a chip together with the active circuitry (such as in the case of the sensor-network processor of Slide 10.60). No special demands are placed on the technology. The ripple on the output voltage is determined by the current drawn (represented by the load resistor R_{load}), the total capacitance in the converter, and the clocking frequency. During the standby mode, the load resistance is large, which means that the clock frequency of the converter can be reduced substantially while keeping the voltage ripple constant. Hence, high levels of efficiency can be maintained for both active and standby modes. The only disadvantage is that the capacitors composing SC converters consume a substantial amount of silicon area. This makes their use prohibitive for applications that draw a substantial amount of current. Advanced packaging technologies can help to offset some of these concerns.

Slide 10.62

Using the SC-converter concept, it is possible to create a fully integrated power train for wireless sensor applications. As was mentioned in the introduction chapter (Chapter 1), distributed sensor

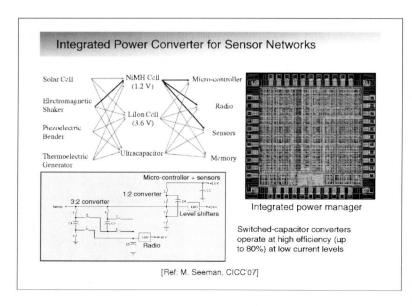

network nodes strive to harvest their energy from the environment to ensure operational longevity. Depending upon the energy source, rectification may be necessary. The scavenged energy is temporarily stored on either a rechargeable battery or a supercapacitor to balance supply and demand times. The sensor node itself requires a variety of voltages. Sensors, for example, tend to require higher operational voltages than digital or mixed-signal hardware. A bank of switched-capacitor converters can be used to provide the necessary voltage levels, all of which need to have the possibility to be ramped down to zero volt or the DRV for standby.

A dedicated integrated power-conversion chip, accomplishing all these functions for a wireless sensor node targeting tire-pressure monitoring applications, is shown in this slide. The IC contains the rectifiers as well as the various level converters. High levels of conversion efficiency are maintained over all operational modes.

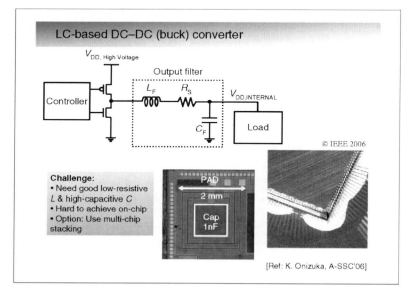

Slide 10.63

Unfortunately, SC voltage converters are only effective at low current and power levels (i.e., at the mA and mW range). Most integrated circuits run at substantially higher current levels, and require more intricate voltage regulators and converters. The most efficient ones are based on resonant LC networks (most often called buck converters), where energy is transferred with minimal losses between an inductor and a capacitor at a well-defined switching rate. An example of such a converter is shown on the slide.

LC-based voltage regulators are generally implemented as stand-alone components. There is a very good reason for this: the required values and quality factors of the inductors and capacitors are hard to accomplish on-chip. Hence, the passives are most often implemented as discrete components. For SoCs with multiple power domains and dynamically varying voltage requirements, there are some compelling reasons to strive for a tighter integration of the passives with the active circuitry. Direct integration of the controller circuitry with the load leads to more precise control, higher efficiencies, and increased flexibility.

Though integrating the Ls and Cs directly on the IC may not be feasible, an alternative approach is to implement the passives on a second die (implemented on a substrate of silicon or some other material such as a plastic/glass interposer), which provides high-quality conductors and isolators, but does not require small feature sizes. The dies can then be connected together using advanced packaging strategies. An example of such a combined inductor/capacitor circuit is shown. Capacitance and inductance values in the nF and nH range, respectively, can be realized in this way. The concept of stacking dies in a 3D fashion is gaining rapid acceptance these days – driven mostly by the size constraints of mobile applications. This trend surely plays in favor of closer integration of the power regulation with the load circuitry, and of distributed power generation and conversion.

Slide 10.64

The possibility of multi-dimensional integration may lead to a complete rethinking of how power is distributed for complex SoCs. In the ICs of the past years, the power distribution network consisted of a large grid of connected Copper (or Al) wires, all of which were set to the nominal supply voltage (e.g., 1 V). The concept of power gating has changed this practice a little: instead of being connected directly to the power grid, modules now connect through switches that allow an idle module to be disconnected.

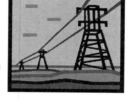

Revisiting Power Distribution Concept

- Current on-chip power distribution: Single metal grid extended with switches for power gating
- Need for: higher voltage distribution and integrated level converters and switches
- Hard to accomplish on standard integrated circuit
- Opportunities offered by stacked dies and 2.5D integration
  - Thicker wires, better inductors and capacitors
- Toward "PGE on a chip"

[Ref: J. Rabaey, MPSOC'04]

If it becomes possible to integrate voltage converters (transformers) more tightly into the network, a totally new approach may arise. This would resemble the way power is distributed in large scale at the metropolitan and national levels: the main grid is operated at high voltage levels, which helps to reduce the current levels and improves the efficiency. When needed, the power is down-converted to lower levels. In addition to the introduction of transformers, switches also can be introduced at multiple levels of the hierarchy.

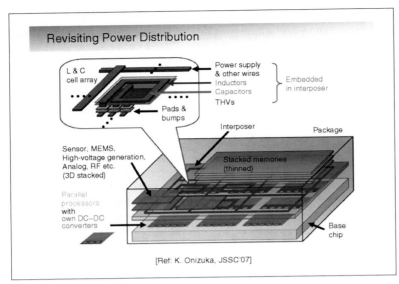

Slide 10.65

A graphical representation of the constructions this vision could lead to is shown. Most standard circuitry is implemented on the base chip. Also included on the die is the control circuitry for the power regulators of the various power domains. However, the power grids of the latter are not connected on the die. The "higher levels" of the power distribution network are implemented on an interposer die, which implements a grid of high-quality inductors and capacitors, as well as a high-voltage power grid. The 2.5D integration strategy also allows for non-traditional technologies such as MEMs, or non-digital technologies such as DRAMs, to be tightly integrated with the computational fabric in a compact package.

Note: The term 2.5D integration relates to a three-dimensional IC technology, where individual dies are stacked on top of each other and interconnected using solder bumps or wire bonding. A true three-dimensional integration strategy, on the other hand, supposes that all active and passive devices are realized as a single artifact by constructively creating a stack of many layers deposited on top of one another. Although this ultimately may be the better solution, a large volume of economical and technological issues make the latter approach quite impractical for the time being.

Slide 10.66

In summary, the combination of variations in activity, process, and environmental conditions is leading to fundamental changes in the way ICs and SoCs are being designed and managed. Rather than relying solely on design-time optimizations, contemporary integrated circuits adjust parameters such as the supply and well voltages on the fly, based on observation of parameters such as the workload, leakage, and temperature. In addition, different parameter sets can be applied to individual regions of the chip called *power domains*.

This design strategy represents a major departure from the methodologies of the past. It challenges the standard design flows – yet does not make them obsolete. Actually, upon further contemplation, we can come to the conclusion that the idea of runtime optimization may make the traditional design strategies more robust in light of the challenges of the nanometer era, at the same time helping to reduce energy substantially.

Slide 10.67–10.69
Some references . . .

Literature

Books, Magazines, Theses

- T. Burd, *Energy-Efficient Processor System Design*," http://bwrc.eecs.berkeley.edu/Publications/2001/THESES/energ_eff_process-sys_des/index.htm, UCB, 2001.
- Numerous authors, *Better than worst case design*, IEEE Computer Magazine, March 2004.
- T. Simunic, "Dynamic Management of Power Consumption", in *Power-Aware Computing*, edited by R. Graybill, R. Melhem, Kluwer Academic Publishers, 2002.
- A. Wang, *Adaptive Techniques for Dynamic Processor Optimization*, Springer, 2008.

Articles

- L. Anghel and M. Nicolaidis, "Cost reduction and evaluation of temporary faults detecting technique," *Proc. DATE 2000*, pp. 591–598, 2000.
- T. Burd, T. Pering, A. Stratakos and R. Brodersen, "A dynamic voltage scaled microprocessor system," *IEEE Journal of Solid-State Circuits*, 35, pp. 1571–1580, Nov. 2000.
- T. Chen and S. Naffziger, "Comparison of adaptive body bias (ABB) and adaptive supply voltage (ASV) for improving delay and leakage under the presence of process variation," *Trans. VLSI Systems*, 11(5), pp. 888–899, Oct. 2003.
- D. Ernst et al., "Razor: A low-power pipeline based on circuit-level timing speculation," *Micro Conference*, Dec. 2003.
- V. Gutnik and A. P. Chandrakasan, "An efficient controller for variable supply voltage low power processing," *IEEE Symposium on VLSI Circuits*, pp. 158–159, June 1996.
- T. Kehl, "Hardware self-tuning and circuit performance monitoring,": *Proceedings ICCD* 1993.
- T. Kobayashi and T. Sakurai, "Self-adjusting threshold-voltage scheme (SATS) for low-voltage high-speed operation," *IEEE Custom Integrated Circuits Conference*, pp. 271–274, May 1994.

References (cont.)

- T. Kuroda et al., "Variable supply-voltage scheme for low-power high-speed CMOS digital design", *IEEE Journal of Solid-State Circuits*, 33(3), pp. 454–462, Mar. 1998.
- W. Liao, J. M. Basile and L. He, "Leakage power modeling and reduction with data retention," in *Proceedings IEEE ICCAD*, pp. 714–719, San Jose, Nov. 2002.
- M. Miyazaki, J. Kao, A. Chandrakasan, "A 175 mV multiply-accumulate unit using an adaptive supply voltage and body bias (ASB) Architecture," *IEEE ISSCC*, pp. 58–59, San Francisco, California, Feb. 2002.
- L. Nielsen and C. Niessen, "Low-power operation using self-timed circuits and adaptive scaling of the supply voltage," *IEEE Transactions on VLSI Systems*, pp. 391–397, Dec. 1994.
- H. Okano, T. Shiota, Y. Kawabe, W. Shibamoto, T. Hashimoto and A. Inoue, "Supply voltage adjustment technique for low power consumption and its application to SOCs with multiple threshold voltage CMOS," *Symp. VLSI Circuits Dig.*, pp. 208–209, June 2006.
- K. Onizuka, H. Kawaguchi, M. Takamiya and T. Sakurai, "Stacked-chip Implementation of on-chip buck converter for power-aware distributed power supply systems," *A-SSCC*, Nov. 2006.
- K. Onizuka, K. Inagaki, H. Kawaguchi, M. Takamiya and T. Sakurai, "Stacked-chip Implementation of on-chip buck-converter for distributed power supply system in SIPS, IEEE JSSC, pp. 2404–2410, Nov. 2007.
- T. Pering, T. Burd and R. Brodersen, "The simulation and evaluation of dynamic voltage scaling algorithms." *Proceedings of International Symposium on Low Power Electronics and Design 1998*, pp. 76–81, June 1998.
- H. Qin, Y. Cao, D. Markovic, A. Vladimirescu and J. Rabaey, "SRAM leakage suppression by minimizing standby supply voltage," *Proceedings of 5th International Symposium on Quality Electronic Design*, 2004, Apr. 2004.
- J. Rabaey, "Power Management in Wireless SoCs," Invited presentation MPSOC 2004, Aix-en-Provence, Sep. 2004; http://www.eecs.berkeley.edu/~jan/Presentations/MPSOC04.pdf

References (cont.)

- T. Sakurai, "Perspectives on power-aware electronics", *IEEE International Solid-State Circuits Conference*, vol. XLVI, pp. 26–29. Feb 2003.
- M. Seeman, S. Sanders and J. Rabaey, "An ultra-low-power power management IC for wireless sensor nodes", *Proceedings CICC 2007*, San Jose, Sep. 2007.
- A. Sinha and A. P. Chandrakasan, "Dynamic voltage scheduling using adaptive filtering of workload traces," *VLSI Design 2001*, pp. 221–226, Bangalore, India, Jan. 2001.
- M. Sheets et al., "A power-managed protocol processor for wireless sensor networks," *Digest of Technical Papers VLSI06*, pp. 212–213, June 2006.
- B. Shim and N. R. Shanbhag, "Energy-efficient soft error-tolerant digital signal processing," *IEEE Transactions on VLSI*, 14(4), 336–348, Apr. 2006.
- J. Tschanz et al., "Adaptive body bias for reducing impacts of die-to-die and within-die parameter variations on microprocessor frequency and leakage," *IEEE International Solid-State Circuits Conference*, vol. XLV, pp. 422–423, Feb. 2002.
- A. Uht, "Achieving typical delays in synchronous systems via timing error toleration," *Technical Report TR-032000-0100*, University of Rhode Island, Mar. 2000.
- G. Varatkar and N. Shanbhag, "Energy-efficient motion estimation using error-tolerance," *Proceedings of ISLPED 06*, pp. 113–118, Oct. 2006.
- F. Worm, P. Ienne, P. Thiran and G. D. Micheli. "An adaptive low-power transmission scheme for on-chip networks," *Proceedings of the International Symposium on System Synthesis (ISSS)*, pp. 92–100, 2002.

Chapter 11
Ultra Low Power/Voltage Design

Ultra Low Power/Voltage Design

Jan M. Rabaey

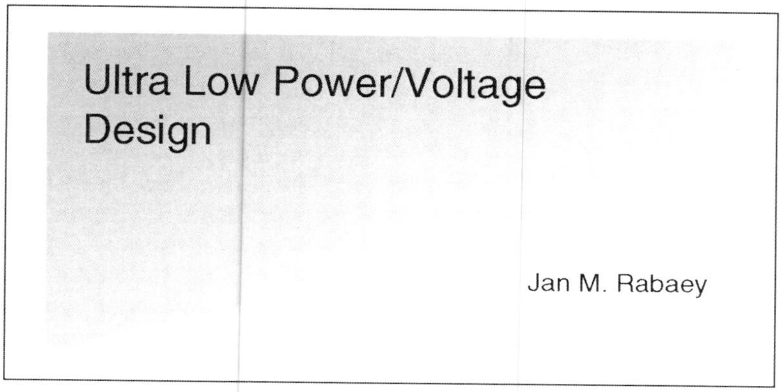

Slide 11.1
In previous chapters, we had established that considering energy in isolation rarely makes sense. Most often, a multi-dimensional optimization process is essential equally considering other metrics such as throughput, area, or reliability. Yet, in a number of applications, minimizing energy (or power) is the singlemost important goal, and all other measures are secondary. Under such conditions, it is worth exploring what the minimum energy is to perform a given task in a given technology. Another interesting question is whether and how this minimum changes with further scaling of the technology. Addressing these questions is the main goal of this chapter.

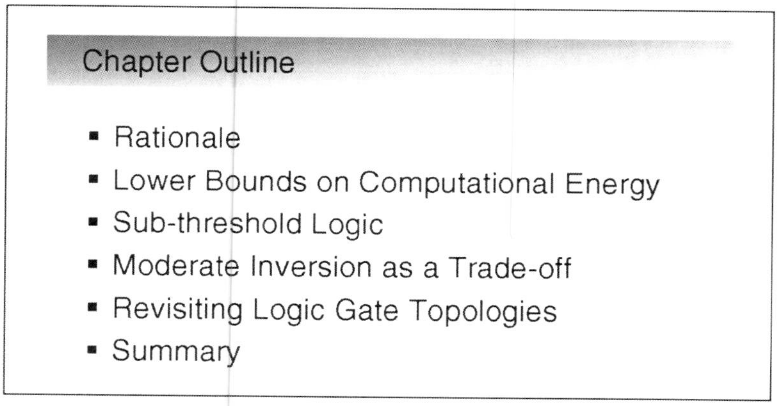

Chapter Outline

- Rationale
- Lower Bounds on Computational Energy
- Sub-threshold Logic
- Moderate Inversion as a Trade-off
- Revisiting Logic Gate Topologies
- Summary

Slide 11.2
The chapter commences with the rationale behind the concept of ultra low power (ULP) design, and the establishment of some firm lower bounds on the minimum energy for a digital operation. From this, it emerges that ULP is quite synonymous to ultra low voltage (ULV) design. Unless we find a way to scale down threshold voltages without dramatically increasing the leakage currents, ULV circuits, by necessity, operate in the sub-threshold region. A sizable fraction of the chapter is hence devoted to the modeling, operation, and optimization of digital logic and memory operating in this mode. We will show that, though this most often leads to a minimum-energy design, it also comes at an

exponentially increasing cost in performance. Hence, backing off just a bit to the moderate-inversion region gives almost identical results in energy, but with a dramatically better performance. An E–D optimization framework that covers all the possible operational regions of a MOS transistor (strong, moderate, and weak inversion) is proposed. Finally, we ponder the question whether other logic families than traditional complementary CMOS might not be better suited to enable ULP at better performance.

Rationale

- Continued increase of computational density must be combined with decrease in energy per operation (EOP)
- Further scaling of supply voltage essential to accomplish that
  - The only other option is to keep on reducing activity
- Some key questions:
  - How far can the supply voltage be scaled?
  - What is the minimum energy per operation that can be obtained theoretically and practically?
  - What to do about the threshold voltage and leakage?
  - How to practically design circuits that approach the minimum energy bounds?

Slide 11.3
Already in the Introductory chapter of this book, it had become apparent that the continuation of technology scaling requires the power density (i.e., the power consumed per unit area) to be constant. The ITRS projects a different trajectory though, with both dynamic and static power densities continuing to increase over the coming decade unless some innovative solutions emerge. One option is to have the circuit do less – reduce the activity, in other words – but this seems hardly attractive. The other more sensible approach is to try continuing the scaling of the energy per operation (EOP). This begs for an answer to the following questions: Is there an absolute lower bound on the EOP? And how far away we are from reaching it?

It turns out that answers to these queries are closely related to the question of the minimum supply voltage at which a digital logic gate can still operate, which turns out to be well-defined. The major topic of this chapter is the exploration of circuit techniques that allow us to approach as close to that minimum as possible.

Opportunities for Ultra-Low Voltage

- Number of applications emerging that do not need high performance, only extremely low power dissipation
- Examples:
  - Standby operation for mobile components
  - Implanted electronics and artificial senses
  - Smart objects, fabrics, and e-textiles
- Need power levels below 1 mW (even µW in certain cases)

Slide 11.4
Although keeping the power density constant is one motivation for the continued search to lower the EOP, another, maybe even more important, reason is the exciting applications that only become feasible at very low energy/power levels. Consider, for instance, the digital wristwatch. The concept, though straightforward, only became attractive once the power dissipation

Ultra Low Power/Voltage Design

was made low enough for a single small battery to last for many years. As such, wristwatches in the early 1980s became the very first application using ultra low power and voltage design technologies.

Today, ULP technology is making it possible for a range of far more complex applications to become reality. Wireless sensor network nodes, combining integrated wireless front ends with signal acquisition and processing, are currently making their way into the market. Further power reductions by one or two orders of magnitude may enable even more futuristic functionality, such as intelligent materials, smart objects that respond to a much broader range of input sense, and the in situ observation of human cells. Each of these requires that the electronics are completely embedded into the object, and operate solely off the ambient energy. To realize this lofty goal, it is essential that power levels for the complete node are at the microwatt level, or below (remember the microwatt nodes described in Chapter 1).

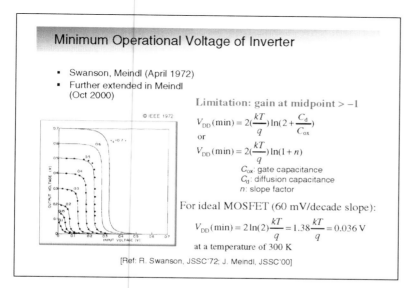

Slide 11.5

The question of the minimum operational voltage of a CMOS inverter was addressed in a landmark paper [Swanson72] in the early 1970s – published even before CMOS integrated circuits came in vogue! For an inverter to be regenerative and to have two distinct steady-state operation points (a "1" and a "0"), it is essential that the absolute value of the gain of the gate in the transient region be larger than 1. Solving for those conditions leads to an expression for V_{min} equal to $2(kT/q)\ln(1+n)$, where n is the slope factor of the transistors. One important observation is that V_{min} is proportional to the operational temperature T. Cooling down a CMOS circuit to temperatures close to absolute zero (e.g., liquid Helium), makes operation at mV levels possible. (Unfortunately, the energy going into the cooling more than often offsets the gains in operational energy.) Also, the closer the MOS transistor operating in sub-threshold mode gets to the ideal bipolar transistor behavior, the lower the minimum voltage. At room temperature, an ideal CMOS inverter (with a slope factor of 1) could marginally operate at as low as 36 mV!

Slide 11.6

Sub-threshold Modeling of CMOS Inverter

- From Chapter 2:

$$I_{DS} = I_S e^{\frac{V_{GS}-V_{TH}}{nkT/q}} \left(1 - e^{\frac{-V_{DS}}{kT/q}}\right) = I_0 e^{\frac{V_{GS}}{nkT/q}} \left(1 - e^{\frac{-V_{DS}}{kT/q}}\right)$$

where

$$I_0 = I_S e^{\frac{-V_{TH}}{nkT/q}}$$

(DIBL can be ignored at low voltages)

Given the importance of this expression, a quick derivation is worth undertaking. We assume that at these low operational voltages, the transistors operate only in the sub-threshold regime, which is often also called the *weak-inversion* mode. The current–voltage relationship for a MOS transistor in sub-threshold mode was presented in Chapter 2, and is repeated here for the sake of clarity. For low values of V_{DS}, the DIBL effect can be ignored.

Slide 11.7

Sub-threshold DC model of CMOS Inverter

Assume NMOS and PMOS are fully symmetrical and all voltages normalized to the thermal voltage $\Phi_T = kT/q$
($x_i = V_i/\Phi_T$; $x_o = V_o/\Phi_T$; $x_D = V_{DD}/\Phi_T$)

The VTC of the inverter for NMOS and PMOS in sub-threshold can be derived:

$$x_o = x_D + \ln\left(\frac{1 - G + \sqrt{(G-1)^2 + 4Ge^{-x_D}}}{2}\right) \quad \text{where } G = e^{(2x_i - x_D)/n}$$

so that

$$A_V = -\frac{2(1 - e^{x_o - x_D} - e^{-x_o} - e^{-x_D})}{n(2e^{-x_D} - e^{x_o - x_D} - e^{-x_o})} \quad \text{and} \quad A_{Vmax} = -(e^{x_D/2} - 1)/n$$

For $|A_{Vmax}| = 1$: $x_D = 2\ln(n+1)$

[Ref: E. Vittoz, CRC'05]

The (static) voltage transfer characteristic (VTC) of the inverter is derived by equating the current through the NMOS and PMOS transistors. The derivation is substantially simplified if we assume that two devices have exactly the same strength when operating in sub-threshold. Also, normalizing all voltages with respect to the thermal voltage Φ_T leads to more elegant expressions. Setting the gain to -1 yields the same expression for the minimum voltage as was derived by Swanson.

Slide 11.8

Using the analytical models derived in the previous slide, we can plot the VTC of the inverter. It becomes clear that, when the normalized supply voltage approaches its minimum value, the VTC degenerates, and the static noise margins are reduced to zero. With no gain in the intermediate region, distinguishing between "0" and "1" becomes impossible, and a flip-flop composed of such inverters would no longer be bi-stable. This presents a boundary condition. For reliable operation, a margin must be provided. As can be seen from the plots, setting the supply voltage at 4 times the

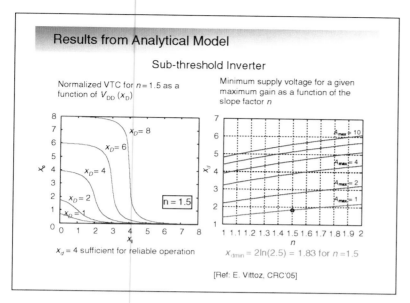

thermal voltage leads to reasonable noise margins (assuming $n = 1.5$). This is approximately equal to 100 mV.

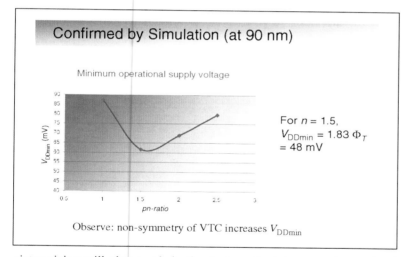

Slide 11.9

Simulations (for a 90 nm technology) confirm these results. When plotting the minimum supply voltage as a function of the PMOS/NMOS ratio, a minimum can be observed when the inverter is completely symmetrical, that is when the PMOS and NMOS transistors have identical drive strengths. Any deviation from the symmetry causes V_{min} to rise. This implies that transistor sizing will play a role in the design of minimum-voltage circuits.

Also worth noticing is that the simulated minimum voltage of 60 mV is slightly higher than the theoretical value of 48 mV. This is mostly owing to the definition of "operational" point. At 48 mV, the inverter is only marginally functional. In the simulation, we assume a small margin of approximately 25%.

Slide 11.10

The condition of symmetry between pull-up and pull-down networks for minimum-voltage operation proves to be important in more complex logic gates. Consider, for instance, a two-input NOR gate. The drive strengths of the pull-up and pull-down networks depend upon the input values.

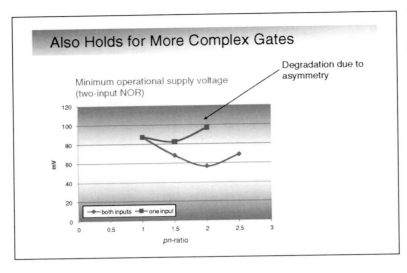

When only one input is switched (and the other fixed to "0"), the built-in asymmetry leads to a higher minimum voltage than when both are switched simultaneously. This leads to a useful design rule-of-thumb: when designing logic networks for minimum-voltage operation, *one should strive to make the gate topology symmetrical over all input conditions.*

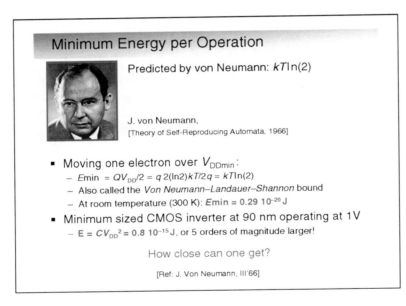

Slide 11.11

Now that the issue of the minimum voltage is settled, the question of the minimum energy per operation (EOP) can be tackled. In a follow-up to the 1972 paper by Swanson, Meindl [Meindl, JSSC'00] [argued that moving a single electron over the minimum voltage requires an energy equal to $kT\ln(2)$. This result is remarkable in a number of ways.

- This expression for the minimum energy for a digital operation was already predicted much earlier by John von Neumann (as reported in [von Neumann, 1966]). Landauer later established that this is only the case for "logically irreversible" operations in a physical computer that dissipate energy by generating a corresponding amount of entropy for each bit of information that then gets irreversibly erased. This bound hence does not hold for reversible computers (if such could be built) [Landauer, 1961].
- This is also exactly the same expression that was obtained in Chapter 6 for the minimum energy it takes to transmit a bit over an interconnect medium. That result was derived from Shannon's theorem, and was based on information-theoretic arguments.

The fact that the same expression is obtained coming from a number of different directions seems to be surprising at first. Upon closer analysis, all the derivations are based on a common

Ultra Low Power/Voltage Design

assumption of white thermal noise with a Gaussian distribution. Under such conditions, for a signal to be distinguishable it has to be at a level of ln(2) above the noise floor of kT. In light of its many origins, $kT\ln(2)$ is often called the *Shannon–von Neumann–Landauer limit*.

A more practical perspective is that a standard 90 nm CMOS inverter (with a 1 V supply) operates at an energy level that is approximately five orders of magnitude higher than the absolute minimum. Given that a margin of 100 above the absolute lower bound is probably necessary for reliable operation, this means that a further reduction in EOP by three orders of magnitude may be possible.

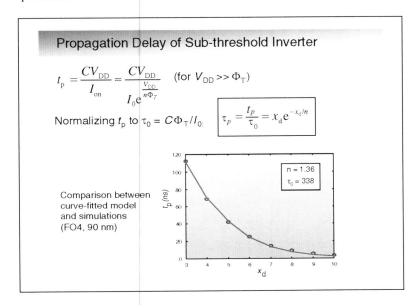

Slide 11.12

The above analysis, though useful in setting absolute bounds, ignores some practical aspects, such as leakage. Hence, lowering the voltage as low as we can may not necessarily be the right answer to minimize energy.

Operating an inverter in the sub-threshold region, however, may be one way to get closer to the minimum-energy bound. Yet, as should be no surprise, this comes at a substantial cost in performance. Following the common practice of this book, we again map the design task as an optimization problem in the E–D space.

One interesting by-product of operating in the sub-threshold region is that the equations are quite simple and are exponentials (as used to be the case for bipolar transistors). Under the earlier assumptions of symmetry, an expression of the inverter delay is readily derived. Observe again that a reduction in supply voltage has an exponential effect on the delay!

Slide 11.13

Given the very low current levels, it should be expected that waveforms would exhibit very slow rise and fall times. Hence, the impact of the input slope on delay might be considerable. Simulations show that the delay indeed rises with the input transition time, but the impact is not unlike what we observed for above-threshold circuits. The question also arises whether short-circuit currents (which were not considered in the expression derived in Slide 11.12) should be included in the delay analysis. This turns out to be not the case as long as the input and output signal slopes are balanced, or the input slope is smaller than t_0.

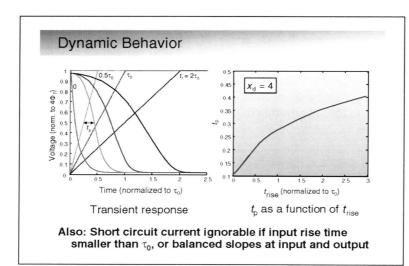

Slide 11.14

An expression for the power dissipation is also readily derived. For $x_d \geq 4$, the logic levels are approximately equal to the supply rails, and the leakage current simply equals I_0. For dynamic and static leakage power to be combined in a single expression, an activity factor α ($= 2t_p/T$) should be introduced, just as we did in Chapter 3.

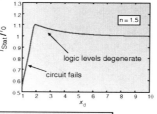

Ultra Low Power/Voltage Design

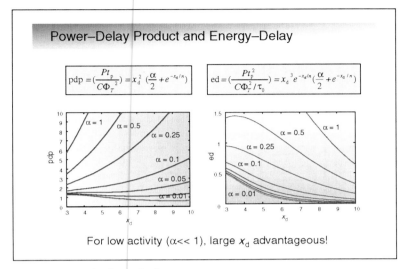

Slide 11.15
We can now plot the normalized power–delay and energy–delay curves as a function of the normalized supply voltage x_d and the activity α. When the activity is very high (α close to 1), the dynamic power dominates, and keeping x_d as low as possible helps. On the other hand, if the activity is low, increasing the supply voltage actually helps, as it decreases the clock period (and hence the amount of energy, leaking away over a single period).

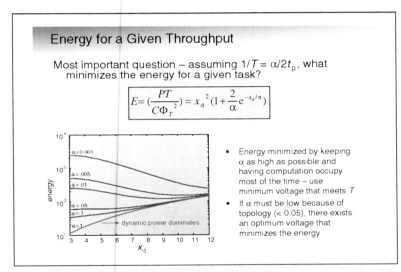

Slide 11.16
The power–delay and energy–delay metrics, as we have indicated earlier, are somewhat meaningless, as they fail to take the energy–delay trade-off into account. A more relevant question is what voltage minimizes the energy for a given task and a given performance requirement. The answer is simple: *keep the circuit as active as possible*, and *reduce the supply voltage to the minimum allowable value*. Low-activity circuits must be operated at higher voltages, which leads to higher energy per operation. This opens the door for some interesting logic- and architecture-level optimizations. Shallow logic is preferable over long complex critical paths. Similarly, increasing the activity using, for instance, time multiplexing is a good idea if minimum energy is the goal.

Slide 11.17
To demonstrate the potential of sub-threshold design and its potential pitfalls, we analyze a couple of case studies. In the first example, an energy-aware Fast Fourier Transform (FFT) module [A. Wang, ISSCC'04] is analyzed. As is typical for FFTs, the majority of the hardware is dedicated to the computation of the "butterfly" function, which involves a complex multiplication. Another

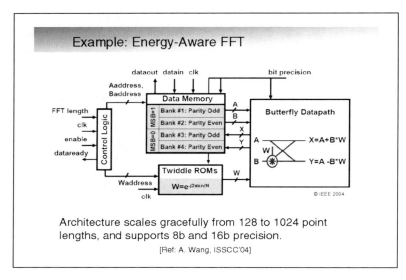

major component of the module is the memory block, storing the data vector. The architecture is parameterized, so that FFTs of lengths ranging from 128 to 1024 points, and data word lengths from 8 to 16 bit can be computed efficiently.

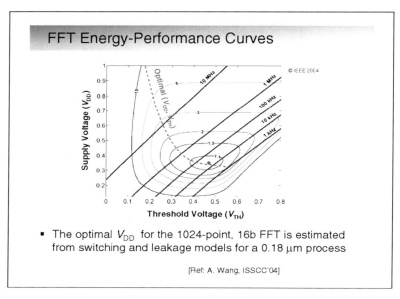

Slide 11.18

Using models in the style we have been doing so far, the energy–delay space of the FFT module is explored. In this study, we are most interested in determining the minimum energy point and its location. The design parameters under consideration are the supply and threshold voltages (for a fixed circuit topology).

The simulated energy–delay plots provide some interesting insights (for a 180 nm technology):

- There is indeed a minimum energy point, situated well in the sub-threshold region (V_{DD} = 0.35 V; V_{TH} = 0.45 V).
- This operation point occurs at a clock frequency of just above 10 kHz – obviously not that stellar ... but performance is not always an issue.
- As usual, an optimal energy–delay curve can be traced (shown in red).
- Lowering the throughput even further by going deeper into sub-threshold does not help, as the leakage power dominates everything, and causes the energy to increase.
- On the other hand, the performance can be increased dramatically (factor 10) if a small increase in energy above the minimum (25%) is allowed. This illustrates again that sub-threshold operation minimizes energy at a large performance cost.

Ultra Low Power/Voltage Design

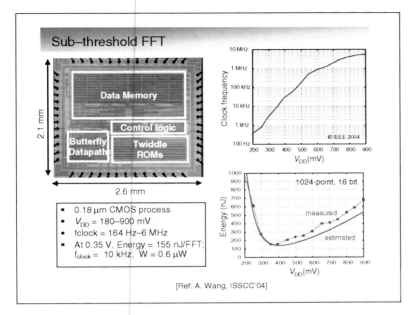

Slide 11.19

The simulation data are confirmed by actual measurements from a 180 nm prototype chip. The design is shown to be functional from 900 mV down to 180 mV. Our earlier analysis would have indicated even lower operational supply voltages — however, as we made apparent in Chapters 7 and 9, SRAM memories tend to be the first points of failure, and probably are setting the lower bound on the supply voltage in this case.

From an energy perspective, simulations and measurements track each other quite well, confirming again the existence of a minimum-energy point in the sub-threshold region.

Slide 11.20

This case study, though confirming the feasibility of operation in the weak-inversion region, also serves to highlight some of the challenges. Careful sizing is essential if correct operation is to be guaranteed. Symmetrical gate structures are of the essence to avoid the impact of data dependencies. Deviations caused by process variations further complicate the design challenge — especially threshold variations have a severe impact in the sub-threshold regime. And as mentioned in the previous slide, ensuring reliable memory operation at these ultra low voltages is not a sinecure. In the following slides, we explore some of these design concerns.

Slide 11.21

The operation in the sub-threshold region boils down to the balancing of the PMOS and NMOS leakage currents against each other. If one of the devices is too strong, it overwhelms the other device under all operational conditions, resulting in faulty operation. For the case of a simple

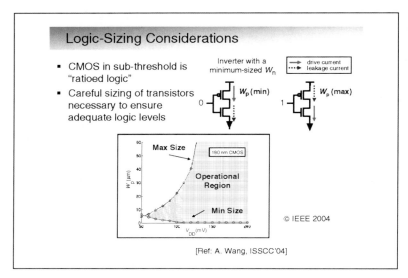

inverter and a given size of the NMOS transistor, the minimum and maximum widths of the PMOS transistor that bound the operational region can be computed, as shown in the graph. From a first glance, the margins seem to be quite comfortable unless the supply voltage drops below 100 mV.

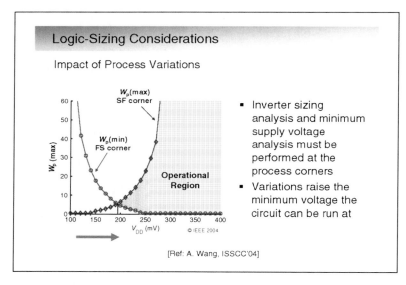

Slide 11.22
However, when process variations are taken into account, those margins shrink substantially. Simulations indicate that guaranteeing correct operation under all conditions is hard for supply voltages under 200 mV. From previous chapters, we know that the designer has a number of options to address this concern. One possibility is to use transistors with larger-than-minimum channel lengths. Unfortunately, this comes again with a performance penalty. Adaptive body biasing (ABB) is another option, and is quite effective if the variations are correlated over the area of interest.

Slide 11.23

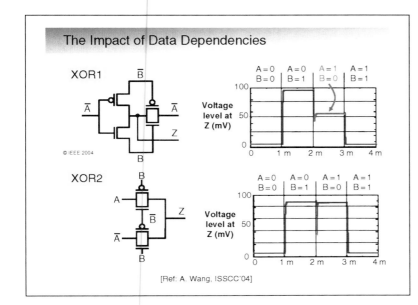

The importance of maintaining symmetry under different data conditions is illustrated in this slide, where the functionality of two different XOR configurations is examined. The popular four-transistor XOR1 topology fails at low voltages for $A = 1$ and $B = 0$. The more complex transmission gate based XOR2 performs fine over all input patterns.

Slide 11.24

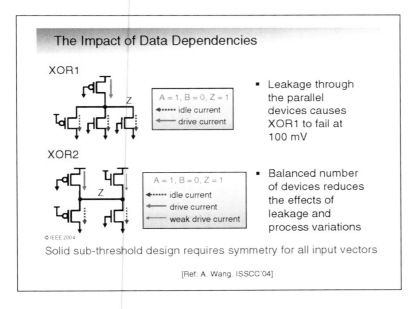

The reasons behind this failure become clear when the circuit is slightly redrawn. The failure of XOR1 for $A = 1$ and $B = 0$ is again caused by asymmetry: under these particular conditions, the leakage of the three "off" transistors overwhelms the single "on" PMOS device, resulting in an undefined output voltage. This is not the case for the XOR2, in which there are always two transistors pulling to V_{DD} and GND, respectively.

Slide 11.25

We have now demonstrated that logic can be safely operated at supply voltages of around 200 mV. This begs the question whether the same holds for another essential component of most digital designs: memory. From our discussions on memory in previous chapters (Chapters 7 and 9), you can probably surmise that this may not be that easy. Scaling down the supply voltage reduces the read, write, and hold static noise margins (SNMs). In addition, it makes the memory more sensitive to process variations, soft errors and erratic failures. Leakage through the cell-access transistors negatively impacts the power.

The Sub-threshold (Low Voltage) Memory Challenge

- **Obstacles that limit functionality at low voltage**
  - SNM
  - Write margin
  - Read current / bitline leakage
  - Soft errors
  - Erratic behavior

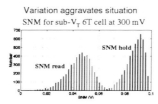
Variation aggravates situation
SNM for sub-V_T 6T cell at 300 mV

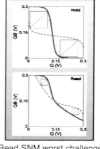
Read SNM worst challenge

When evaluating the potential impact of the different factors, it becomes apparent that the degradation of the read SNM is probably of primary concern. The bottom plot shows the distribution of the read and hold SNMs of a 90 nm 6T cell operated at 300 mV. As expected, the average value of the hold SNM (96 mV) is substantially higher than the read SNM (45 mV). In addition, the distribution is substantially wider, extending all the way to 0 mV.

Solutions to Enable Sub-V_{TH} Memory

- Standard 6T way of doing business won't work
- Voltage scaling versus transistor sizing
  - Current depends exponentially on voltages in sub-threshold
  - Use voltages (not sizing) to combat problems
- New bitcells
  - Buffer output to remove read SNM
  - Lower BL leakage
- Complemented with architectural strategies
  - ECC, interleaving, SRAM refresh, redundancy

Slide 11.26
This indicates that business as usual will not work for sub-threshold memory cells. One approach to combat the impact of variations is to increase the sizes of the transistors. Unfortunately, this leads, by necessity, to increased leakage, and may offset most of the power gains resulting from the voltage scaling.

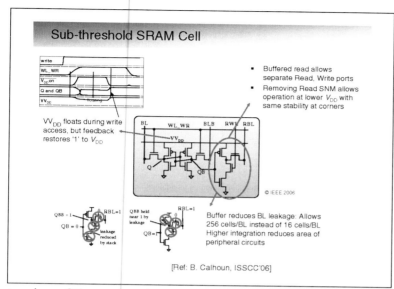

Slide 11.27
This slide presents a bit-cell that has proven to operate reliably down to 300 mV. It specifically addresses the susceptibility to read-SNM failures at low supply voltages. By using a separate read-access buffer, the read process is addressed separately from the cell optimization. Hence, the supply voltage can be reduced without jeopardizing the read operation. Obviously, this comes at a cost of three extra transistors. This is only one possible cell topology. Other authors have proposed a number of similar approaches (e.g., [Chen'06]).

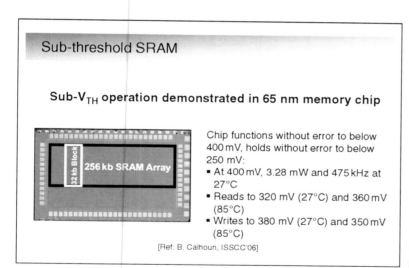

Slide 11.28
A 256 kb SRAM memory was designed and tested using the cell of the previous slide. Reliable operation down to 400 mV was demonstrated.

Though this represents great progress, it is clear that addressing further voltage scaling of SRAMs (or any other memory type) is essential if ultra low power design is to be successful. (sorry if this starts to sound like a broken record!)

Slide 11.29
Sub-threshold digital operation is also effective for the implementation of ultra low energy embedded microprocessors or microcontrollers. A sub-threshold processor for sensor network applications was developed by researchers at the University of Michigan. Implemented in a 130 nm CMOS technology with a 400 mV threshold (for $V_{DS} = 50$ mV), a minimum energy per instruction of 3.5 pJ is reached for a supply voltage of 350 mV. This is by far the lowest recorded energy efficiency for microprocessors or signal processors. At lower supply voltages, leakage power starts to dominate, and energy per instruction creeps upward again. If only the processor core is considered (ignoring memories and register files), the optimal-energy voltage shifts to lower levels

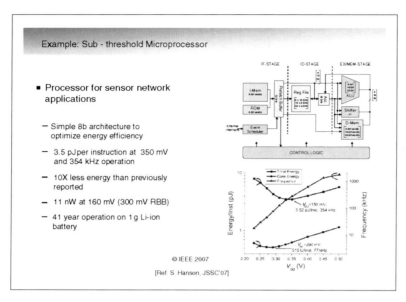

(i.e., 290 mV). Further energy reductions can be obtained by reverse biasing – at the expense of performance, for sure.

As was established earlier in this chapter, the minimum voltage operation coincides with the operational point where the driving strengths of NMOS and PMOS transistors under sub-threshold conditions are matched. Given the strong impact of process variations on the sub-threshold operation, accomplishing this match using design-time techniques only is hard. The use of runtime body-biasing of NMOS and PMOS transistors helps to compensate for the differences in driving strengths. In [Hanson'07], it was established that optimal application of substrate biasing helps to reduce the minimum operational voltage by 30–150 mV.

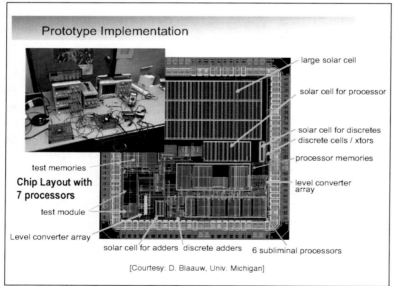

Slide 11.30

A prototype implementation of the sub-threshold processor is shown in this slide [Hanson07]. With the power dissipation of the processor low enough, it actually becomes possible to power the digital logic and memories from solar cells integrated on the same die. This probably represents one of the first ever totally energy self-contained processor dies. In addition to a number of processors, the chip also contains various test circuits as well as voltage regulators.

Note: Just recently, the same group published another incarnation of their ULP processor (now called the Phoenix) [Seok'08]. Implemented in a 180 nm CMOS process, the processor consumes only 29.6 pW in sleep mode, and 2.8 pJ/cycle in active mode at $V_{DD} = 0.5$ V for a clock frequency of 106 kHz. Observe that the authors deliberately chose to use an older-technology node to combat the leakage problems associated with newer processes. This turns out to be a big plus in reducing the standby power.

Slide 11.31

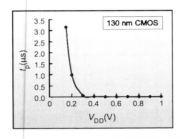

Is Sub-threshold the Way to Go?

- Achieves lowest possible energy dissipation
- But … at a dramatic cost in performance

The preceding discussion clearly demonstrated that sub-threshold operation is a good way to go if operating at minimum energy is the primary design goal. Although this was well-known in the ultra low power design community for a while (dating back to the late 1970s), it has caught the attention of the broader design community only over the past few years. This is inspiring some very interesting efforts in academia and industry, which will be worth-tracking over the coming years.

Yet, you should be fully aware that sub-threshold design comes with an enormous performance penalty. This is apparent in this chart, which plots the propagation delay as a function of the supply voltage (on a linear scale). As we had established earlier, an exponential degradation of the delay occurs once the circuit drops into the sub-threshold region. For applications that happily operate at clock rates of 100s of kHz, this is not an issue. However, not many applications fall in this class.

Slide 11.32

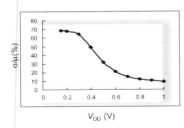

In Addition: Huge Timing Variance

- Normalized timing variance increases dramatically with V_{DD} reduction
- Design for yield means huge overhead at low voltages:
  - Worst-case design at 300 mV : >200% overkill

In addition, the design becomes a lot more vulnerable and sensitive to process variations. This is clearly demonstrated in this chart, which plots the variance of the delay as a function of the supply voltage. When operating under worst-case design conditions, this results in a huge delay penalty.

Slide 11.33

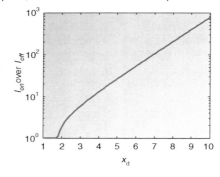

Increased Sensitivity to Variations

- Sub-threshold circuits operate at low I_{on}/I_{off} ratios, from about a 1000 to less than 10 (at $x_d = 4$)
- Small variations in device parameters can have a large impact, and threaten the circuit operation

Beyond the impact on performance, process variations may also threaten the reliability of the sub-threshold circuits. This is illustrated magnificently with a plot of the I_{on}/I_{off} ratio of an inverter as a function of the supply voltage (using the expressions derived earlier in the chapter). With ratios in the small tens, variations can easily cause the circuit to fail, or force an increase in the supply voltage. Proper transistor sizing can help to mitigate some of these vulnerabilities (as we had demonstrated in the design of the FFT processor). Yet, variability clearly puts a limit on how far sub-threshold design can be driven.

Slide 11.34

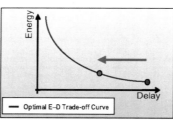

One Solution: Back Off a Bit ...

- The performance cost of minimum energy is exponentially high.
- Operating slightly above the threshold voltage improves performance dramatically while having small impact on energy

The Challenge: Modeling in the moderate-inversion region

A very attractive option to deliver both low energy as well as reasonable performance is obtained by practicing one of our established rules of thumb: that is, *operating at the extremities of the energy–delay space is rarely worth it*. Backing off just a bit gets you very close at a much smaller cost. Operating a circuit just above the threshold voltages avoids the exponential performance penalty.

One of the reasons designers avoided this region for a long time was the (mostly correct) perception that transistor models in this region are inaccurate. Manual analysis and optimization in this region is considered hard as well, as the simple performance models (such as the -law) do not apply any longer. However, the combination of reducing supply and constant threshold voltages has made the moderate-inversion region increasingly attractive. Even better, transistor and performance models are now available that cover all the operational regions of the MOS transistor equally well.

Ultra Low Power/Voltage Design

Slide 11.35

In 1995, Enz and Vittoz [Enz95] introduced the EKV MOS transistor model, which covers all operational regions of the transistor in a seamless manner. A crucial parameter in the model is the *Inversion Coefficient* (*IC*), which measures the degree of inversion of the device. Large values of *IC* indicate strong inversion, whereas values substantially below mean that the transistor operates in weak inversion (or sub-threshold). The transistor operates in moderate inversion when *IC* is between 1 and 10. *IC* is a direct function of the supply and threshold voltages.

Modeling over All Regions of Interest

- The EKV Model covers strong, moderate, and weak inversion regions

$$t_p = k \frac{CV_{DD}}{IC \cdot I_S}$$

where *k* is a fit factor, I_S the specific current and *IC* the inversion coefficient.

- Inversion Coefficient IC measures the degree of saturation

$$IC = \frac{I_{DS}}{I_S} = \frac{I_{DS}}{2n\mu C_{ox}(\frac{W}{L})\Phi_T^2}$$

and is related directly to V_{DD}

$$V_{DD} = \frac{V_{TH} + 2n\Phi_T \ln(e^{\sqrt{IC}} - 1)}{1 + \lambda_d}$$

[Ref: C. Enz, Analog'95]

Slide 11.36

The relationship between supply voltage and inversion coefficient *IC* is plotted in this chart (for a given threshold voltage). Moving from weak to moderate and strong inversion requires an increasing amount of supply voltage investment.

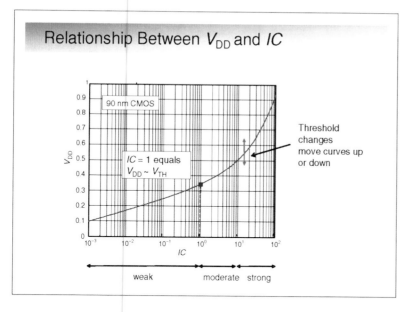

Slide 11.37

The EKV-based delay equation presented in Slide 11.33 provides an excellent match over a broad operation range (after fitting the model parameters *k* and I_S to the simulation data). Though the delay model performs very well in the weak- and moderate-inversion regions, some deviation can be observed for stronger inversion, caused by the fact that the model does not handle velocity saturation very well. However, this should not be an issue in most contemporary designs, in which

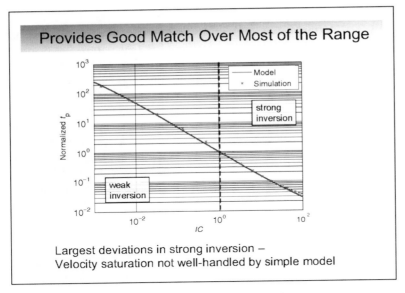

the *IC* rarely exceeds 100. Also, more recent incarnations of the model have resolved this issue. This analysis demonstrates that there should be no fear in operating a design at the boundary of the moderate and weak inversion.

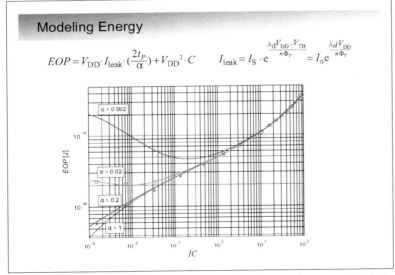

Slide 11.38

An energy model based on *IC* can be constructed as well. Again, the activity parameter α (= $2t_p/T$) is used to combine the active and static energy components. The capacitance *C* is the *effective capacitance*, which is the average switching capacitance per operation. The energy-per-operation (EOP) is plotted as a function of *IC* and α. The full lines represent the model, whereas the dots represent simulation results. As can be expected, the EOP is minimized for all operational regimes when α = 1. Under that condition, reducing the supply voltage as much as possible and operating the circuit at very low values of *IC* make sense. However, for less-active circuits, the minimum energy point shifts to larger values of *IC*, and eventually even moves to the moderate threshold region.

Slide 11.39

We can now combine the EKV delay and energy models, and plot the energy and delay planes as a function of V_{DD} and V_{TH}, with activity as an independent parameter. For an activity level of 0.02, we see that the minimum EOP (red dots) moves from the weak- to the moderate-inversion region

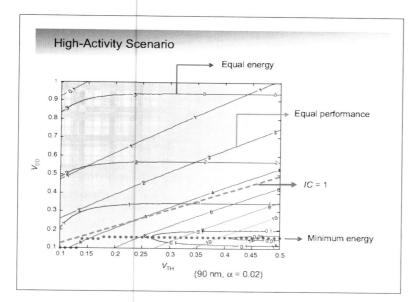

for increasing performance values. It is interesting to notice that the increased performance is obtained almost completely from a reduction of the threshold voltage at a fixed supply voltage. Also, observe how the equal-energy curves are almost horizontal, which illustrates how little impact the increased performance has on the EOP.

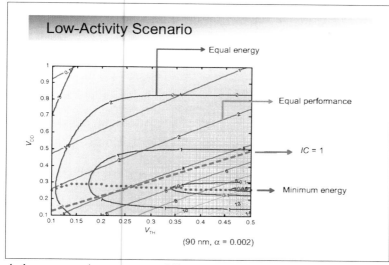

Slide 11.40

When the activity is further reduced ($\alpha = 0.002$), the minimum-energy point occurs at a higher supply voltage. Hence, the moderate-inversion region becomes attractive even earlier.

This set of simple-analysis examples shows that the EKV model is an effective tool in analyzing the energy–delay performance over a broad range of operation regimes. Even from this simple example, we can deduce some important guiding principles (most of which we had already mentioned when discussing sub-threshold behavior):

- From an energy perspective, a circuit performs better if the activity (i.e., duty cycle) is as high as possible.
- Sometimes this is not possible – for instance, if the required throughput is substantially below what the technology can deliver, the optimum operation point is not within the reliable

operation range of the technology, or the operation conditions vary over a wide range. In such cases, there are a number of possible solutions to consider:

1. Change the architecture to increase activity (e.g., time multiplexing).
2. Run the circuit at a high activity level, and put it in standby when finished.
3. Find the best possible operation condition given the lower activity level. This will often be at a higher supply voltage.

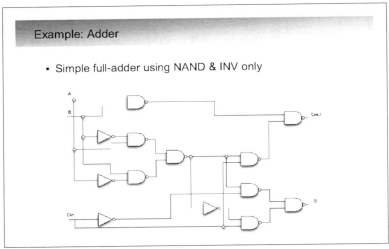

Slide 11.41
Whereas the example used in the previous slides is quite indicative in illustrating the potential of the EKV model, it is quite simple as well. To demonstrate that the modeling and optimization strategy also works for more complex examples, let us consider the case of a full adder cell. The cell we are considering is implemented using only two-input NAND gates and inverters.

The EKV model for the 90 nm CMOS technology-of-choice was derived in the earlier slides. The logical-effort model parameters are also available. For a small effort, we can now derive the delay and energy models of the complete adder.

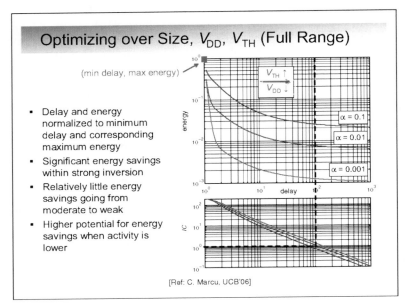

Slide 11.42
The resulting energy–delay curves as a function of activity (and the inversion coefficient IC) are plotted here. Observe that these analytical results now span all operational regions. As expected, the largest energy savings are obtained when migrating from the strong- to the moderate-inversion region. Going from moderate to weak inversion has only a minor impact on the EOP. Evident also is that the lowest delay comes at a huge penalty in energy. In addition, the

graph shows that a larger reduction in energy consumption is achievable for lower activity factors. This can largely be explained by the increased importance of leakage currents at low activities.

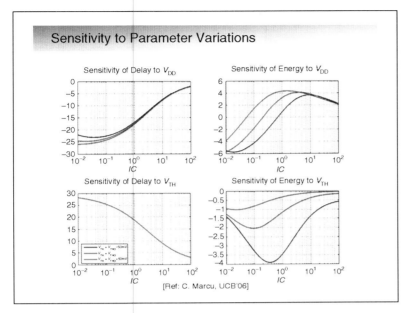

Slide 11.43
Another advantage of having analytical delay and energy expressions available is that sensitivities to parameter variations can be easily computed. In this slide, we plot the normalized sensitivities of delay and energy with respect to variations in the supply and threshold voltages (for different values of V_{TH}). Please observe that these sensitivities hold only for small deviations of the corresponding variable, as they represent the gradients of energy and delay.

The plots on the left show a major increase in delay sensitivity with respect to both V_{DD} and V_{TH} when going into weak inversion (more or less independent of the chosen V_{TH}). The sensitivity of energy with respect to V_{DD} is positive when dynamic energy dominates, and negative when leakage takes the largest share. This makes intuitive sense: an increase in V_{DD} in strong inversion mainly affects the energy consumption; in weak inversion, it increases the performance substantially, reducing the time the circuit is leaking. The absolute sensitivity for energy with respect to V_{TH} turns out to be equal for all values of V_{TH}. However, owing to the different absolute values of the energy for different V_{TH}, a maximum in the normalized sensitivity occurs at the point where the absolute energy is at its minimum.

Slide 11.44
So far, we have investigated how ultra low energy operation can be obtained by further lowering of the supply voltage, while keeping the threshold voltage more or less constant. Though this enables substantial energy reductions, the ultimate savings are limited. Moreover, the Von Neumann–Landauer–Shannon bound stays far out of reach. This becomes more than apparent in the graph on this slide, which shows the energy–delay curves for an inverter for technologies ranging from 90 nm to 22 nm. The graphs are obtained from simulations of a 423-stage ring-oscillator using the predictive technology models (PTM), operating at the nominal threshold voltages. Though it is clear from the simulations that the minimum EOP will continue to scale down, the projected reduction is quite small – no more than a factor of four, over five technology generations. This, of course, does not reflect potential technology innovations that are not included in the models. Even when assuming that the ultimate energy bound is a factor of 500 above $kT\ln(2)$ (for reliability reasons), the 22 nm solution still is at least a factor of 40 above that!

Hence, we cannot refrain from wondering what is needed to reduce the energy even further. Aside from adopting completely different technologies, it seems that the only plausible option is to

Moving the Minimum Energy Point

- Having the minimum-energy point in the sub-threshold region is unattractive
  - Sub-threshold energy savings are small and expensive
  - Further technology scaling not offering much relief
- Can it be moved upward?
- Or equivalently... Can we lower the threshold?

Remember the stack effect ...

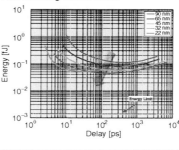

find ways to further reduce the threshold voltages while avoiding substantial increases in leakage current. The net effect of this would be to move the minimum-energy point out of weak into moderate inversion. A number of ways to accomplish this can be envisioned:

- **Adoption of switching devices with steep sub-threshold slopes (< 60 mV/dec)** – Such transistors, if available, would definitely provide a major boost to energy-efficient logic. Some promising technologies have recently been proposed by the research community. Unfortunately, reliable production of any of those is still a long distance away. It is definitely worthwhile keeping an eye on the developments in this area, anyhow.
- **Adoption of different logic styles** – One of the main disadvantages of complementary CMOS logic is that leakage control is not easy. Other logic styles offer more precise control of the off-current, and hence may be better candidates for ultra low voltage/power logic.

Complex Versus Simple Gates

- Example (from Chapter 4)

  Fan-in(4) versus Fan-in(2)

Complex gates improve the I_{on}/I_{off} ratio!

Slide 11.45
In Chapter 4, we analyzed the *stack effect*, which causes the off-current to grow slower than the on-current as a function of fan-in in a cascode connection of transistors. Exploiting this effect may help to create logic structures with reasonable I_{on}/I_{off} ratios, even while reducing the threshold voltages. The minimum-energy point for large fan-in (complex) gates hence should be located at lower threshold voltages. It should be repeated that, as an additional benefit, complex gates typically come with lower parasitic capacitances.

Ultra Low Power/Voltage Design

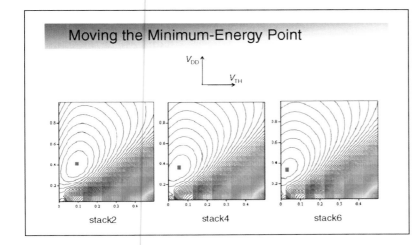

Slide 11.46
The simulations shown in this slide, in which the equal-energy curves are plotted as a function of V_{DD} and V_{TH} for different stack depths, confirm this. A clear shift to lower threshold values for larger stack depths can be observed.

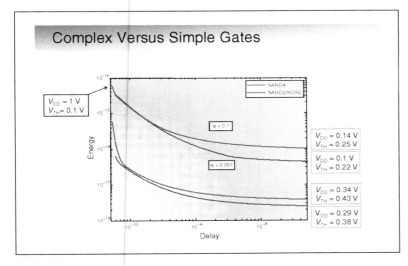

Slide 11.47
This observation is confirmed by comparing the energy–delay plots of a four-input NAND and an equivalent NAND–NOR implementation using only fan-in of 2 gates. Independent of the activity levels, the NAND4 implementation consumes less energy for the same performance.

Slide 11.48
Though this seems to be an encouraging line of thought, the overall impact is limited. Additional modifications in the gate topology are necessary for more dramatic improvements. One promising option to do so is offered by the pass-transistor logic (PTL) approach. Over the years, PTL has often been considered as a premier candidate for low-energy computation, mainly owing to its simple structure and small overhead. Even in the original papers on CPL (complementary pass-transistor logic) [Yano'90], it was suggested that the threshold voltages of the pass-transistors could be reduced to almost zero to improve performance with little impact on energy. At that time, standby power was not considered to be a major issue yet, and it turns out that sneak paths between inputs make this approach non-effective from a leakage perspective.

Yet, a basic observation stands: networks of switches – regardless of their complexity – on their own do not add any meaningful leakage, as they do not feature any direct resistive connections to either V_{DD} or GND. The only leakage of such a network occurs through the parasitic diodes, and is

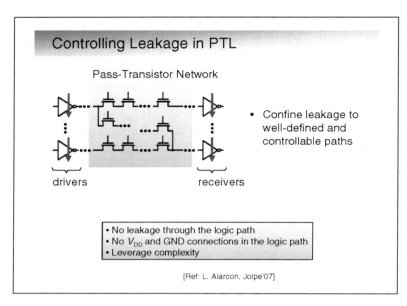

small. The switches do nothing more (or less) than steering the currents between the input and output nodes.

Hence, a clever connection and control of the input and output sources could help to combine to attractive features: complex logic with very low thresholds and controllable on- and off-currents.

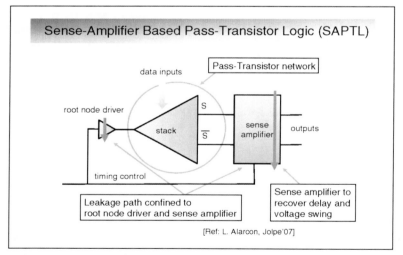

Slide 11.49

The sense amplifier based PTL (SAPTL) family represents an example of such a scheme [Alarcon'07]. In evaluation mode, current is injected into the root of a pass-transistor tree. The network routes the current to either the S or the $\bar{S}$ node. A high-impedance sense amplifier, turned on only when the voltage difference between S and $\bar{S}$ has built up sufficiently, is used to restore voltage levels and to speed up the computation. As a result, the gate can function correctly even at very low I_{on}/I_{off} ratios. Observe that leakage current only occurs in the driver and sense amplifiers, and hence can be carefully controlled, independent of the logic complexity.

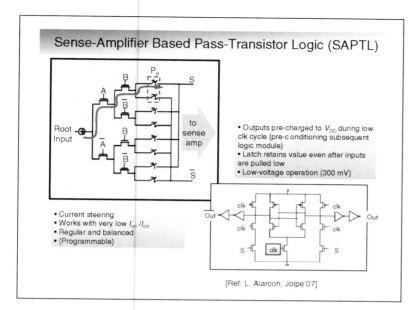

Slide 11.50

A more detailed transistor diagram is shown in this slide. The pass-transistor network is implemented as an inverse tree, and can be made programmable (just like an FPGA Look-up table or LUT) by adding an extra layer of switches. The sense amplifier output nodes are pre-charged high, and get conditionally pulled low during evaluation, effectively latching the result. Correct operation down to 300 mV has been demonstrated.

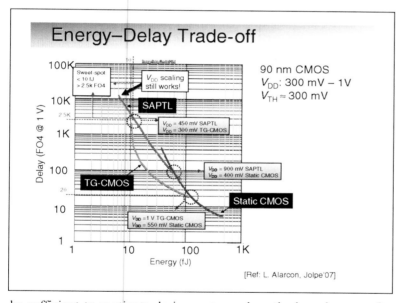

Slide 11.51

The energy–delay curves of equivalent functions in CMOS, transmission-gate logic, and SAPTL are plotted in this chart. The graph clearly demonstrates how each logic style has its own "sweet spot". Most importantly though, it shows that pass-transistor based logic families extend the operational range of MOS logic to energy levels that are substantially below what can be accomplished with traditional complementary CMOS. This observation alone should be sufficient to motivate designers to explore the broad space of opportunities offered by alternative logic styles. We contend that the exploration space will expand even further with the emergence of new switching devices.

Slide 11.52

In summary, further scaling of the EOP is essential if further miniaturization of computing is to continue. There is definitely room for improvement, as indicated by the Von Neumann–Landauer–Shannon bound on digital switching. Operating the transistor in the weak- or moderate-inversion

Summary

- To continue scaling, a reduction in energy per operation is necessary
- This is complicated by the perceived lower limit on the supply voltage
- Design techniques such as circuits operating in weak or moderate inversion, combined with innovative logic styles, are essential if voltage scaling is to continue
- Ultimately the deterministic Boolean model of computation may have to be abandoned

regions helps somewhat to get closer to this far-out bound, but does not make huge inroads. The only way to accomplish the latter is to profoundly revise the ways we design logic functions. Moreover, it will most probably require that we revisit the way we do logic altogether. In today's design world, the outcome of logic is supposed to be fully deterministic. When we approach the physical limits of scaling, it may be unwise to stick to this model – when all processes are statistical, maybe computation should be as well.

Slides 11.53–11.54
Some references . . .

References

Books and Book Chapters
- J. von Neumann, "*Theory of self-reproducing automata,*" A.W. Burks, Ed., University of Illinois Press, Urbana, 1966.
- E. Vittoz, "Weak Inversion for Ultimate Low-Power Logic," in C. Piguet, Ed., *Low-Power Electronics Design*, Ch. 16, CRC Press, 2005.
- A. Wang and A. Chandrakasan, *Sub-Threshold Design for Ultra Low-Power Systems*, Springer, 2006.

Articles
- L. Alarcon, T.T. Liu, M. Pierson and J. Rabaey, "Exploring very low-energy logic: A case study," *Journal of Low Power Electronics*, 3(3), Dec. 2007.
- B. Calhoun and A. Chandrakasan, "A 256kb Sub-threshold SRAM in 65nm CMOS," Digest of Technical Papers, ISSCC 2006, pp. 2592–2601, San Francisco, Feb. 2006.
- J. Chen et al., "An ultra-low_power memory with a subthreshold power supply voltage," *IEEE Journal of Solid-State Circuits*, 41(10), pp. 2344-2353, Oct. 2006.
- C. Enz, F. Krummenacher and E. Vittoz, "An analytical MOS transistor model valid in all regions of operation and Dedicated to low-voltage and low-current applications," *Analog Integrated Circuits and Signal Proc.*, 8, pp. 83–114, July 1995.
- S. Hanson et al., "Exploring variability and performance in a sub-200-mV Processor," *Journal of Solid State Circuits*, 43(4), pp. 881–891, Apr. 2008.
- R. Landauer, "Irreversibility and heat generation in the computing process," *IBM Journal of Research and Development*, 5:183–191, 1961.
- C. Marcu, M. Mark and J. Richmond, "Energy-performance optimization considerations in all regions of MOSFET operation with Emphasis on IC=1", Project Report EE241, UC Berkeley, Spring 2006.
- J.D. Meindl and J. Davis, "The fundamental limit on binary switching energy for tera scale integration (TSI)", *IEEE Journal of Solid-State Circuits*, 35(10), pp. 1515–1516, Oct. 2000.
- M. Seok et al., "The phoenix processor: A 30 pW platform for sensor applications," *Proceedings VLSI Symposium*, Honolulu, June 2008.

References (contd.)

- R. Swanson and J. Meindl, "Ion-implanted complementary MOS transistors in low-voltage Circuits," *IEEE Journal of Solid-State Circuits*, SC-7, pp. 146–153, Apr. 1972.
- E. Vittoz and J. Fellrath, "CMOS analog integrated circuits based on weak-inversion operation," *IEEE Journal of Solid-State Circuits*, SC-12, pp. 224–231, June 1977.
- A. Wang and A. Chandrakasan, "A 180mV FFT processor using subthreshold circuit techniques", Digest of Technical Papers, ISSCC 2004, pp. 292–293, San Francisco, Feb. 2004.
- K. Yano et al., "A 3.8 ns CMOS 16 × 16 Multiplier using complementary pass-transistor logic," *IEEE Journal of Solid-State Circuits*, SC-25(2), pp. 388–395, Apr. 1990.

Chapter 12
Low Power Design Methodologies and Flows

**Low Power
Design Methodologies and Flows**

Jerry Frenkil
Jan M. Rabaey

Slide 12.1
The goal of this chapter is to describe the methodologies and flows currently used in low-power design. It is one thing to understand a technique for achieving low power; it is another to understand how to efficiently and effectively implement that technique. Previous chapters have focused upon particular techniques and how they achieve energy efficiency. This chapter explores methodologies for implementing those techniques along with issues and trade-offs associated with those methodologies.

Slide 12.2
There is more to the story of low-power design than power minimization. A substantial amount of time and effort are needed to achieve that sought-after energy efficiency. One of the main challenges is that, in effect, the task of low-power design itself is a multi-variable optimization problem. As we will later see, optimizing for power in one mode of operation can actually cause increased power in other modes and, if one is not careful, the time and effort spent on various power-related issues can swell to match and even exceed that spent on basic functionality. Hence, the motivations for an effective low-power design methodology include *minimizing time and effort* in addition to minimizing power.

The first decade of power-efficient design was mostly devoted to the development of design technologies – that is, techniques to reduce power. With power efficiency becoming a prime design metric, at the same level of importance as area and speed, electronic design automation (EDA) companies are gradually getting engaged in the development of integrated design flows (and the supporting tools) for low power. This proves to be advantageous, as many design techniques, described in this book, are now becoming an integral part of the design flow, and, as such, are available to a broader audience. This is testified by the

Low Power Design Methodology – Motivations

- Minimize power
  - Reduce power in various modes of device operation
  - Dynamic power, leakage power, or total power

- Minimize time
  - Reduce power quickly
    - Complete the design in as little time as possible
  - Prevent downstream issues caused by LPD techniques
    - Avoid complicating timing and functional verification

- Minimize effort
  - Reduce power efficiently
    - Complete the design with as few resources as possible
  - Prevent downstream issues caused by LPD techniques
    - Avoid complicating timing and functional verification

detailed low-power design methodologies, advertised by each EDA company (e.g., [Cadence, Sequence, Synopsys]).

Methodology Issues

- **Power Characterization and Modeling**
  - How to generate macro-model power data?
  - Model accuracy
- **Power Analysis**
  - When to analyze?
  - Which modes to analyze?
  - How to use the data?
- **Power Reduction**
  - Logical modes of operation
    - For which modes should power be reduced?
  - Dynamic power versus leakage power
  - Physical design implications
  - Functional and timing verification
  - Return on Investment
    - How much power is reduced for the extra effort? Extra logic? Extra area?
- **Power Integrity**
  - Peak instantaneous power
  - Electromigration
  - Impact on timing

Slide 12.3

There are a variety of issues to consider when developing a low-power design methodology.

The first issue pertains to *characterization and modeling*. In cell-based system-on-chip (SoC) design flows, each cell must have a power model in addition to the usual functional and timing models. Without a full set of sufficiently accurate models, the methodology's effectiveness will be compromised.

The methodology must define the points in the design process at which *power analysis* will be performed; that is, is it only a major milestone check or something that is performed regularly and with great frequency, or something in between? The answer may depend upon the project's goals and the complexity of the device being designed. Similarly, what will the results of the analysis be used for – is it simply to check whether the overall design is on target to meet a basic specification or to check against multiple operational modes, or to use as a guidance on how to reduce power further? Or perhaps, is the goal not power reduction at all, but instead verification of the power delivery network (also known as *power integrity*)? The answers to these questions will determine the frequency of and the level of detail needed from power analysis.

Similarly, *power reduction* efforts should be driven by specific project objectives. How many different power targets have been specified? What are the priorities of those targets relative to other project parameters such as die size, performance, and design time? The answers to these questions help determine not only the particular power reduction techniques to be employed but also the methodologies used to implement them.

Some Methodology Reflections

- Generate required models to support chosen methodology
- Analyze power early and often
- Employ (only) as many LPD techniques as needed to reach the power spec
  - Some techniques are used at only one abstraction level; others are used at several
    - Clock Gating: multiple levels
    - Timing-slack redistribution: only physical level
- Methodology particulars dependent upon choice of techniques
  - Power gating versus Clock gating
    - Very different methodologies
- No free lunch
  - Most LPD techniques complicate the design flow
  - Methodology must avoid or mitigate the complications

Slide 12.4
Despite the various questions raised in contemplating the aforementioned issues, several truisms hold. Models are needed and can be generated during the earliest stages of the project, often well before usage is actually required. Power should be analyzed early and often so as to keep a close watch on it and to prevent surprises. The number of reduction techniques should be limited to only those that are needed to meet project power targets as each technique usually complicates other design tasks.

Power Characterization and Modeling

- **Objective: Build models to support low-power design methodology**
  - Power consumption models
  - Current waveform models
  - Voltage-sensitive timing models
- **Issues**
  - Model formats, structures, and complexity
    - Example: Liberty-power
  - Run times
  - Accuracy

[Ref: Liberty]

Slide 12.5
In cell-based SoC design flows, power models are needed for each cell used in the design. So an enabling task is to build the models required for the chosen methodology. For example, to calculate leakage power consumption, leakage currents must be characterized and modeled. To analyze voltage drop, basic average-power models can be used, but for best accuracy, current waveform models should be created for each cell. Similarly, if the methodology calls for checking the effects of voltage drop on timing (more on this later), voltage-sensitive timing models need to be built.

Whichever models are built, they must be compatible with the tools to be used in the methodology. There are several standards, along with proprietary formats, for modeling power, and usually various modeling options exist for each format. The most widely used format for modeling power at the time of this writing is *Liberty power*. It provides for pin-based and path-based dynamic power modeling. In the former case, power is consumed whenever a particular pin on that cell transitions, whereas in the latter a full path – from an input transition to an output transition – must be stimulated to consume power. In either case, power is represented as a single value, for a particular input transition time and output load pair, representing the total power consumed for the entire event. A recent extension to Liberty power, known as CCS (composite current source [Liberty Modeling Standard]) uses a time-varying current waveform to model power, instead of a single value.

Both formats support both state-dependent and state-independent leakage models. As the name implies, state-independent models use a single value to represent the leakage of the cell, independent of whatever state the cell might be in, whereas state dependent models contain a different leakage value for each state. The trade-off here is model complexity versus accuracy and evaluation time. A state-independent model sacrifices accuracy for fast evaluations and a compact model, whereas a state-dependent model provides the best accuracy. In most cases, for standard-cell type primitives, full state dependent models are preferred. State-independent models or limited state dependency models are often used for more complex cells or for those cells with more than about eight inputs.

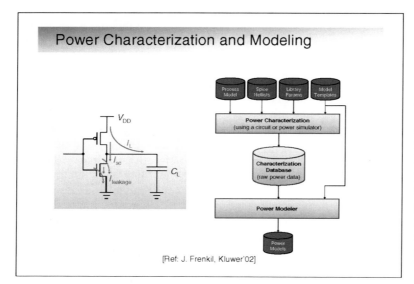

Slide 12.6
A common flow for generating power models is shown here. This flow is almost identical to that for generating timing models. In effect, multiple SPICE, or SPICE-like, simulations are run on the transistor-level netlist for each cell primitive: one simulation for each input-to-output path/input transition time/output load combination, monitoring the current drawn from the supply. The resulting data, I_L, I_{sc}, and $I_{leakage}$ (in this case), are collected and formatted into a particular model structure. This flow is usually encapsulated into a characterization tool that automatically runs SPICE based upon user-supplied parameters such as the characterization conditions (PVT: process, voltage, temperature), data model table sizes, and types of characterization to perform (power, timing, noise, etc.). The simulation stimulus is automatically generated by exhaustively stimulating every potential input-to-output path. However, the ease of use of this approach is mitigated by its lack of scalability, as the number of simulations grows exponentially $O(2^n)$.

Note that this type of characterization can be performed on larger objects or functional blocks, such as memories, controllers, or even processors. In such applications, the characterization engine might be a high-capacity SPICE simulator, or a gate or RTL (register transfer level) power analysis tool. For objects that have a large number of states, a target model template is essential. Such a template specifies the specific logical conditions under which to characterize the target block. This utilizes the block designer's knowledge to avoid the 2^n simulation runs.

Slide 12.7

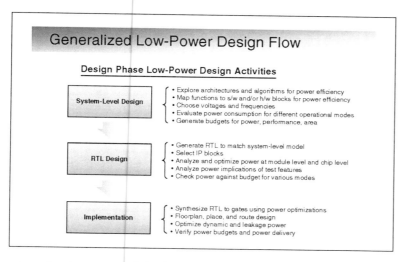

A generalized low-power design flow involves both analysis and reduction at multiple points in the development cycle, although the precise motivations for each analysis and reduction activity will differ. In this abstracted view of a low-power design flow, the overall effort is divided into three phases: system-level design, RTL design, and implementation. The system-level design (SLD) phase is when most of the large design decisions are made, especially in regard to how particular algorithms will be implemented. As such, this is also the phase in which designers have the largest opportunities to influence power consumption.

The RTL design phase is when the decisions made during the system-level design phase are codified into executable RTL form. The RTL may be manually coded to match the system conceived in the earlier phase, or it may be directly synthesized from a system-level description in C, C++, or SystemVerilog.

The implementation phase includes both the synthesis of the RTL description into a netlist of logic gates and the physical implementation of that netlist. It also includes tasks associated with final signoff such as timing and power closure and power grid verification.

Slide 12.8

Power analysis has two different, albeit related, motivations. The first is to determine whether the power consumption characteristics meet the desired specification; the second is to identify opportunities, or to enable methods, for reducing the power if the specification is not met.

The *technique* for analyzing power is to simulate the design using a power simulator or create an estimate using an estimation tool. The *methodology* is to do so regularly and automatically by creating a set of power regression tests as soon as possible in the overall development. Such a setup raises the awareness not just of the design's overall power characteristics but also of the impact of individual design decisions. Having the power data always available and regularly updated is a major help in accomplishing such analysis.

Slide 12.9

This "early and often" methodology raises some issues requiring consideration. A general trade-off exists between the amount of detail used to generate a power estimate and its accuracy – the earlier the estimate, the less detail is available and the less accuracy is possible. On the other hand, the later in the design process the design is analyzed, the more detailed the information (sized gates, extracted parasitics) available and the greater the accuracy. But this increased accuracy comes at the expense of longer run times and increasing difficulty in using the results to find power reduction opportunities, as the greater detail can obscure the big picture – a "can't see the forest for the trees" situation. In addition, the later in the design process, the harder it is to introduce major changes with large power impact. A useful concept to keep in mind is the following: *the granularity of the analysis should be on par with the impact of the design decisions that are being made.*

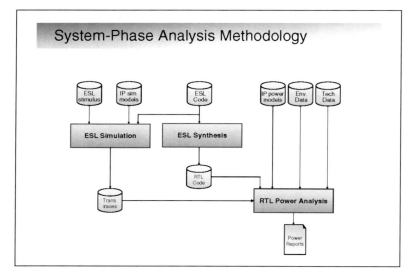

Slide 12.10

A power analysis flow during the system-level design phase involves binding different parts of the system description to particular hardware elements. This slide shows a flow in which the binding is accomplished by synthesizing the ESL (electronic system level) description to RTL code. A simulation of the ESL description produces a set of activity data, often in the form of *transaction traces*. An RTL power estimator will read the RTL code and the transaction traces to calculate the power. Note that several other inputs are required as well: environmental data (such as power supply voltages and external load capacitances), technology data (representing the target fabrication technology – this is often encapsulated in the power models for the cell primitives) and power models for any non-synthesizable IP blocks.

These "additional" inputs are also required to estimate power during other phases and at other abstraction levels as well.

It is worth observing that this picture presents an ideal scenario. Today, most SoC designs rarely use a fully-automated analysis flow in the early phases of the design process. Power analysis is often performed using spreadsheets, combining activity estimates for the different operational modes with design data obtained from previous designs, IP vendors, or published data.

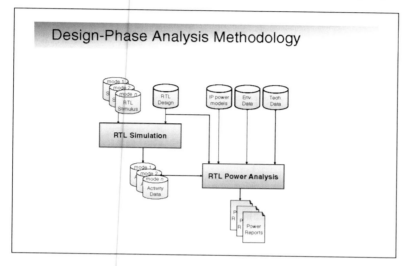

Slide 12.11

This flow is similar to that shown for the *system phase*, but with several significant differences. First, the design is simulated and analyzed at a single abstraction level: RTL. Second, the power analysis uses activity data (nodal transition changes) instead of transaction data. These two differences result in longer run times for a given simulation period (or, equivalently, a shorter simulation period for the same run time) but provide better calculation accuracy. A third difference is that this flow shows several RTL simulations (resulting in several activity files and power reports); this represents the idea that modal operation will be explored during this phase.

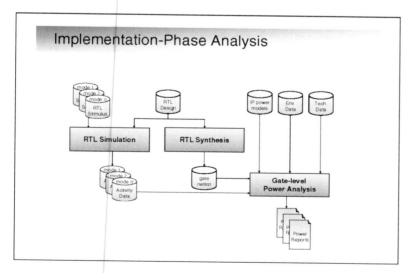

Slide 12.12

This flow bears great resemblance to the flow shown for the *design phase*, the primary difference being that a *gate-level netlist* is used for analysis instead of the RTL design. Note that the simulation is still performed on the RTL design. Although it is certainly possible (and even simpler in terms of design automation technology) to obtain activities from a gate-level simulation, in practice the gate-level simulations are often difficult to set up and too time-consuming to run. Using RTL simulation data, as shown here, enables power analysis over longer simulation periods with only a slight compromise in accuracy. Activity data for the nodes in the gate-level netlist that are not present in the RTL simulation are calculated probabilistically.

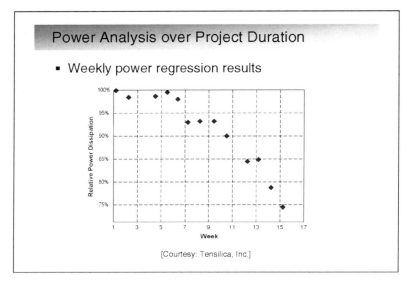

Slide 12.13
Shown here are the power analysis results for a microprocessor design, tracked over the course of the RTL design phase. In this particular case, power regressions were set up to run automatically every weekend enabling the design team to easily see and understand where they were relative to the project goals. This type of attention is needed to cultivate an environment of power awareness, enabling the team – designers and managers alike – to efficiently pursue their power objectives. Two particular items should be noted. The power goal of this project was to achieve a 25% power reduction from the previous design; it can be seen that this goal was achieved by week 15. Second, during week 13, a change occurred that resulted in power actually increasing from the previous week's results, promptly alerting the team that corrective action was in order. It can also be seen that the corrective action occurred by week 14.

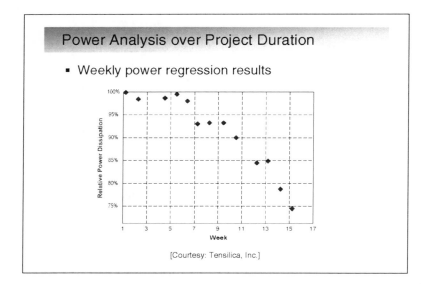

Slide 12.14
For some projects, it is important only to know whether a particular power target is achieved or not. For others, the goal is to reduce power as much as possible. For the latter cases it is especially important to begin the process of low-power design during the system phase.

The primary low-power design objectives during the system phase are the minimization of f_{eff} (i.e., the *effective switching frequency*, which is the

Low Power Design Methodologies and Flows

product of the clock frequency and the average switching activity) and V_{DD}. These can be accomplished by various means, but the main idea is to explore different options and alternatives. One of the key techniques for the reduction of f_{eff} is the use of modes, whereas V_{DD} reduction can often be achieved using parallelism and/or pipelining to increase the raw throughput. In either case, the essential methodology challenge is the evaluation of the different alternatives in terms of their power consumption.

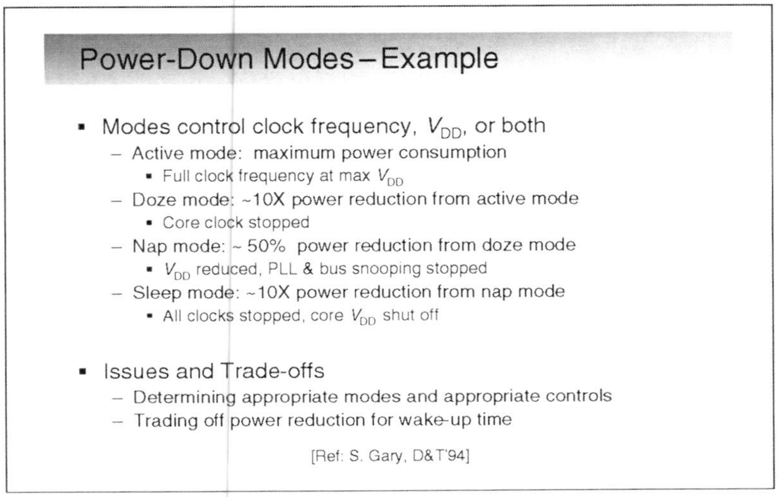

Slide 12.15

One of the best examples of a system-level low-power design technique is the use of different operating modes in a microprocessor (See also Chapter 10). Clearly, if a processor is not fully active it should not be consuming the full power budget. Hence the clock can be slowed down or even stopped completely. If the clock is slowed down, the supply voltage can also be reduced as the logic need not operate at the highest clock frequency, a technique known as *dynamic voltage and frequency scaling* (DVFS) (see earlier chapters). Multiple levels of power-down are possible, ranging from the four coarse levels shown here to an almost continuous range of levels possible with DVFS, with voltage steps as small as 20 mV.

However, controlling the power-down modes can be quite complicated involving software policies and protocols. In addition, there is often a trade-off between the amount of power reduction and the time needed to emerge from the power-down mode, known as the wake-up time. These and other trade-offs should be explored early, preferably in the system phase and no later than the design phase, to avoid late discoveries of critical parameters not meeting specifications. This need for early investigation is, of course, not unique to power, but it is perhaps more prevalent with power parameters as there is usually no single-large factor to address. Instead, the power consumption problem is the amalgam of millions of tiny power consumers.

Power-Down Modes – Example

- Modes control clock frequency, V_{DD}, or both
  - Active mode: maximum power consumption
    - Full clock frequency at max V_{DD}
  - Doze mode: ~10X power reduction from active mode
    - Core clock stopped
  - Nap mode: ~50% power reduction from doze mode
    - V_{DD} reduced, PLL & bus snooping stopped
  - Sleep mode: ~10X power reduction from nap mode
    - All clocks stopped, core V_{DD} shut off

- Issues and Trade-offs
  - Determining appropriate modes and appropriate controls
  - Trading off power reduction for wake-up time

[Ref: S. Gary, D&T'94]

Slide 12.16

Parallelism and Pipelining – Example

- Concept: maintain performance with reduced V_{DD}
  - Total area increases but each data path works less in each cycle
    - V_{DD} can be reduced such that the work requires the full cycle time
    - Cycle time remains the same, but with reduced V_{DD}
  - Pipelining a data path
    - Power can be reduced by 50% or more
    - Modest area overhead due to additional registers
  - Paralleling a data path
    - Power can be reduced by 50% or more
    - Significant area overhead due to paralleled logic
  - Multiple CPU cores
    - Enables multi-threaded performance gains with a constrained V_{DD}
- Issues and Trade-offs
  - Application: can it be paralleled or threaded?
  - Area: what is the area increase for the power reduction?
  - Latency: how much can be tolerated?

[Ref: A. Chandrakasan, JSSC'92]

Parallelism and pipelining can be employed to reduce power consumption, but at the expense of increased area (Chapter 5). Parallelism can be deployed at a relatively low level, such as paralleling a data path, or at a much higher level, such as using multiple processor cores in a single design, an example of which is the Intel Penryn™ multi-core processor. In either case, the delay increases resulting from the reduced supply voltages are mitigated by the enhanced throughput of the replicated logic.

Slide 12.17

System-Phase Low-Power Design Flow

Create design in C / C++ → Simulate C / C++ under typical work loads → Create /vsynthesize different versions → Evaluate power of each version → Choose lowest power version

Transmitter Design (IFFT Block)	Area (mm²)	Symbol Latency (cycles)	Throughput (cycle/symbol)	Min. Freq to Achieve Req. Rate	Avg. Power (mW)
Combinational	4.91	10	4	1.0 MHz	3.99
Pipelined	5.25	12	4	1.0 MHz	4.92
Folded (16 Bfly4s)	3.97	12	4	1.0 MHz	7.27
Folded (8 Bfly4s)	3.69	15	6	1.5 MHz	10.9
Folded (4 Bfly4s)	2.45	21	12	3.0 MHz	14.4
Folded (2 Bfly4s)	1.84	33	24	6.0 MHz	21.1
Folded (1 Bfly4)	1.52	57	48	12.0 MHz	34.6

Example: Exploration of IFFT block for 802.11a transmitter using BlueSpecSystemVerilog

[Ref: N. Dave, Memocode'06]

The generalized low-power design flow during the *system phase* begins with the creation of the design, or at least a model of the design, in a high-level language such as C, C++, or SystemVerilog. The design is simulated under typical workloads to generate activity data, usually in the form of transactions, for use in power analysis. Several different versions of the model are created so as to evaluate and compare the power characteristics of each, thus enabling the designer to choose the lowest-power version or the best alternative. As an example of this flow, seven different versions of an 802.11a transmitter were synthesized from a SystemVerilog description at MIT [Dave'06]. Power ranged from 4 to 35 mW with a corresponding range in area of 5–1.5 mm².

Slide 12.18

Design-Phase Low-Power Design

- Primary objective: minimize f_{eff}
- Clock gating
  - Reduces / inhibits unnecessary clocking
    - Registers need not be clocked if data input hasn't changed
- Data gating
  - Prevents nets from toggling when results won't be used
    - Reduces wasted operations
- Memory system design
  - Reduces the activity internal to a memory
    - Cost (power) of each access is minimized

The key low-power design objective in the *design phase* is the minimization of f_{eff}. Here again this does not mean reducing the clock frequency, although it often involves a reduction in how often the clock toggles by implementing clock gating. Another technique of reducing f_{eff} is the use of data gating. Similarly, a reduction in the number of accesses to a memory is an effective design-phase technique for reducing power.

Slide 12.19

Clock Gating

- Power is reduced by two mechanisms
  - Clock net toggles less frequently, reducing f_{eff}
  - Registers' internal clock buffering switches less often

Local Gating / Global Gating

Clock gating is the single-most popular technique for reducing dynamic power. It conserves power by reducing f_{eff}, which in turn reduces two different power components. The first is the power consumed in charging the load capacitance seen by the clock drivers, and the second is the power consumed by the switching of each register's internal clock buffers. (virtually all common standard-cell registers have buffered clock inputs so as to produce the true and complement clock signals needed by the basic latching structure.)

Two different flavors of clock gating are commonly used. Local clock gating involves gating individual registers, or banks of registers, whereas global clock gating is used to gate all the registers within a block of logic. Such a block can be relatively small, such as perhaps a few hundred instances, or it can be an entire functional unit consisting of millions of instances.

Slide 12.20

Several different methods are used to insert clock gating into designs. The easiest method is to use the logic synthesizer to insert the clock gating circuitry wherever it finds a logical structure involving a feedback multiplexer, as shown on the previous slide. This can work well for local gating, but is not applicable to global clock gating, as synthesizers generally cannot recognize or

> **Clock-Gating Insertion**
>
> - Local clock gating: Three methods
>   - Logic synthesizer finds and implements local gating opportunities
>   - RTL code explicitly specifies clock gating
>   - Clock-gating cell explicitly instantiated in RTL
> - Global clock gating: Two methods
>   - RTL code explicitly specifies clock gating
>   - Clock-gating cell explicitly instantiated in RTL

find those conditions. On the other hand, global clock-gating conditions are often easily recognizable by designers who will insert the clock-gating logic explicitly into the RTL code either by adding code that will get synthesized into the clock-gating logic or by instantiating clock-gating logic cells.

Although using a synthesizer to insert clock gating wherever possible is attractive owing to the methodological ease, it is usually desirable to avoid such a simplistic approach for two reasons. The first is that clock gating does not always reduce power, as the additional logic inserted to gate the clock will consume extra power when the gated clock leaf node toggles. Thus, this may result in higher maximum sustained power – for instance, if the design has a mode in which few of the clocks are simultaneously disabled. Even the long-term time-averaged power can be higher with clock gating. If the *enable* is active for a high percentage of time, the power consumed by the extra logic may exceed the amount of power saved when the clock is disabled. The second reason to avoid "wherever-possible" clock gating is to prevent complicating the clock tree synthesis, which occurs later in the implementation phase. Excessive numbers (exceeding roughly a hundred) of gated clock leaves can make it very difficult to achieve low timing skew between the leaf nodes.

> **Clock Gating Verilog Code**
>
> - Conventional RTL Code
>
> ```
> //always clock the register
> always @ (posedge clk) begin    // form the flip-flop
>     if (enable) q = din;
> end
> ```
>
> - Low-Power Clock-Gated RTL Code
>
> ```
> //only clock the register when enable is true
> assign gclk = enable && clk;    // gate the clock
> always @ (posedge gclk) begin   // form the flip-flop
>     q = din;
> end
> ```
>
> - Instantiated Clock-Gating Cell
>
> ```
> //instantiate a clock-gating cell from the target library
> clkgx1 i1 .en(enable), .cp(clk), .gclk_out(gclk);
> always @ (posedge gclk) begin   // form the flip-flop
>     q = din;
> end
> ```

Slide 12.21
The first snippet of Verilog RTL code shows a canonically enabled register that will become a register with a feedback multiplexer when synthesized without low-power optimizations, or a clock-gated register when synthesized using low-power synthesis features. However, low-power synthesizers usually clock gate all recognized opportunities, which is usually undesirable, for the reasons mentioned on the previous slide. One method of controlling which registers get synthesized with gated clocks is to explicitly define the gating in the RTL code, as shown in the middle snippet, and to turn automatic clock gating off during synthesis. Another method of explicitly defining the clock gating is to instantiate an integrated clock-gating cell from the target cell library as shown in the third code snippet.

Alternately, a synthesizer can be provided with constraints explicitly specifying which registers should be clock gated and which should not.

Clock Gating: Glitchfree Verilog

- Add a Latch to Prevent Clock Glitching

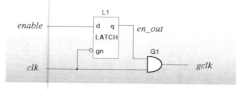

- Clock-Gating Code with Glitch Prevention Latch

```
always @ (enable or clk) begin
   if !clk then en_out = enable  // build latch
end
assign gclk = en_out && clk;     // gate the clock
```

Slide 12.22
Most implementations of clock gating employ some version of the logic shown here to prevent glitches on the gated clock output. The RTL code, when synthesized, will produce a gated clock with a glitch prevention latch as shown. This logic is often implemented together in a single library cell known as an *integrated clock gating cell*.

Data Gating

- **Objective**
  - Reduce wasted operations → reduce f_{eff}

- **Example**
  - Multiplier whose inputs change every cycle, whose output conditionally feeds an ALU

- **Low-Power Version**
  - Inputs are prevented from rippling through multiplier, if multiplier output is not selected

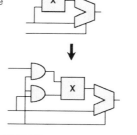

Slide 12.23
Data gating is another method of reducing f_{eff}. Whereas clock gating reduces f_{eff} for clock signals, data gating focuses upon non-clock signals. The general concept is to prevent signal toggles from propagating through downstream logic if those toggles result in unused computations. The particular example shown here is known as *operator isolation*; the multiplication operator is isolated so that it will not be called upon to operate unless its results will be selected to pass through the downstream multiplexer. Power is saved by preventing unused operations.

Slide 12.24

The data-gating insertion methodology is similar to that for clock-gating insertion – the synthesizer can do the insertion, or the RTL designer can explicitly specify it. Similar to clock gating, some applications of operator isolation can cause power to go up, depending upon how often the operator's results are used. Unlike clock gating, however, operator isolation requires additional area for the gating cells, and the gating cells introduce additional delays into the timing path, which may be undesirable depending upon timing requirements. For these reasons, it is sometimes desirable to explicitly embed the data gating into the RTL code.

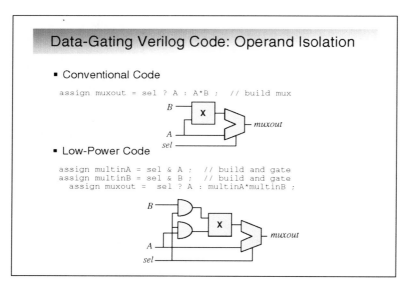

Slide 12.25

Shown here is an example of conventional RTL code of a data-gating opportunity. An operator isolation capable synthesizer recognizes this gating opportunity and automatically inserts the isolation logic. The low-power code snippet shows an example of RTL code that describes the data gating explicitly, obviating the need for an operator isolation capable synthesizer. Observe that activity would be reduced further if the inputs of the multiplier are latched – keeping the old data intact and thus avoiding unnecessary activity.

Slide 12.26

Memory systems often represent significant opportunities for power reduction, as they usually consume considerable portions of a design's overall power budget. It is often possible to minimize this power by implementing some form of memory banking or splitting. This technique involves splitting the memory into several sections, or banks, so as to minimize the extent of the memory that must be activated for a particular access, thus reducing the power consumed for any particular access.

Memory System Design

- Primary objectives: minimize f_{eff} and C_{eff}
  - Reduce number of accesses or (power) cost of an access
- Power Reduction Methods
  - Memory banking / splitting
  - Minimization of number of memory accesses
- Challenges and Trade-offs
  - Dependency upon access patterns
  - Placement and routing

Another technique for lowering memory system power is to reduce the number of accesses through such techniques as storing intermediate results in a local register instead of writing to memory or restructuring algorithms such that fewer memory accesses are needed, although this latter technique is best addressed during the system phase.

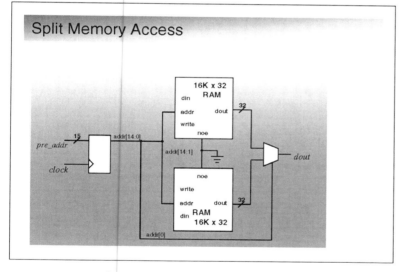

Slide 12.27

Shown here is a banked memory, nominally 32 K words deep by 32 bits wide but implemented as two banks, each 16 K words deep. When an upper address bit changes, a read is initiated to both memory banks in parallel, but when the least significant address bit changes a new read is not initiated; instead, the output multiplexer simply flips so as to select the other memory bank. This saves power because instead of two accesses to a $32\,K \times 32$ memory, we now have two accesses to a $16\,K \times 32$ memory, performed in parallel, with a later multiplexer toggle.

Note that this particular banking strategy saves power only for an application that sequentially accesses memory. For different access patterns, different banking techniques are needed. In terms of methodology, the key issue here is to analyze the access patterns; the knowledge gained from that effort will illuminate opportunities to restructure the memory system to reduce its power consumption.

Slide 12.28

Multiple techniques exist for reducing power during the *implementation phase*; however, they are generally limited in terms of how much total power can be saved as so much of the design is fixed by

Implementation Phase Low-Power Design

Primary objective: minimize power consumed by individual instances

- Low-power synthesis
  - Dynamic power reduction via local clock gating insertion, pin-swapping
- Slack redistribution
  - Reduces dynamic and/or leakage power
- Power gating
  - Largest reductions in leakage power
- Multiple supply voltages
  - The implementation of earlier choices
- Power integrity design
  - Ensures adequate and reliable power delivery to logic

this time. Nevertheless, efforts at this level are important in an overall low-power design flow, especially for leakage reduction.

Note that this phase also includes tasks focused on power integrity. Although, strictly speaking, these efforts are not focused on power reduction, they are nonetheless essential components in an overall methodology, as the use of very low supply voltages can easily result in many Amperes coursing through the power grid. One must be careful to verify that the chip will function correctly with such large currents, and rapid changes in those currents, occurring.

Slack Redistribution

- **Objective**
  - Reduce dynamic Power or leakage power or both by trading off positive timing slack
  - Physical-level optimization
    - Best optimized post-route
    - Must be noise-aware
- Dynamic power reduction by cell resizing
  - Cells along non-speed critical path resized
    - Usually downsized, sometimes upsized
  - Power reduction of 10–15%
- Leakage power reduction by V_{TH} assignment
  - Cells along non-speed critical path set to High V_{TH}
  - Leakage reduction of 20–60%
- Dynamic & leakage power can be optimized independently or together

[Ref: Q. Wang, TCAD'02]

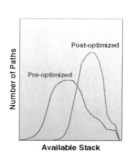

Slide 12.29

Slack redistribution is a particularly common, and conceptually straightforward, implementation phase power reduction technique, involving the trade-off between positive timing-slack and power. As synthesizers generally work to speed up all timing paths, not just the critical paths, many instances on the non-critical paths end up being faster than they need to be. Slack redistribution works by slowing down the instances off the critical path, but only to the extent that the changes do not produce new timing violations (see Chapters 4 and 7). The changes produce a new slack timing histogram, as shown in the chart, hence the moniker. Typical changes reduce the drive strength of a particular instance (usually saving dynamic power) or increase the instance's threshold voltage (reducing leakage power).

Slide 12.30

An example of slack redistribution, as applied to dynamic power reduction, is shown here. This optimization can be performed either before or after routing, but is usually performed after routing as actual extracted parasitics can be used for delay calculation to provide the most accurate timing analysis and hence the best optimization results. Note that after this optimization, the design must

Low Power Design Methodologies and Flows

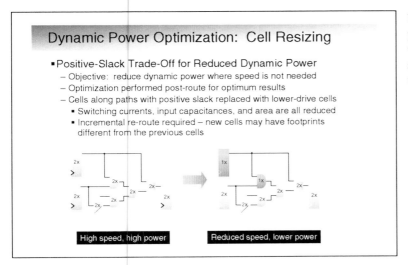

Slide 12.31

Another example of slack redistribution targets leakage power reduction. Also best-performed after routing, this optimization replaces instances off the critical path with same-sized instances but with higher threshold voltages, thus reducing the leakage power with each replacement.

It should be noted that one side effect of replacing a low-V_{TH} cell with a high-V_{TH} cell (or replacing a high-drive cell with a low-drive version) is that the net driven by the replaced cell becomes more susceptible to signal integrity (SI) noise effects, such as coupling delays and glitches. Thus, it is essential to verify noise immunity after each slack redistribution optimization.

Slide 12.32

Two slightly different slack redistribution flows are commonly employed. The left-hand flow illustrates a sequential flow, in that timing, noise, and power are each verified in sequence. This flow works, but tends to require multiple iterations through the entire flow as noise fixes and power optimizations tend to "fight" each other – changes made by the noise fixer tend to be reversed by the power optimizer, and vice versa. For example, a common noise fix is to increase the drive strength of a particular instance so that it is less sensitive to coupling delays. However, when the power optimizer sees this, it may conclude that particular instance to be oversized and either downsizes it or raises its threshold voltage leading to a potential flow convergence issue.

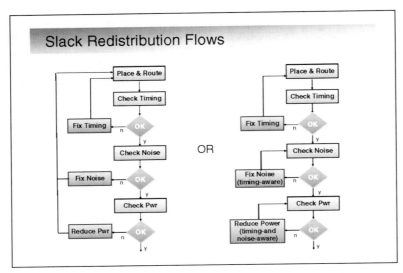

The right-hand flow avoids this issue through the use of concurrent optimizing tools, software that concurrently checks multiple parameters during each optimization. In this case, the noise optimizer is timing-aware – any changes it makes to fix noise problems will not break timing – and the power optimizer is both noise- and timing-aware – any changes it makes to reduce power will not break timing or introduce any new noise problems. These tools tend to be more complex than those employed in the sequential flow, but result in faster timing/noise/power closure as the number of iterations through the flow is greatly reduced.

Slide 12.33

Slack redistribution flows are very popular as they are fully automatic, but they introduce some issues of their own. Perhaps the most significant of these is that it is difficult to predict in advance just how much power will be saved, as the results are so design-dependent. In addition, the resulting slack distribution is narrower with more paths becoming critical, which makes the design more vulnerable to the impact of process variations. For these reasons, slack redistribution is rarely employed as the only low-power design method. Instead, it is usually used in addition to several other techniques and methods employed earlier in the design process.

Power Gating

- Objective
  - Reduce leakage currents by inserting a switch transistor (usually high-V_{TH}) into the logic stack (usually low-V_{TH})
    - Switch transistors change the bias points (V_{SB}) of the logic transistors
- Most effective for systems with standby operational modes
  - 1 to 3 orders of magnitude leakage reduction possible
  - But switches add many complications

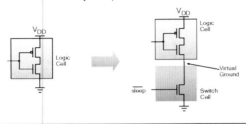

Slide 12.34
Power gating is perhaps the most effective technique for reducing leakage power, resulting in savings ranging from 1 to 3 orders of magnitude (see Chapter 7). Conceptually very simple, power gating adds many complications to the design process. It should also be noted that power gating is not appropriate for all applications. To be effective, power gating should only be used on those designs that contain blocks of logic that are inactive for significant portions of time.

Power Gating: Physical Design

- Switch placement
  - In each cell?
    - Very large area overhead, but placement and routing is easy
  - Grid of switches?
    - Area-efficient, but a third global rail must be routed
  - Ring of switches?
    - Useful for hard layout blocks, but area overhead can be significant

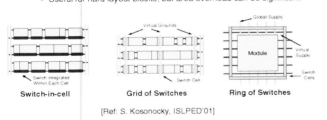

[Ref: S. Kosonocky, ISLPED'01]

Slide 12.35
One of the major decisions to be made regarding power-gating implementation concerns switch placement. Shown here are three different switch placement styles: *switch-in-cell*, *grid-of-switches*, and *ring-of-switches*.

The switch-in-cell implementation uses a switch transistor in each standard cell. In practice, the standard cells are designed up front, each containing a switch, so that each individual instance is power-gated. This is sometimes referred to as fine-grained power gating. This has the advantage of greatly simplifying physical design, but the area overhead is substantial, almost equaling that of the non-power gated logic area.

The grid-of-switches implementation style involves placing the switches in an array across the power-gated block. This generally results in three rails being routed through the logic block: power, ground, and the virtual rail. For this reason, this is often the desired style when state must be retained in registers within the block, as state retention registers need access to the real rails as well as the virtual rails.

The ring-of-switches implementation style places, as the name implies, a ring of switches around the power-gating block; the switches "break" the connection between the external real rail and the internal virtual rail. This style is often utilized for legacy designs for which the original physical design should not be disturbed.

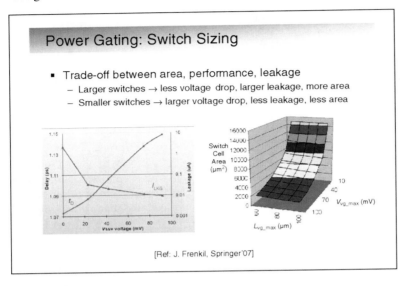

[Ref: J. Frenkil, Springer'07]

Slide 12.36
Perhaps the biggest challenge in designing power-gated circuits is *switch sizing*, owing to the relationship between switch size, leakage reduction, and performance and reliability impact. On the one hand, smaller switches are desirable as they occupy less area. However, smaller switches result in larger switch resistance, which in turn produces a larger voltage drop across them. This larger voltage drop is undesirable as it results in increased switching times and degraded signal integrity. On the other hand, though larger switches impact performance less, they occupy more area. In addition, leakage reduction is smaller with larger switches, as the smaller voltage drop reduces the body effect on the logic transistors. This relationship between virtual-rail voltage drop (a proxy for switch size), delay degradation, and leakage reduction is clearly shown in the left-hand chart, while the relationship between switch area and voltage drop is shown on the right.

Slide 12.37
Numerous additional issues must also be addressed in the design of power-gated circuitry. Some of these are planning issues (the choice of headers or footers, which registers must retain state, the mechanism by which state will be retained), some others are true implementation issues (inserting isolation cells to prevent the power-gated logic outputs from floating), and the rest are verification-oriented (verifying power-up and power-down sequencing, verifying voltage drop and wake-up time limits).

Low Power Design Methodologies and Flows

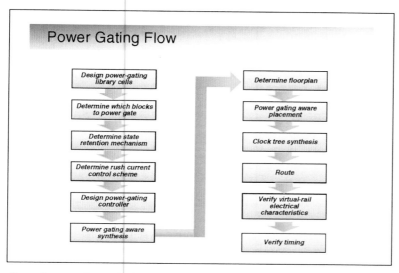

Slide 12.38
A generalized flow for implementing power gating is shown here. Note that the choice of state retention mechanism dictates, to a certain extent, the design of the cell library – if the desired mechanism is to use *state retention flip-flops* (SRFFs), then those primitives must be present in the library. Further, this choice can influence whether the ring-of-switches or grid-of-switches floorplan is chosen. Alternatively, the choice of floorplan can force the issue. For example, the use of a ring-of-switches floorplan makes it awkward to use SRFFs, and a scan-out (for retention) and scan-in (for state restoration) approach may be preferable. In any event, the key concept is that many of these issues are intimately inter-related, thus requiring careful consideration up-front prior to proceeding in earnest on physical implementation.

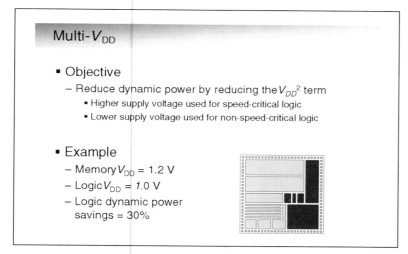

Slide 12.39
Whereas power gating addresses leakage power reduction, multi-V_{DD} (or multiple supply voltages – MSV) addresses dynamic power reduction. Shown here is a plain vanilla example of a design with two voltage domains, but several variants are possible. One variant simply uses more – three or four – voltage domains. Another variant, known as dynamic voltage scaling (DVS) uses time-varying voltages such that the supply is kept at a high value when maximum performance is needed, but reduces the supply voltage at other times. A more complex variant, dynamic voltage and frequency scaling (DVFS) adjusts the clock frequency along with the supply voltage. As one might imagine, while reducing power, these techniques increase the overall complexity of design and verification.

Slide 12.40

Similar to power gating, the use of multiple supply voltages poses a variety of issues, which can be categorized as planning-, implementation-, and verification-oriented. While most of the complications occur during the implementation phase, the choice of which particular voltages to use, and on which blocks, should be made during the system phase. During implementation, the design will be floorplanned, placed, and routed taking into consideration the multitude of unique supplies (ground is usually shared amongst the different power domains). Level shifters must be inserted to translate voltage levels between domains and, depending on the level shifter design, may have placement constraints regarding which domain – driving or receiving – they can be placed in.

Verification becomes much more complicated, as timing analysis must be performed for all PVT corner cases involving signals crossing across power islands.

Slide 12.41

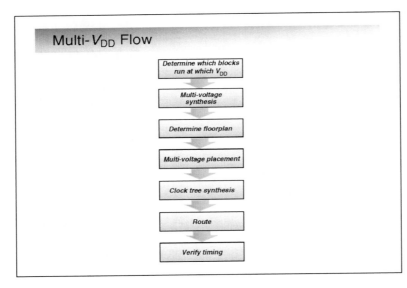

A generalized flow for implementing a multi-V_{DD} design is shown here. The overall flow is very similar to a standard design flow, but several tasks deserve particular attention. First, and perhaps foremost, planning is needed both in terms of logically determining which blocks will run at which voltage and in terms of how the power rails will be routed and the physical blocks will be placed. Multi-voltage synthesis will logically insert level shifters, as appropriate, and the placement of those level shifters must be considered during the subsequent multi-voltage placement. Clock tree synthesis must be multi-voltage aware; that is, it must understand that a clock buffer placed in one voltage domain will have different timing characteristics than the same buffer placed in a different voltage domain, and it should use that knowledge for managing latency and skew. Finally, after routing, timing must be verified using timing models characterized at the particular operating voltages for each different power domain.

Slide 12.42

Power integrity is a concern in all integrated circuits, but it is an especially heightened concern in power-optimized devices, as many of the reduction techniques tend to stress the power delivery network and reduce the noise margins of the logic and memory. For example, power-gating switches insert additional resistance in the supply networks, making them particularly susceptible to excessive voltage drop. Clock gating, with its cycle-by-cycle enabling, often leads to large current differentials from one cycle to the next. Even basic operation at low supply voltages results in huge on-chip currents. Consider a 3 GHz Intel Xeon™ processor, consuming 130 W while running on a 1.2 V power supply – the on-chip supply currents exceed 100 A!

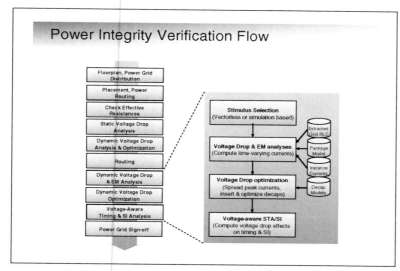

Slide 12.43

The basic idea behind power integrity verification is to analyze the equation $V(t) = I(t)*R + C*dv/dt *R + L*di/dt$, and to determine its effects upon the overall operation of the circuit.

A power integrity verification flow consists of several critical steps, employed in a successive-refinement approach. First, the grid is evaluated for basic structural integrity by checking the resistance from each instance to its power source (*effective resistance check*). Second, static currents (also known as average, or effective dc, currents) are introduced into the grid to check for basic voltage drop (*static voltage drop analysis*). Next, time-dependent behavior of the grid is analyzed (*dynamic voltage drop analysis*). Finally, the effects of voltage drop upon the behavior of the circuit are verified (*voltage-aware timing and SI analysis*). Along the way, if any of the analyses indicate problems, repairs or optimizations can be made. Prior to routing, the repairs are usually implemented by several mechanisms such as strap resizing, instance movement to spread peak currents, and decoupling-capacitor insertion. After routing, the latter two techniques are used most frequently.

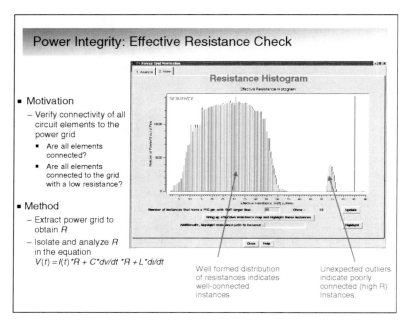

Slide 12.44

The first step in the flow is to verify the connectivity of each instance to the grid by computing its effective resistance to the power source. In effect, this isolates the R term in the voltage drop equation so that it can be checked, individually, for every instance. The result of an effective resistance analysis is a histogram indicating the number of instances with a particular effective resistance. Note the outliers in this chart, as they indicate a much higher resistance than all the others, and hence highlight problems such as missing viae or especially narrow rail connections. A well formed, errorfree power delivery network will produce a resistance histogram with a well-behaved distribution without any outliers.

This analysis, run without the need for any stimuli or current calculations, is computationally fast and, as it covers all instances, can easily highlight grid connectivity issues that are difficult to find by other methods.

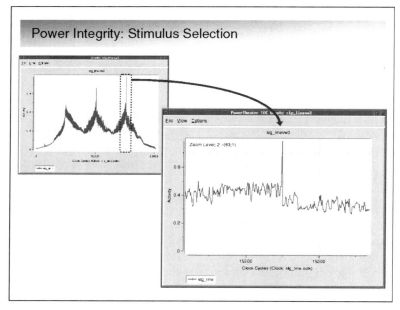

Slide 12.45

After verifying grid connectivity, the next step is to analyze voltage drop. However, to do so the circuit must be stimulated, and, as power is a strong function of activity, a high-activity cycle should be chosen for analysis. This requires a careful selection from a simulation trace as indicated here.

Vectorless stimulation can be an attractive alternative to simulation-based stimuli as it obviates the need for lengthy simulations and vector selection. However vectorless analysis presents its own set of issues, such as choosing the vectorless set-up conditions like total power consumption targets or activity percentages. Also, the analysis may be overly-pessimistic, resulting in over-sizing (and potentially an increase in leakage power).

Slide 12.46

Power Integrity: Static Voltage Drop

- Motivation
  - Verify first-order voltage drop
    - Is grid sufficient to handle average current flows?
    - Static voltage drop should only be a few % of the supply voltage
- Method
  - Extract power grid to obtain R
  - Select stimulus
  - Compute time-averaged power consumption for a typical operation to obtain I
  - Compute: $V = IR$
    - Non time-varying

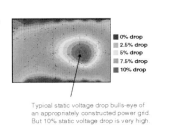

- 0% drop
- 2.5% drop
- 5% drop
- 7.5% drop
- 10% drop

Typical static voltage drop bulls-eye of an appropriately constructed power grid. But 10% static voltage drop is very high.

Once an appropriate cycle has been chosen for analysis, time-averaged currents are computed for use in the $V = IR$ calculation, producing a time-averaged voltage gradient across the chip as shown in the color coded layout. This analysis is used to indicate any grid sensitivities, current crowding, or high-current instances. Note that a static voltage drop analysis is not an effective substitute for effective-resistance analysis, as the former does not check the connections of every instance (because realistic stimuli do not activate all instances).

A useful metric to consider is the effective voltage, or the difference between V_{DD} and V_{SS}, seen by individual instances. The effective voltage, considering static voltage drop on both the V_{DD} and V_{SS} rails, should not deviate from the ideal values by more than 2–3%.

Slide 12.47

Power Integrity: Dynamic Voltage Drop

- Motivation
  - Verify dynamic voltage drop
    - Are current and voltage transients within spec?
    - Can chip function as expected in external RLC environment?
- Method
  - Extract power grid to obtain on-chip R and C
  - Include RLC model of the package and bond wires
  - Select stimulus
  - Compute time-varying power for specific operation to obtain $I(t)$
  - Compute $V(t) = I(t)*R + C*dv/dt*R + L*di/dt$

Time step 1 @ 20 ps Time step 2 @ 40 ps Time step 3 @ 60 ps Time step 4 @ 80 ps

The best measure of power integrity is *dynamic voltage drop* as it accounts for the time-dependent contributions of capacitances and inductances. The method for determining dynamic voltage drop is identical to that for static voltage drop with two exceptions: parasitic models and calculation method. The parasitic model for static voltage drop analysis need only contain resistors resulting from the effective-dc calculations, whereas for dynamic voltage drop analysis full RLC models should be used. Instead of a single solve of the $V(t)$ equation, multiple solves are performed, one for each specified time step, much like SPICE, to compute the time-varying voltage waveforms for all power grid nodes. The four plots on the bottom are from successive time steps of the same dynamic voltage drop computation – the progression illustrates typical voltage drop variations over time.

A maximum effective dynamic voltage drop of 10% is often considered acceptable, although less is obviously better owing to the deleterious effects of voltage drop upon delays.

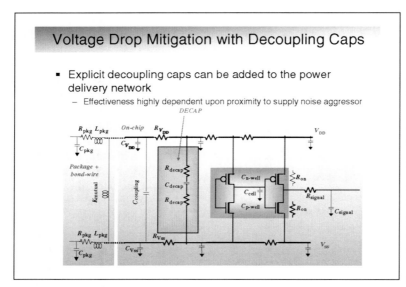

Slide 12.48
A typical and effective voltage drop mitigation technique involves the use of explicit on-chip decoupling capacitors. *Decaps* can be inserted before routing, as a preventative technique, or after routing, as a mitigation technique, or both. Decaps function as local charge reservoirs with a relatively fast time constant, providing energy for high transient-current demand. Shown here is the RC model of a decoupling capacitor, as inserted into the parasitic netlist, consisting of the RC rail models and an RLC package model.

Decaps are usually formed by either of two structures: a metal-to-metal capacitor or a thin gate-oxide capacitor. The latter is the prevalent construction as the capacitance per unit area is much higher than that for metal-to-metal capacitors. However, beginning at 90 nm and becoming substantially worse at 65 nm, gate leakage through the thin gate-oxide capacitors has become a concern. This has prompted the avoidance of decap "fill", which filled all otherwise-unused silicon area with decoupling capacitors, and instead has been a motivation for optimization of the number of decaps inserted and where they are placed.

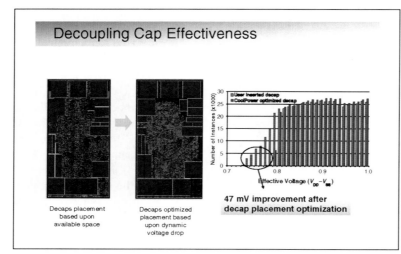

Slide 12.49
A second motivation for decap optimization is that decap effectiveness is a strong function of proximity to the aggressor. For a decoupling capacitor to prevent a transient current-demand event from disturbing the power grid, the decap's response time must be faster than that of the event itself. That is, the resistance through which the current will flow from the decap to the aggressor must be sufficiently low so as to provide the charge at the demanded di/dt rate. From a physical perspective, the decap must be close to the aggressor. This can be seen in the chart which compares the results of two voltage drop analyses, one before placement optimization and one after. The post-optimization results used fewer decaps than the pre-optimization results, yet effectively

improved the worst voltage drop. The use of fewer decaps means that the leakage current due to all the decaps will also be reduced, assuming the use of thin gate-oxide decaps.

Decoupling capacitors can be inserted before or after routing; however, in either case a voltage drop analysis is needed to determine the location and decoupling capacitance requirements of the voltage drop aggressors.

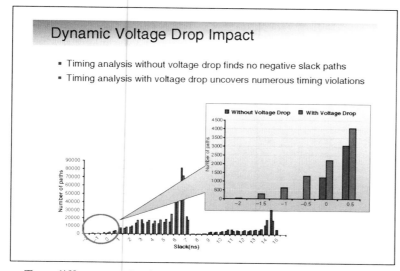

Slide 12.50
The final step, after ensuring that all power parameters have been met for all modes, is to verify timing. This verification step is, of course, independent of whether low-power design techniques have been employed or not. However, owing to the low voltages and the impact upon timing of even small supply voltage deviations, the validation of the impact of voltage drop is especially important.

Two different methods are used to check the impact of voltage drop. The first, and most commonly used, is to check whether the worst-case voltage drop is within the library timing characterization conditions; that is, the actual voltage drop value is of little concern as long as it is within the voltage drop budget assumed during the up-front timing characterization.

The second method is to run static timing analysis using delays computed with actual voltages resulting from a dynamic voltage drop analysis. Such a voltage-sensitive timing analysis can illuminate timing problems that may have otherwise been uncovered, as illustrated in this chart.

Slide 12.51
Low-power design is more than running a power optimization tool. It especially requires designers' creativity in crafting power-efficient algorithms, architectures, and implementations. As the design progresses through the various design phases, increasing levels of power-aware design automation can aid the pursuit of particular power goals. The effectiveness of such automation is dependent upon the chosen methodologies, but several truisms pretty much hold for all SoC design projects. Library cells must be characterized for power. Power should be analyzed early and often. Power consumption can be reduced at all abstraction levels and during all phases, but the best opportunities for major reductions occur during the earliest design phases. Finally, voltage drop concerns must be addressed before tape-out, particularly with regard to effects upon timing and noise characteristics.

As mentioned in the beginning of the chapter, the emergence of power efficiency as a primary design concern has caused the electronic design automation industry to increasingly focus on design flows and tools incorporating power as a primary criterion. Many of the outcomes have been described in this chapter. Yet, it is clear that the effort still falls short with respect to incorporating many of the aggressive low-power design technologies described in this book. A close collaboration between design and design automation engineers and researchers is essential to accomplish truly-automated low-power design.

Slide 12.52
Some references . . .

Summary – Low Power Methodology Review

- Characterization and modeling for power
  - Required for SoC cell-based design flows

- Power analysis
  - Run early and often, during all design phases

- Power reduction
  - Multiple techniques and opportunities during all phases
  - Most effective opportunities occur during the early design phases

- Power integrity
  - Voltage drop analysis is a critical verification step
  - Consider the impact of voltage drop upon timing and noise

Some Useful References

Books and Book Chapters
- A. Chandrakasan and R. Brodersen, *Low Power Digital CMOS Design*, Kluwer Academic Publishers, 1995
- D. Chinnery and K. Keutzer, *Closing the Power Gap Between ASIC and Custom*, Springer, 2007
- J. Frenkil, "Tools and Methodologies for Power Sensitive Design", in *Power Aware Design Methodologies*, M. Pedram and J. Rabaey, Kluwer, 2002.
- J. Frenkil and S. Venkatraman, "Power Gating Design Automation", in *Closing the power crap Between ASIC and custom*, Chapter 10, Springer'2007
- M. Keating et al., *Low Power Methodology Manual -For System-on-Chip Design*, Springer, 2007
- C. Piguet, Ed., *Low-Power Electronics Design*, Ch. 38–42, CRC Press, 2005

Articles and Web Sites
- Cadence Power Forward Initiative, http://www.cadence.com/partners/power_forward/index.aspx
- A. Chandrakasan, S. Sheng and R. W. Brodersen, "Low-power digital CMOS design," *IEEE Journal of Solid-State Circuits*, pp. 473–484, Apr. 1992
- N. Dave, M. Pellauer and S. Gerding, Arvind, "802.11a transmitter: A case study in microarchitectural exploration", MEMOCODE, 2006.
- S. Gary, P. Ippolito, G. Gerosa, C. Dietz, J. Eno and H., Sanchez, "PowerPC603, a microprocessor for portable computers", *IEEE Design and Test of Computers*, 11(4), pp. 14–23, Winter 1994.
- S. Kosonocky, et. al., "Enhanced multi-threshold (MTCMOS) circuits using variable well bias", ISLPED Proceedings, pp. 165–169, 2001.
- Liberty Modeling Standard, http://www.opensourceliberty.org/resources_ccs.html#1
- Sequence PowerTheater, http://www.sequencedesign.com/solutions/powertheater.php
- Sequence CoolTime, http://www.sequencedesign.com/solutions/coolproducts.php
- Synopsys Galaxy Power Environment, http://www.synopsys.com/products/solutions/galaxy/power/power.html
- Q. Wang and S. Vrudhula, "Algorithms for minimizing standby power in deep submicrometer, dual-Vt CMOS circuits," *IEEE Transactions on Computer-Aided Design of Integrated Circuits and Systems*, 21(3), pp 306–318, Mar 2002.

Chapter 13
Summary and Perspectives

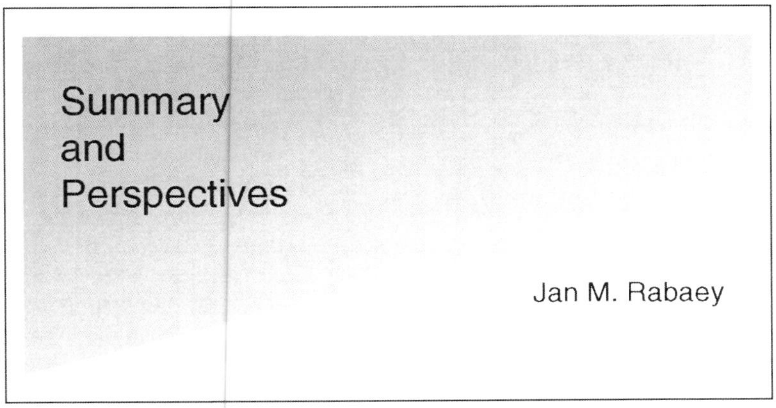
Summary
and
Perspectives
Jan M. Rabaey

Slide 13.1
In this book, we have been exploring a broad range of technologies to address the power and energy challenges in digital integrated circuit design. Many of these techniques have only been developed over the past decade or so. Yet, one cannot help wondering where the future may lead us. In this last chapter, we present some short summary of the developments in low-power design over the past years, where the state-of-the-art is today, and what may be in store for tomorrow.

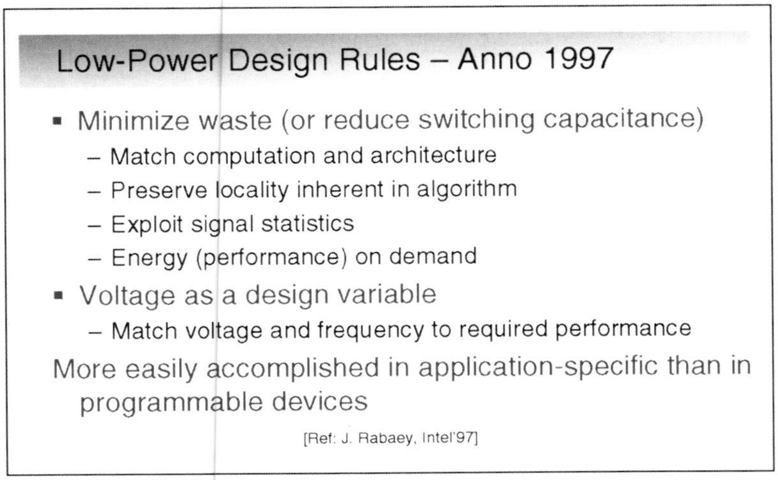
Low-Power Design Rules – Anno 1997
▪ Minimize waste (or reduce switching capacitance)
– Match computation and architecture
– Preserve locality inherent in algorithm
– Exploit signal statistics
– Energy (performance) on demand
▪ Voltage as a design variable
– Match voltage and frequency to required performance
More easily accomplished in application-specific than in programmable devices
[Ref: J. Rabaey, Intel'97]

Slide 13.2
It is fair to state that the main developments in power reduction in the early 1990s can be classified under two headers: *cutting the fat*, and *turning the supply voltage into a design variable*.

Indeed, before power became an issue, energy waste was rampant (Does this not seem to bear an eerie resemblance to the world at large today?). Circuits were oversized, idle modules were still clocked, and architecture design was totally driven by performance. The supply voltage and its distribution

J. Rabaey, *Low Power Design Essentials*, Series on Integrated Circuits and Systems,
DOI 10.1007/978-0-387-71713-5_13, © Springer Science+Business Media, LLC 2009

grid were considered sacred and untouchable. And the idea that general-purpose computing was "somewhat" (three orders of magnitude!) inefficient was a shocker.

Both of the above-mentioned concepts are now main-stream. Most designs are doing away with any excess energy consumption quite effectively, and multiple and, sometimes, variable supply voltages are broadly accepted.

Slide 13.3

Adding Leakage to the Equation

- The emergence of power domains
- Leakage not necessarily a bad thing
  - Optimal designs have high leakage ($E_{Lk}/E_{Sw} \approx 0.5$)
- Leakage management requires runtime optimization
  - Activity sets dynamic/static power ratio
- Memories dominate standby power
  - Logic blocks should not consume power in standby

[Emerged in late 1990s]

The emergence of leakage as a substantial source of power dissipation came as somewhat of a surprise in the late 1990s, more specifically with the introduction of the 130 and 90 nm technology nodes. Few roadmaps had foreseen this. Solutions were put forward and adopted swiftly.

The most important one was the adoption of *power domains*. They are essentially an extension of the voltage domains that were introduced earlier, and, as such, made the concept of dynamic supply and body voltage management a reality in most SoC and general-purpose designs. In a sense, this is the root of the idea of runtime optimization, a technique that is now becoming one of the most prevalent tools to further improve energy efficiency.

Designers also learned to live with leakage, and to use it to their advantage. For instance, when a circuit is active, allowing a healthy dose of leakage is actually advantageous, as was predicted by Kirsten Svensson and his group in Linkoping in 1993 [Liu'93]. Finally, the realization that memories are the prime consumers of standby power (which is a major issue in mobile devices) pushed the issue of energy-efficient memory design to the forefront.

Slide 13.4

With all the low-hanging fruit picked, it became clear that further improvements in energy efficiency could only be accomplished through novel and often-disruptive design solutions.

Although the concept of using *concurrency* to improve energy efficiency (or to improve performance within a constant power budget) was put forward in the early 1990s [Chandrakasan92], it was not until the beginning of the 2000s that the idea was fully adopted in the general-purpose computing world – it most often takes a major disaster for disruptive ideas to find inroads. Once the dam bursts, there is no holding back however. Expect large (and even huge) amounts of concurrency to be the dominating theme in the architectural design community for the foreseeable future.

A second contemporary evolution is that the concept of *runtime optimization* is now coming to full fruition. If energy efficiency is the goal, it is essential that circuits and systems are always functioning at the optimal point in the energy–delay space, taking into account the varying environmental conditions, activity level, and design variations. In an extreme form, this means that the traditional "worst-case" design strategy is no longer appropriate. Selecting the operating

Summary and Perspectives   347

Low-Power Design Rules – Anno 2007

- Concurrency Galore
  - Many simple things are way better than one complex thing
- Always-Optimal Design
  - Aware of operational, manufacturing, and environmental variations
- Better-than-worst-case Design
  - Go beyond the acceptable and recoup
- The Continuation of Voltage Scaling
  - Descending into ultra low voltages
  - How close can we get to the limits?
- Explore the Unknown

[Ref: J. Rabaey, SOC'07]

---

condition of a circuit for the worst case comes with major overhead. Allowing occasional failures to happen is ok, if the system can recuperate from them.

It also seems obvious that we continuously need to explore ways of further scaling the supply voltage, the most effective knob in reducing energy consumption. *Ultra low voltage design* will continue to be a compelling topic over the next few decades.

Finally, we should realize that further reduction in energy per operation will ultimately challenge the way we have been doing digital computation for the past six or seven decades. Based on the Boolean–von Neumann–Turing principles, we require our digital engines to execute computational models that are entirely deterministic. When signal-to-noise ratios get extremely small and variances large, it is worth exploring the opportunities offered by *statistical computational models*. Though these may not fit every possible application domain, there are plenty of cases where it is very effective, as is amply demonstrated in nature.

The latter observation is worth some further consideration. Design tools and technologies have always been constructed such that the design statistics were hidden or could be ignored. However, of late this is less and less the case. Statistical design is definitely becoming of essence, but, unfortunately, our design tools are ill-equipped to deal with it. Analyzing, modeling, abstracting, composing, and synthesizing distributions should be at the core of any low-power design environment.

Some Concepts Worth Watching

- Novel switching devices
- Adiabatic logic and energy recovery
- Self-timed and asynchronous design
- Embracing non-conventional computational paradigms
  - Toward massive parallelism?

Slide 13.5

In the remainder of this "perspectives" chapter, we briefly discuss some concepts worth watching, none of which were discussed so far.

Slide 13.6

From an energy perspective, the ideal switching device would be a transistor or a switch that has an infinite sub-threshold slope, and a fully deterministic threshold voltage. Under such conditions, we could continue to scale supply and threshold voltages, maintaining performance while eliminating static power consumption. Trying to accomplish this with semiconductor switches is most probably a futile exercise. In Chapter 2, we introduced a couple of device structures that may be able to reduce the sub-threshold slope below 60 mV/decade.

Hence, it is worth considering whether some of the device structures that have been explored within the realm of nanotechnology may offer relief. Researchers have been proposing and examining a wide array of novel devices operating on concepts that are vastly different from that of the semiconductor transistors of today. Although this indeed has shown some interesting potential (e.g., spintronics operating on magnetism rather than electrostatics), the long-term prospective of most of these devices with respect to building digital computational engines remains unclear. The next digital switch may very well emerge from a totally unexpected side.

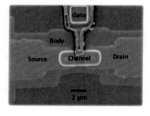 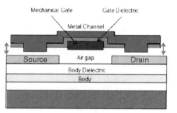

Slide 13.7

This is illustrated by the following example. Recently, a number of researchers have been exploring the idea of using a mechanical switch. This indeed seems to be a throwback to the beginning of the 20th century, when relays were the preferred way of implementing digital logic. Technology scaling has not only benefited transistors, but has also made it possible to reliably manufacture a broad range of micro- and nano-electromechanical systems (MEMS and NEMS, respectively) of increasing complexity. A micro-mechanical switch, which reliably opens and closes trillions of times, would bring us closer the ideal switch (at least for a while ...). The main advantage of the mechanical switch is that its on-resistance is really small (on the order of 1 Ω) compared to that of a CMOS transistor (which is on the order of 10 kΩ for a minimum-size device), whereas the off-current also is very small (air is the best dielectric available). The challenge, however, is to produce a device that can reliably switch trillions of times. At the nano-scale level, atomic forces start to play, and stiction can cause a device to be stuck in the closed position. Successful industrial MEMS products, such as the Texas Instruments' DLP displays, have demonstrated that these concerns can be overcome.

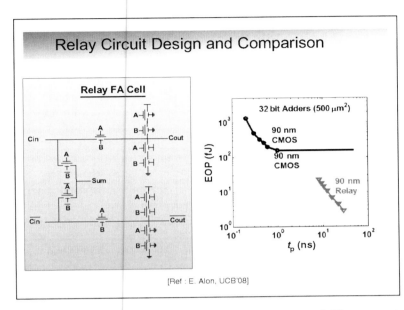

Slide 13.8
The availability of a device with very low on-resistance has a profound impact on how logic can and should be built. Instead of the shallow-logic structures of today, complex deep logic becomes a lot more attractive. The (in)famous "fan-out of four" rule that governs logic design today no longer holds. But most importantly, it may help to move the energy–delay bounds into new territory. This is illustrated in this slide, where the energy–delay optimal curves of 32-bit adders, implemented in 90 nm CMOS and 90 nm relay logic, are compared. The concept of relay logic has been around for ages – however, until now, it was confined to the domain of bulky electromagnetic relays. With the advent of MEMS switches, it may rise to the foreground again. And, as is evident from the chart, it has better potential of getting us closer to the minimum energy bound than any other device we have seen so far. Only one "minor" detail: we have to figure out how to make it work in a scalable and reliable way.

This example shows only one possible way of how novel devices may disruptively change the low-energy design scene. We are sure that other exciting options will emerge from the large cauldron of nanotechnology in the coming decades.

Slide 13.9
One recurring topic in the low-energy research community is the search for more efficient ways of charging and discharging a capacitor. In fact, if we could do this without any heat dissipation, we may even be capable of designing a logic that consumes virtually no energy (!?). A clue on how this might be done was given in Chapter 3, when we discussed adiabatic charging: If a capacitor is charged extremely slowly, virtually no energy is dissipated in the switch (see also Chapter 6). This idea created a lot of excitement in the 1990s and led to a flurry of papers on adiabatic logic and reversible computing. The latter idea postulates that using adiabatic charging and discharging, all charge taken from the supply during a computation can be put back into the supply afterward by reversing the computation [Svensson'05].

None of these ideas ever made it into real use. Mark Horowitz and his students [Indermauer'94] rightly argued that if one is willing to give up performance, one might as well lower the supply voltage of a traditional CMOS circuit rather than using the more complex adiabatic charging. And the overhead of reversing a computation proved to be substantial and not really practical.

Yet, with voltage scaling becoming harder, the idea of adiabatic charging is gaining some following again. An excellent example of this is shown in this slide, where adiabatic charging is

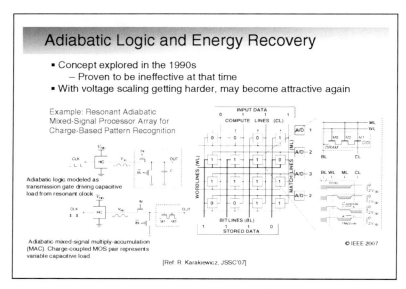

used to implement an energy-efficient processor array for pattern recognition. The (albeit mixed-signal) processor realizes 380 GMACs/mW (10^9 multiply-accumulates per second), which is 25 times more efficient than what would be accomplished with static CMOS drivers. It demonstrates that the creative use of resonant adiabatic structures can lead to substantial energy savings.

At the time of writing, it is hard to see whether the adiabatic approach is limited to niche circuits, or if it may lead to something more substantial.

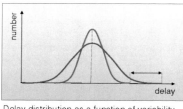

Slide 13.10

One topic that always floats to the foreground when discussing low-power technology of the future is the potential of asynchronous-timing or self-timing strategies. The belief that self-timing may help to reduce power dissipation is correct, but the reasoning behind it is often misguided. The common understanding is that the power savings come from the elimination of the global clock. In reality, that is not the case. A well-thought-out synchronous-clocking strategy using hierarchical clock gating can be just as efficient, and eliminates the overhead of completion signal generation and protocol signaling. More meaningful is that self-timed methodologies inherently support the "better-than-worst-case" design strategy we are advocating. Variations in activity and implementation platforms are automatically accounted for. In fact, asynchronous design, when implemented in full, realizes an "average case" scenario. In a design world where variations are prominent, the small overheads of the self-timing are easily offset by the efficiency gains.

Summary and Perspectives 351

Yet, this does not mean that all designs of tomorrow will be asynchronous. For smaller modules, the efficiency of a clocked strategy is still hard to beat. This explains the current popularity of the GALS (globally asynchronous locally synchronous) approach, in which islands of synchronicity communicate over asynchronous networks [Chapiro84]. Over time, the size of the islands is bound to shrink gradually.

Another often-voiced concern is that asynchronous design is complex and not compatible with contemporary design flows. Again, this is only a partial truth. The design technology is well-understood, and has been published at length in the literature. The real challenge is to convince the EDA companies and the large design houses that asynchronous design is a viable alternative, and that the effort needed to incorporate the concept into the traditional environments is relatively small. This rather falls into the domains of policy and economic decision-making rather than technology development.

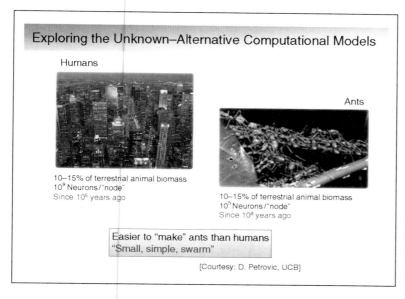

Slide 13.11

We would like to devote the last pages of this book to some far-out speculation. In the text, we have advocated a number of strategies such as concurrency, better-than-worst-case design and aggressive deployment. Although these techniques can be accommodated within the traditional computational models of today, it is clear that doing so is not trivial. To state it simply, these models were not built to deal effectively with statistics.

There are other computational systems that do this much better, famously, those that we encounter in biology and nature. Look, for instance, at the brain, which performs amazingly well under very low signal-to-noise conditions and adapts effectively to failure and changing conditions. Maybe some of the techniques that nature uses to perform computation and/or communication could help us to make the integrated systems of tomorrow work better and more efficiently.

Let us take the case of concurrency, for example. We have discussed earlier how multi- and many-core systems are being adopted as a means to improve performance of SoCs while keeping energy efficiency constant. We can take this one step further. What is it that keeps us from envisioning chips with millions of processors, each of them very simple? This model is indeed working very well in nature – think again about the brain, or alternatively a colony of ants. Instead of building a system based on a small number of very reliable and complex components, a complex reliable system can emerge from the communication between huge numbers of simple nodes. The motto is "small, simple, swarm".

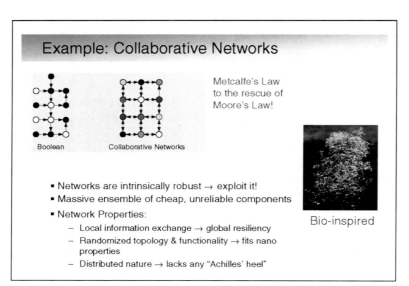

Slide 13.12

The advantage of these "collaborative" networks is that they avoid the Achilles' heels of traditional computing, as redundancy is inherent, and, as such, the networks are naturally robust. This may allow for computation and communication components to operate much closer to the von Neumann and Shannon bounds.

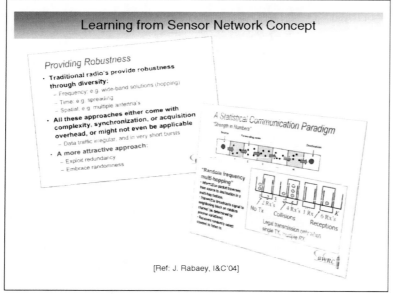

Slide 13.13

These ideas are definitely not new. Cellular automata are an example of systems that are built on similar ideas. The "neural networks" concept of the late 1980s is another one – however, that concept was doomed by the limitations of the computational model, the attempt to transplant computational models between incompatible platforms, and the hype. Hence, it has proven to be useful in only a very limited way.

Yet, the idea of building complex electronic systems by combining many simple and non-ideal components found its true calling in the concept of wireless sensor networks, which emerged in the late 1990s. There it was realized that the statistical nature of the network actually helped to create robust and reliable systems, even if individual nodes failed or ran out of power and in the presence of communication failures. In one landmark paper, it was shown how the combination of pure stochastic communications and network coding leads to absolutely reliable systems if the number of nodes is large enough.

Summary and Perspectives

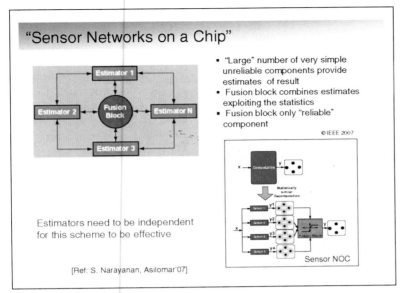

[Ref: S. Narayanan, Asilomar'07]

Slide 13.14

Based on these observations, a number of researchers have started to explore the idea of bringing similar ideas to the chip level – the "sensor-network-on-a-chip (SNOC)" approach [Narayanan'07]. Instead of having a single unit perform a computation (such as filtering or coding), why not let N simple units simultaneously estimate the result and have a single fusion block combine these estimations into a final output. Each of the "sensors" only works on a subset of data and uses a simplified estimation algorithm. The technique obviously only works effectively if the estimations are non-correlated, yet have the same mean value.

This technique is inherently robust – failure in one or more of the sensors does not doom the final results, it just impacts the signal-to-noise ratio or the QOS (quality-of-service). As such, aggressive low-energy computation and communication techniques can be used. The only block that needs to be failproof is the fusion module (and even that can be avoided).

An attentive reader may notice that the SNOC approach is nothing less than an extension of the "aggressive deployment" approach advocated in Chapter 10. Consider, for instance, the example of Slide 10.53, where a computational block was supplemented by an estimator to catch and correct the occasional errors. The SNOC simply eliminates the computational block altogether, and uses only estimators, realizing that the computational block is nothing more than a complex estimator itself.

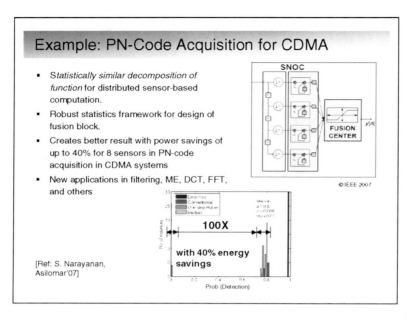

Slide 13.15

An example of an SNOC is shown in this slide. The application is *pn*-code acquisition, an essential function in any wideband-CDMA wireless receiver. Its main task is to correlate an incoming data stream with a long pseudo-random code. The traditional approach is to use a single correlator – which obviously is vulnerable to failures. As shown in the chart, a single failure dooms the hardware. The SNOC-architecture divides the function over many (16–256) simple correlators, each operating on a sub-sampled data stream and a sub-set of the *pn*-code. The fusion block either accumulates over the different sensors or performs a median filtering. As can be observed, the algorithm performs very well even in the presence of large number of failures. In addition, owing to aggressive deployment, the energy efficiency is improved by 40%.

One can envision many examples that fall into the same category. In fact, any application in the RMS (recognition, mining, synthesis) class is potentially amenable to the ideas presented. With the growing importance of these applications, it is clear that the opportunities for innovation are huge.

Slide 13.16

The goal of this book was to present low-power design in a methodological and structured fashion. It is our hope that, by doing so, we have offered you the tools to engage effectively in state-of-the-art low-energy design, and, furthermore, to contribute actively to the field in the future.

Slide 13.17
Some references . . .

Interesting References for Further Contemplation

Books and Book Chapters
- L. Svensson, "Adiabatic and Clock-Powered Circuits," in C. Piguet, *Low-Power Electronics Design*, Ch. 15, CRC Press, 2005.
- R. Wasser (Ed.), *Nanoelectronics and Information Technology*, Wiley-CVH, 2003.

Articles
- E. Alon et al., "Integrated circuit design with NEM relays," *UC Berkeley Technical Report*, 2008.
- A.P. Chandrakasan, S. Sheng and R.W. Brodersen, "Low-power CMOS digital design," *IEEE Journal of Solid-State Circuits*, 27, pp. 473–484, Apr 1992.
- D.M. Chapiro, "Globally asynchronous locally synchronous Systems," PhD thesis, Stanford University, 1984.
- Digital Light Processing (DLP), http://www.dlp.com
- Handshake Solutions, "Timeless Designs," http://www.handshakesolutions.com
- T. Indermaur and M. Horowitz, "Evaluation of charge recovery circuits and adiabatic switching for low power CMOS design," *Symposium on Low Power Electronics*, pp. 102–103, Oct.1994.
- H. Kam, E. Alon and T.J. King, "Generalized scaling theory for electro-mechanical switches.," UC Berkeley, 2008.
- R. Karakiewicz, R. Genov and G. Cauwenberghs, "480-GMACS/mW resonant adiabatic mixed-signal processor array for charge-based pattern recognition," *IEEE Journal of Solid-State Circuits*, 42, pp. 2573–2584, Nov. 2007.
- D. Liu and C. Svensson, "Trading speed for low power by choice of supply and threshold voltages," *IEEE Journal of Solid-State Circuits*, 28, pp. 10–17, Jan 1993.
- S. Narayanan, G.V. Varatkar, D.L. Jones and N. Shanbhag, "Sensor networks-inspired low-power robust PN code acquisition", *Proceedings of Asilomar Conference on Signals, Systems, and Computers*, pp. 1397–1401, Oct. 2007.
- J. Rabaey, "Power dissipation, a cause for a paradigm shift?", Invited Presentation, Intel Designers Conference, Phoenix, 1997.
- J. Rabaey, "Embracing randomness – a roadmap to truly disappearing electronics," Keynote Presentation, I&C Research Day, http://www.eecs.berkeley.edu/~jan/presentations/randomness.pdf, EPFL Lausanne, July 2004
- J. Rabaey, "Scaling the power wall", Keynote Presentation SOC 2007, http://www.eecs.berkeley.edu/~jan/presentations/PowerWallSOC07.pdf, Tampere, Nov. 2007.

Summary and Perspectives

Index

A

Abstraction design methodology, 79–80
Abstraction levels, optimizations at, 114
Accelerator approach, 143
Active (dynamic) power, 54–55
Active deskew, 212
Adaptive body bias (ABB), in runtime optimization, 265–268
    advantage at low V_{DD}/V_{TH}, 269
Adiabatic charging approach, 163–164
Adiabatic logic, 350
Aggressive deployment (AD), in runtime optimization, 272
    algorithmic-level AD, effectiveness, 279
    components, 273
        error correction, 273
        error detection, 273
        voltage-setting mechanism, 273
Air conditioning system, power issues in, 3
Algebraic transformations, 101
Algorithmic BTWC, 278
Alpha power law model, nanometer transistors, 29
Alpha-power based delay model, 84
Amdahl's law, 145–146
Application-specific instruction processors (ASIPs), 136, 141
    advantage, 141
Application-specific integrated circuit (ASIC) design, 97–98
Architecture, algorithms and systems level optimizations @ design time, 113–148
    in 1990s, 121–122
    architectural choices, 137
    concurrency exploitation, 116–118
        See also Concurrency
    (re)configurable processors, 143
    design abstraction stack, 115
    domain of video processing, 141–142
    embedded applications, 146
    energy–delay space mapping, 119
    extensible-processor approach, 141–142
    flexibility, 135
        quantifying flexibility, 135
        trade-off between energy efficiency and, 136–137
    hardware accelerators, 142–144
    locality of reference, 132
    matching computation to architecture issue, 128–129
    to minimize activity, 133–134
    multi-antenna techniques, 129–130

    parallel implementation, 117
    pipelining, 117–118
    programming in space versus time, 144
    simple versus complex processors, 138
    software optimizations, 133
    time-multiplexing, 120
    word-length optimization, 129–131
    See also Platform-based design strategy; Singular-value decomposition (SVD)
Asynchronous logic, 350–351
Asynchronous signaling, 166
Automated optimization, 81

B

Back-gated (BG) MOSFET, 203–204
Battery technology
    battery storage, as limiting factor, 6
    energy storage, calculation, 9
    evolution, 7
    fuel cells, 8–9
    higher energy density, need for, 8
    Lithium-ion, 7
    micro batteries, 9
    saturating, 7
    supercapacitor, 10
Better-than-worst-case (BTWC) design, 272
Biological machinery, 13
Bipolar systems, heat flux in, 14–15
6T Bitcells, 202
Bitline leakage, SRAM
    during read access, 196
    solutions, 197
Body biasing
    body bias circuitry, 226
        central bias generator (CBG), 226
        local bias generators (LBGs), 226
    nanometer transistors, 35
        body-effect parameter γ, 30
        forward, 30–31
        reverse, 30–31
    in standy mode leakage control, 224–225
    in standby mode leakage reduction of embedded SRAM, 244
        body biasing and voltage scaling, combining, 245

Body biasing (*cont.*)
    forward body bias (FBB), 245
    raised V_{SS} and RBB, combining, 245
    reverse body biasing (RBB), 244
    See also Adaptive body bias (ABB)
Boolean–von Neumann–Turing principles, 347
Boosted-gate MOS (BGMOS), in standy mode leakage control, 218–219
Boosted-sleep MOS, in standy mode leakage control, 219
Bus protocols and energy, 172
Bus-invert coding (BIC), 168–169

C

Canary-based feedback mechanism, 243
Capacitors
    See also Charging capacitors
Carbon-nanotube (CNT) transistors, 51
Cell writability, SRAM, 191
Central bias generator (CBG), 226
Channel length impact on nanometer transistors threshold voltages, 31
Channel-coding techniques, 167
Charge recycling, 164–165
Charging capacitors, 55, 57
    driving from a constant current source, 57
    driving using sine wave, 58
Chip architecture and power density, 4
Circuit optimization, 83–84, 114–115
Circuit with dc bias currents, 70
    power management, 70
    trade off performance for current, 70
Circuit-level techniques, optimizing power @ design time, 77–111
    abstraction methodology, 79–80
    algebraic transformations, 101
    alpha-power based delay model, 84
    ALU for 64-bit microprocessor, 96
    circuit optimization framework, 83
        generic network, 84
    complex gates, 108
    delay modeling, 84
    'design-time' design techniques, 78
    dual-V_{TH} domino, 106
        for high-performance design, 107
    dynamic energy, 85
    dynamic-power optimization, 78
    energy–delay trade-off optimization, framework, 78–79
    hierarchy methodology, 79–80
    inverter chain, 86–88
    layered approach, 80
    leakage at design time, 102
        reducing the voltage, 103
        reducing using higher thresholds, 103
        reducing using longer transistors, 103
        reducing, 103–104
    for leakage reduction, 199
    level-converting flip-flops (LCFFs), 95
    logical effort formulation, 85
    longer channels, 104
    low-swing bus and level converter, 96
    multiple thresholds voltages, 104–106
    optimal energy–delay curve, 79, 82
    optimization model, refining, 103
        leakage energy, 103
        switching energy, 103
    reducing active energy @ design time, 82
    Return on Investment (ROI), optimizing, 86
    'run-time' optimizations, 78
    shared *n*-well, 94
    sizing, transistor, 97–98
        continuous, 98
        discrete, 98
    static power optimization, 78
    technology mapping, 98–100
        logical optimizations, 100
    variables to adjust, 79
        continuous variables, 79, 80–81
        discrete variables, 79
    See also Multiple supply voltages
Circuits and systems, optimizing power @ runtime, 249–288
    adaptive body bias, 267–268
    aggressive deployment at the algorithm level, 278
    clock frequency, adjusting, 252
        disadvantages, 252
    converter loop sets V_{DD}, f_{clk}, 260
    delay sensitivity, 265
    dynamic frequency scaling (DFS), 253
    dynamic logic, 263
    dynamic voltage scaling (DVS), 253–254
    energy–performance characteristics, 277
    error rate versus supply voltage, 274
    generalized self-adapting approach, 271
    high-performance processor at low energy, 261
    relative timing variation, 264
    static CMOS logic, 263
    stream-based processing and voltage dithering, 256
    threshold variability and performance, 266
    timing, managing, 281
    using discrete voltage levels, 255
    variable workload
        adapting to, 252
        in general-purpose computing, 251
        in media processing, 251
    V_{DD} and f_{clk}, relating, 257
        on-line speed estimation, 257
        self-timed, 257
        synchronous approach, 257
        table look-up, 257
    V_{DD} and throughput, 253
    voltage scheduling impact, 260
    voltage/frequency scheduling, impact, 259
    See also Aggressive deployment (AD)
Circuits and systems, optimizing power @ standby, 207–230
    See also Standby mode
Circuits with reduced swing, 56
Circuit-switched versus packet-based network, 175
Clock distribution network, optimizing power @ design time, 178–180
    advantages, 178

Index

reduced-swing clock distribution, 178–179
    transmission line based, 179
Clock frequency, adjustment, 252
Clock gating, 209–210
    implementing, 210
    low power design flow, 327
        clock-gating insertion, 328
        conventional RTL code, 328
        data gating, 329
        glitchfree verilog, 329
        global clock gating, 328
        instantiated clock-gating cell, 328
        local clock gating, 328
        low power clock-gated RTL code, 328
        verilog code, 328
    reducing power, 210
Clock hierarchy, 211
Closed-loop feedback approach, for DRV, 242
Clustered voltage scaling (CVS), 94
CMOS systems
    heat flux in, 14–15, 17
    power dissipation in, 54
        active (dynamic) power, 54
        static (leakage) power, 54
        static currents, 54
    reducing SRAM power, 203–204
Code transformations, 125–126
Code-division multiple access (CDMA), 136–137
Code-division multiplexing (CDM), 171
Coding strategies, 167
    activity reduction through, 168
    bus-invert coding (BIC), 168–169
    channel-coding techniques, 167
    error-correcting coding (ECC), 167, 170
    source codes, 168
    transition coding techniques, 169–170
Collaborative networks, 352
Communication infrastructure, power consumption in, 2–3
Complex gates, 108
    complex-versus-simple gate trade-off, 109
Complex logic, power dissipation evaluation in, 62
Composite current source (CCS), 320
Computation, power consumption in, 2–3
Computational architecture, 128–129
Computational efficiency, improving, 128
Computing density, 19
Concerns over power consumption, 1–2
Concurrency, optimization using, 116–118, 346
    in 2000s, 122–123
    alternative topologies, 126
    concurrent compilers to pick up the pace, 125
    fixed EOP, 119
    fixed performance, 119
    manipulating through transformations, 124
    and multiplexing combined, 120
    quest for, 123
Conditional probability, 61
(re)configurable processors, 143
Constant-current (CC) technique, 30
Constraints, design, *see* Design constraints
Consumer and computing devices, 5–6

'Microwatt nodes', 6
'Milliwatt nodes', 5
'Watt nodes', 5–6
Continuous design parameters, 79–81
Cooling issues
    chip cooling system, 3–4
    computing server rack, 3
'Custom voltage assignment' approach, 93

D

Data gating, low power design flow, 329
    data-gating insertion, 330
        logic synthesizer, 330
        RTL code, 330
    data-gating verilog code, operand isolation, 330
        conventional code, 330
        low power code, 330
    low power version, 329
Data link/media access layer, 167
Data retention voltage (DRV), in embedded SRAM, 236–237
    approaching, 242
        closed-loop feedback approach, 242
        open-loop approach, 242
    lowering using error-correcting (ECC) strategies, 240
    power savings of, 237
    process balance impact, 238
    process variations impact on, 238–239
    reducing, 240
        by lowering voltage below worst-case value, 240
        optimization, 240
    statistical distribution, 239–240
    and transistor sizes, 237
Decoupling caps
    effectiveness, 342
    placement, 220
    voltage drop mitigation with, 342
Delay (s), 54
Delay modeling, 84
Delay sensitivity, 265
Design abstraction stack, 115
Design constraints, 2–3, 5
    communication infrastructure, growth, 2–3
    computation, growth, 2–3
    cooling issues, 3
    mobile electronics emergence, 5
    'zero-power electronics' emergence, 10–11
Design phase analysis methodology, low power design, 323
Design phase low power design, 327
    clock gating, 327
    data gating, 327
    f_{eff} minimizing, 327
    memory system design, 327
'Design time' design techniques, 78
Design time techniques, in standby mode leakage reduction of embedded SRAM, 235
Differential logic networks, 61
Differential signaling, 162
Digital frequency dividers (DFDs), 211
Digital signal processors (DSPs), 132, 139–140

Digital signal processors (DSPs) (*cont.*)
  advantages, 140
  performance of, 140
Diode reverse-bias currents, 69
Direct-oxide tunneling currents, nanometer MOS transistor, 39
Disappearing electronics, *see* Zero-power electronics
Discrete design parameters, 79–81
Discrete voltage levels, 255
Dissipation, power, 1
Dithering, 255
Double-gated (DG) MOSFET, 203–204
Double-gated fully depleted SOI, 49
Drain induced barrier lowering (DIBL), 29, 32, 35
Drain leakage, 66
Drain-induced barrier lowering (DIBL) effect, 235
Dual voltage approach, 93
Dynamic body biasing (DBB), 266
  dynamics of, 225–226
  effectiveness of, 227
  in standy mode leakage control, 224–227
Dynamic energy, 85
Dynamic frequency scaling (DFS), 253
Dynamic hazards, 63
Dynamic logic networks, 60–61, 263
Dynamic power, 19, 53
  consumption, 58
  dissipation, 70
  reduction by cell resizing, 332–333
  in standby, 208
Dynamic RAM (DRAM), 183–205
Dynamic voltage and frequency scaling (DVFS), 325
Dynamic voltage drop, 339, 341, 343
Dynamic voltage scaling (DVS), in runtime optimization, 253–254
  ABB and, combining, 269–271
  in general-purpose processing, 259
  verification challenge, 262
  workload estimation, 255
Dynamic-power optimization, 78

E
Edge-triggered flip-flop, 95
Effective capacitance, 59
Effective resistance check, 339
Embedded SRAM, 234
  standby leakage reduction, 234
    body biasing, 244
    canary-based feedback mechanism, 243
    data retention voltage (DRV), 236–237
    design-time techniques, 235
    leakage current reduction, 235
    periphery leakage breakdown, 246
    by raising V_{SS}, 243–244
    voltage knobs, 235
    voltage reduction, 235
    voltage scaling approach, 235–236
    voltage scaling in and around the bit-cell, 246
Energy (Joule), 54
Energy efficiency of brain, 13
Energy per operation (EOP), 116–117, 290
  minimum EOP, 294–295
Energy recovery, 350
Energy scavenging, 12
Energy storage technologies, 9–10
  See also Battery technology
Energy-area-delay tradeoff in SVD, 131
Energy–delay (E–D) trade-off optimization framework, 78–79
  optimal energy–delay curve, 79
Energy–Delay space, 54, 73–74, 119
Equivalent oxide thickness (EOT), 40
Error correction, aggressive deployment (AD), 273
Error detection, aggressive deployment (AD), 273
Error rate versus supply voltage, runtime optimization, 274
Error-correcting coding (ECC) strategies, 167, 170
  combining cell optimization and, 241–242
  DRV lowering using, 240

F
Factoring, 101
Fast Fourier Transform (FFT) module, 297–298
  energy-performance curves, 298
  sub–threshold FFT, 299
Fine-grained power gating, 335
FinFETs, 49
  backgated FinFET, 50
Fixed deskew, 212
Fixed-voltage scaling model, 16
Flit-routing, 176
Forced transistor stacking, in standy mode leakage control, 215–216
Fowler–Nordheim (FN) tunneling, 38
Frequency scaling model, 72
Frequency-division multiplexing (FDM), 171
Fuel cells, 8–9
  miniature fuel cells, 9
Full-depleted SOI (FD-SOI), 19, 48

G
Gate leakage, 66
  nanometer MOS transistor, 37–38
    gate leakage current density limit versus simulated gate leakage, 41
    gate-induced drain leakage (GIDL), 36–37
    mechanisms, 38
Gate tunneling, 69
Gate-induced drain leakage (GIDL), 235
Gate-level trade-offs for power, 99–100
Generalized low power design flow, 321
  design phase, 321
    implementation, 321
    RTL design, 321
    system-level design (SLD), 321
Glitchfree verilog code, clock gating
  low power design flow, 329
    prevention latch, 329
Glitching

Index

occurrence, 64–65
in static CMOS, 63
Global clock gating, 328
Globally asynchronous locally synchronous (GALS) methodology, 166, 351

H
Hardware accelerators, 142–144
Heterogeneous networking topology, 174
Hetero-junction devices, 47
Hierarchical bitlines, SRAM, 195
Hierarchical optimization, challenge in, 114
Hierarchical wordline architecture, SRAM, 195
Hierarchy design methodology, 79–80
High-performance microprocessors, 3–4
Homogeneous networking topology, 174
    binary tree network, 174
    crossbar, 174
    mesh, 174
Human brain, power consumption by, 13

I
Idealized wire-scaling model, 153
Implementation phase low power design, 323, 331–332
    low power synthesis, 332
    multiple supply voltages, 332
    power gating, 332
    power integrity design, 332
    slack redistribution, 332
Instruction loop buffer (ILB), 132
Integrated chips (ICs), memory role in, 184
Iintegrated clock gating cell, 329
Integrated power converter for sensor networks, 284
Interconnect network/Interconnect-optimization, @ design time, 151–180
    charge recycling, 164
    circuit-switched versus packet-based, 175
    communication dominant part of power budget, 153
    data link/media access layer, 167
    idealized wire-scaling model, 153
    increasing impact of, 152
    interconnect scaling, 156
    ITRS projections, 152
    layered approach, 157
        physical layer, 158
        repeater insertion, 158
    logic scaling, 155
    multi-dimensional optimization, 160
    networking topologies, 174
        binary tree network, 174
        crossbar, 174
        exploration, 175
        heterogeneous, 174
        hierarchy, 174
        homogeneous, 174
        mesh, 174
    network trade-offs, 173
    OSI approach, 151
    OSI protocol stack, 157–158
    quasi-adiabatic charging, 164
    reduced swing, 160
    reducing interconnect power/energy, 157
    research in, 154–155
        dielectrics with lower permittivity, 154
        interconnect materials with lower resistance, 154
        novel interconnect media, 15
        shorter wire lengths, 155
    signaling protocols, 166
    wire energy delay trade-off, 159–160
    See also Clock distribution network
Interconnect scaling, 156
Inverted-clustering, 175
Inverter chain, 86–87
    gate sizing, 87
    V_{DD} optimization, 87

J
Junction leakage, 66

K
Kogge–Stone tree adder, 88–89, 109, 274, 276
    sizing vs. dual-V_{DD} optimization, 89

L
Latch-retaining state during sleep, 222–223
Layered design methodology, 80
LC-based DC–DC (buck) converter, 284
Leakage, 66, 346
    advantages, 102
    components, nanometer MOS transistor, 33
    drain leakage, 66
    at design time, 102
    effects/concerns, 18, 20–21, 37
    gate leakage, 66
    in standby mode, 214
        See also under Standby mode
    junction leakage, 66
    mechanisms, memory and, 192
        See also under Memory
    power reduction by V_{TH} assignment, 332–333
    reduced threshold voltages, 34
    sub-threshold leakage, 33, 35, 67
Level-converting flip-flops (LCFFs), 95
    pulsed LCFF, dynamic realization, 95
    pulsed precharge (PPR) LCFF, 95
Liberty power, 319
Lithium-ion battery technology, 7–8
Local bias generators (LBGs), 226
Local clock gating, 328
Locality of reference, 132
Logic function, 59
Logic networks
    activity as a function of topology, 60
    differential logic networks, 61
    See also Complex logic; Dynamic logic
Logic scaling, 155
Logical effort based design optimization methodology, 71

Logical optimizations, 100
    logic restructuring, 100
Logical effort formulation, 85
Logic-sizing considerations, sub-threshold design, 300
Loop transformations, 124
Low power design methodologies and flows, 317–344
    clock gating, 327
    design phase, 327
        analysis methodology, 323
        dynamic voltage drop impact, 343
        in implementation phase, 323, 331–332
        methodology issues, 318
            power analysis, 318
            power integrity, 318
            power reduction, 318
        motivations, 318
            minimize effort, 318
            minimize power, 318
            minimize time, 318
    power analysis methodology, 321
        issues, 322
        method, 321
        motivation, 321
        over project duration, 324
    power characterization and modeling issue, 318–320
        SPICE-like simulations, 320
        state-dependent leakage models, 320
        state-independent leakage models, 320
    power-down modes, 325
    power integrity methodologies, 339
    slack redistribution, 332
    system phase analysis methodology, 322, 324–326
        challenges, 324
        f_{eff} minimization, 324
        modes, 324
        parallelism, 324
        pipelining, 324
        V_{DD} minimization, 324
    voltage drop mitigation with decoupling caps, 342
    See also Clock gating; Generalized low power design flow; Memory system design; Multi-V_{DD}; Power gating

M

Magnetoresistive RAM (MRAM), 183–205
MATLAB program, 81
Media access control (MAC), 167, 170–171
Memory, optimizing power @ design time, 183–205
    cell array power, 192
    leakage and, 192
        multiple threshold voltages reducing, 193
        multiple voltages, 194
        sub-threshold leakage, 192
        threshold voltage to reduce, 192
    low-swing write, 201
    processor area dominated by, 184
    role in ICs, 184
    structures, 185
        power for read access, 185
        power for write access, 185
        power in the cell array, 185
    See also Static random access memory (SRAM)
Memory, optimizing power @ standby, 233–247
    processor area dominated by, 234
    See also Embedded SRAM
Memory system design, low power, 330–331
    objectives
        C_{eff} minimization, 331
        challenges, 331
        f_{eff} minimization, 331
        power reduction methods, 331
        trade-offs, 331
    split memory access, 331
Metrics, 54
    delay (s), 54
    energy (Joule), 54
    energy delay, 54
    power (Watt), 54
    propagation delay, 54
Micro batteries, 9
Micro-electromechanical systems (MEMS), 51, 348
'Microwatt nodes', 6
'Milliwatt nodes', 5
Minimum energy point, moving, 313
    from sub-threshold to moderate inversion, 312–313
        using different logic styles, 312
        using switching devices, 312
Mobile electronics emergence, as design constraint, 5
Mobile functionality limited by energy budget, 6
Moore's law, 6, 18
Motes, 10
MTCMOS derivatives, in standby mode state loss prevention, 223
Multi-dimensional optimization, interconnect, 160
Multiple supply voltages, 90–92
    block-level supply assignment, 90–91
    conventional, 93
    distributing, 93
    multiple supplies inside a block, 90–91
Multiple threshold voltages
    in power optimization, 104
    reducing SRAM leakage, 193
Multiple-input and multiple-output (MIMO) communications, SVD processor for, 129–130
Multi-V_{DD}, low power design, 337–338
    flow, 338
    issues, 338
        level shifters, 338
        partitioning, 338
        physical design, 338
        timing verification, 338
        voltages, 338

N

NAND gates, 60, 64
Nano-electromechanical systems (NEMS), 348
Nanometer transistors, 25–52
    advantages, 26

Index

alpha power law model, 29
behavior, 25–26
body bias, forward and reverse, 30
challenging low power design, 26
device and technology innovations, 45–46
    strained silicon, 46–47
DIBL effect, 32, 35
drain current under velocity saturation, 27
FinFETs, 49–50
leakage components, 33
    direct-oxide tunneling currents, 39
    high-k dielectrics, 40
    high-k gate dielectric, 39
    leakage effects, 37
    reduced threshold voltages impact on, 34
    sub-threshold current, 35
    sub-threshold leakage, 33, 35
    temperature sensitivity and, 41
    See also Gate leakage
output resistance, 29
65 nm bulk NMOS transistor, I_D versus V_{DS} for, 27
Silicon-on-Insulator (SOI), 48
    See also individual entry
sub-100 nm transistor, 26–27
their models and, 25–52
    sub-100 nm CMOS transistors
threshold control, evolution, 31
threshold voltages, channel length impact on, 31–32
thresholds and sub-threshold current, 30
variability impact, 42–43
    environmental source, 43
    physical source, 43
    process variations, 44
    threshold variations, 45
    variability sources and their time scales, 43
    variability sources, 43
Nanotechnology, 348
    nano-mechanical relays, 348
Need for power, 2
Negative bias temperature instability (NBTI), 44
Network trade-offs, 173
Network-on-a-chip (NoC), 166, 172–173, 176
Neumann von and Shannon bounds, 352
Neural networks concept, 352
Non-traditional bit-cells, 202
Novel switching devices, 347
20 nm technology, 21
65 nm bulk NMOS transistor, I_D versus V_{DS} for, 27

O

Off-chip inductors, 163–164
On-chip leakage sensor, 266
Open-loop approach, for DRV, 242
Operator isolation, 329
Optimization methodology, CMOS, 71
    logical effort based, 71
Oracle scheduler, 259
OSI stack, 177
Output resistance, nanometer transistors, 29

P

Packet-switched networks, 175–176
Parallelism, 117, 324, 326
Pareto-optimal curve, 82
Partially-depleted (PD-SOI), 48
Pass-transistor logic (PTL) approach, 313
    leakage control in, 314
Periphery leakage breakdown, embedded SRAM, 246
Physical layer of interconnect stack, 158
Pipelining, 117–118, 324, 326
Platform-based design strategy, 146
    heterogeneous platform, 147
    NXP Nexperia™ platform, 146
    OMAP platform™ for Wireless, 147
Pleiades network, 177
PMOS transistors, 65
PN-code acquisition for CDMA, 354
Power (Watt), 54
    and delay, relationship between, 73
    dissipation, 61–62
        in CMOS, 54
    distribution, 285–286
    power density, 18
    versus energy, 14
Power domains (PDs), 280, 346
    challenges, 280
    interfacing between, 282
    in sensor network processor, 282
Power gating
    low power design, 335
        fine-grained power gating, 335
        flow, 337
        issues, 336
        physical design, 335
        switch placement, 335
        switch sizing, 336
    in standy mode leakage control, 217–218
Power integrity methodologies, 339
    dynamic voltage drop, 341
    resistance check, 340
    static voltage drop, 341
    stimulus selection, 340
    verification flow, 339
Power manager (PM), 280
Power-down modes, 325
    clock frequency control, 325
    issues, 325
    trade-offs, 325
    V_{DD} control, 325
Power-limited technology scaling, 22
Processors, power trends for, 15–17
Propagation delay, 54
    of sub-threshold inverter, 295
Pull-up and pull-down networks, for minimum-voltage operation, 293
Pulsed LCFF, dynamic realization, 95
Pulsed precharge (PPR) LCFF, 95

Q

Quasi-adiabatic charging, 164

R

Random doping fluctuation (RDF), 186
RAZOR, in runtime optimization, 275
　distributed pipeline recovery, 276
　voltage setting mechanism, 276
Read-assist techniques, for leakage reduction, 199
Read-power reduction techniques, 201–202
Reconvergent fan-out, 61
Reduced-swing circuits, 160–162
　issues, 163
Reduced-swing clock distribution, 178–179
Register-transfer level (RTL) code, 209
Relay circuit design and comparison, 349
Return on Investment (ROI), optimizing, 86
Rules, low power design, 345–347
Runtime optimization, 346
'Run-time' optimizations, 78

S

Scaling/Scaling model, 17
　direct consequences of, 152–153
　fixed-voltage scaling model, 16
　frequency, 72
　idealized wire-scaling model, 153
　interconnect scaling, 156
　logic scaling, 155
　supply voltage scaling, 17
　traditional, 72
Scavenging, energy, 12
Self-adapting approach, 271
Self-adjusting threshold voltage scheme (SATS), 266–267
Self-timing strategies, 350–351
Sense amplifier based pass-transistor logic (SAPTL), 314–315
Sensor network concept, 352
Sensor network-on-a-chip (SNOC), 353
Sensor networks, integrated power converter for, 284
Shannon theorem, 156
　Shannon–von Neumann–Landauer limit, 295, 311
Shared n-well, 94
Shared-well technique, 96
Short circuit currents, 63, 65–66
　modeling, 66
　　as capacitor, 66
Silicon-on-Insulator (SOI), 48
　double-gated fully depleted SOI, 49
　fully-depleted (FD-SOI), 48
　partially-depleted (PD-SOI), 48
　types of, 48
Simple versus complex processors, 138
Singular-value decomposition (SVD) processor
　energy-area-delay tradeoff in, 131
　for MIMO, 129–130
　optimization techniques, 130–131
　power/area optimal 4x4 SVD chip, 131
Sizing, transistor, 97–98
　continuous, 98
　discrete, 98
Slack redistribution, low power design, 332–333
　dynamic & leakage power optimization, 332
　dynamic power reduction by cell resizing, 332–333
　leakage power reduction by V_{TH} assignment, 332–333
　objective, 332
　slack redistribution flows, 334
　　issues, 334
　　trade-offs, 334
Sleep mode management, *see* Standby mode
Software optimizations, 133
Source codes, 168
Spatial programming, 129
Split memory access, low power design methodology, 331
6T SRAM cell, 204
Stack effect, 68
Standby mode, optimizing power @ standby, 207–230
　concern over, 209
　decoupling capacitor placement, 220
　design exploration space, 213
　dynamic power in standby, 208
　energy consumption in, 209
　　See also Clock gating
　impacting performance, 220–221
　in µprocessors and µcontrollers, 213
　latch-retaining state during, 222–223
　leakage challenge in, 214
　　control techniques, 215
　leakage control
　　boosted-gate MOS (BGMOS), 218–219
　　boosted-sleep MOS, 219
　　dynamic body biasing (DBB), 224–225
　　forced transistor stacking, 215–216
　　power-gating technique, 217–218
　　supply voltage ramping (SVR), 227–228
　　transistor stacking, 215
　　virtual supplies, 219
　　See also under Embedded SRAM
　MTCMOS derivatives preventing state loss, 223
　preserving state, 222
　reaching, 219–220
　sizing, 221
　sleep modes and sleep time, trade-off between, 212
　sleep transistor layout, 224
　sleep transistor placement, 223
　standard cell layout methodology, integration in, 229
　versus active delay, 216
State retention flip-flops (SRFFs), 337
Static (leakage) power, 54–55
Static CMOS logic, 263
　glitching in, 63
Static currents, 54–55
Static noise margin (SNM), SRAM, 188–189
　cells with pseudo-static SNM removal, 203
　lower precharge voltage improving, 198
　with scaling, 190
Static power, 53
　dissipation, 69–70
　drawbacks, 18
Static random access memory (SRAM), 183–205
　bit-cell array, 187
　　power consumption within, 191
　BL leakage during read access, 196
　cell writability, 191

Index

data retention voltage (DRV), 188
embedded SRAM, 234
functionality constraint in, 205
hierarchical bitlines, 195
hierarchical wordline architecture, 195
metrics, 186
    area, 186
    functionality, 186
    hold, functionality metric, 188
    power, 186
    read, functionality metric, 188
    write, functionality metric, 188
power breakdown during read, 194–195
power consumption in, 186–187
process scaling degrade, 189
6T SRAM cell, 204
static noise margin (SNM), 188–189
sub-threshold SRAM, 303
topology, 187
V_{DD} scaling, 198
voltage transfer characteristic (VTC) curves, 189
write margin, 190
write, power breakdown during, 199–200
    alternative bit-cell reducing, 200
    charge recycling to reduce write power, 200–201
Static routing, 175
Static voltage drop, 339, 341
Static power optimization, 78
Statistical computational models, 347
Strained silicon concept, 46–47
Stream-based processing, 256
Sub-100 nm CMOS transistors, 26–27
    models for, 28
    simplification in, 28
Substrate current body effect (SCBE), 29
Sub-threshold current, nanometer MOS transistor, 30, 35
    as a function of V_{DS}, 36
Sub-threshold design
    challenges in, 299
        data dependencies impact, 301
        erratic behavior, 302
        logic-sizing considerations, 300
        modeling in moderate-inversion region, 306–307
        process variations impact, 300
        read current/bitline leakage, 302
        SNM, 302
        soft errors, 302
        timing variance, 305
        write margin, 302
    CMOS inverter, 292
    microprocessor, 304
    moving away the minimum energy point from, 312
    power dissipation of, 296
    propagation delay of, 295
    prototype implementation of, 304
    SRAM cell, 303
    sub-threshold FFT, 299
    sub-V_{TH} memory, 302
Sub-threshold leakage, 17, 67, 192
    nanometer MOS transistor, 33
Supercapacitor, 10

Supply and threshold voltage trends, 20
Supply voltage ramping (SVR), in standy mode leakage control, 227–228
    impact of, 228
Supply voltage scaling, 17, 82–83
Suspended gate MOSFET (SG-MOS), 51
Switch sizing, power gating, 336
Switched-capacitor (SC) converter concept, 283–284
Synchronous approach, 257
System-level design (SLD), in generalized low power design flow, 321
System-on-a-chip (SoC), 165
System phase analysis methodology, 322, 324–325
    See also under Low power design
System phase low power design flow, 326
Systems-on-a-Chip (SoC), 3
    complications, 20

T

Technology generations, power evolution over, 15
Technology mapping, in optimizations, 98–100
Temperature gradients and performance, 4
Temperature influence, nanometer transistors leakage, 41
Temporal correlations, 63–64
Thermal voltage, 34
Threshold control, nanometer transistors, 31
Threshold current, nanometer transistors, 30
Threshold variations, nanometer transistors, 45
Threshold voltages
    exploitation, 161
    nanometer transistors, channel length impact on, 31–32
Time-multiplexing, 120
Timing, managing, 281
    basic scheduling schemes, 281
    metrics, 281
Trade-off, 78–79
Transistor stacking, in standy mode leakage control, 215
    8T transistor, 203
Transition probabilities for basic gates, 60
Transition coding techniques, 169–170
Transmission line based clock distribution network, 179
Tree adder, 88–89
    in energy–delay space, 89–90
    multi-dimensional search, 90
    See also Kogge–Stone tree adder

U

Ultra low power (ULP)/voltage design, 289–316, 347
    complex versus simple gates, 312–313
    dynamic behavior, 296
    EKV model, 309–310
    energy–delay trade-off, 315
    high-activity scenario, 309
    low-activity scenario, 309
    minimum energy per operation, 294–295
    minimum operational voltage of inverter, 291
        pull-up and pull-down networks for, 293
    modeling energy, 308
    opportunities for, 290–291

Ultra low power (ULP)/voltage design (*cont.*)
  power–delay product and energy–delay, 297
  PTL, leakage control in, 314
  sense amplifier based pass-transistor logic (SAPTL), 314–315
  size, optimizing over, 310
  V_{DD}, optimizing over, 310
  V_{TH}, optimizing over, 310
  See also Fast Fourier Transform (FFT) module; Sub-threshold design
Ultracapacitor, 10

V
Variability impacting nanometer transistors leakage, 42
  See also under Nanometer transistors
V_{DD} scaling, 198–199
Velocity saturation effect, 27
  drain current under, 27
Verilog code, clock gating, 328
  low power design flow
    conventional RTL code, 328
    instantiated clock-gating cell, 328
    low power clock-gated RTL code, 328
Video, optimizing energy in, 141–142
Virtual Mobile Engine (VME), 145
Virtual supplies, in standy mode leakage control, 219
'Virtual' tapering, 87
Voltage dithering, 255–256

Voltage hopping, 255
Voltage knobs
  in standby mode leakage reduction of embedded SRAM, 235
Voltage transfer characteristics (VTC) of inverter, 292–293
  analytical model, 292–293
    simulation confirming, 293
Voltage-setting mechanism, aggressive deployment (AD), 273
Von–Neumann style processor, 128

W
'Watt nodes', 5–6
Wave-division multiplexing (WDM), 171
Weak inversion mode, 292
Wire energy delay trade-off, interconnect, 159
Wireless sensor networks (WSN), 10–11
Word length optimization, 129–131
Write margin, SRAM, 190
Write power saving approaches, 201–202

X
XOR gates, 60

Z
'Zero-power electronics' emergence, 10–11

Continued from page ii

Routing Congestion in VLSI Circuits: Estimation and Optimization
Prashant Saxena, Rupesh S. Shelar, Sachin Sapatnekar
ISBN 978-0-387-30037-5, 2007

Ultra-Low Power Wireless Technologies for Sensor Networks
Brian Otis and Jan Rabaey
ISBN 978-0-387-30930-9, 2007

Sub-Threshold Design for Ultra Low-Power Systems
Alice Wang, Benton H. Calhoun and Anantha Chandrakasan
ISBN 978-0-387-33515-5, 2006

High Performance Energy Efficient Microprocessor Design
Vojin Oklibdzija and Ram Krishnamurthy (Eds.)
ISBN 978-0-387-28594-8, 2006

Abstraction Refinement for Large Scale Model Checking
Chao Wang, Gary D. Hachtel, and Fabio Somenzi
ISBN 978-0-387-28594-2, 2006

A Practical Introduction to PSL
Cindy Eisner and Dana Fisman
ISBN 978-0-387-35313-5, 2006

Thermal and Power Management of Integrated Systems
Arman Vassighi and Manoj Sachdev
ISBN 978-0-387-25762-4, 2006

Leakage in Nanometer CMOS Technologies
Siva G. Narendra and Anantha Chandrakasan
ISBN 978-0-387-25737-2, 2005

Statistical Analysis and Optimization for VLSI: Timing and Power
Ashish Srivastava, Dennis Sylvester, and David Blaauw
ISBN 978-0-387-26049-9, 2005

Printed by Publishers' Graphics LLC
AMZ20121230.19.18.14